ENGINEERING TECHNOLOGY AND INDUSTRIAL CHEMISTRY WITH APPLICATIONS

Innovations in Physical Chemistry: Monograph Series

ENGINEERING TECHNOLOGY AND INDUSTRIAL CHEMISTRY WITH APPLICATIONS

Edited by

Reza K. Haghi, PhD
Francisco Torrens, PhD

APPLE ACADEMIC PRESS

Apple Academic Press Inc.   Apple Academic Press Inc.
3333 Mistwell Crescent   9 Spinnaker Way
Oakville, ON L6L 0A2   Waretown, NJ 08758
Canada   USA

© 2019 by Apple Academic Press, Inc.

First issued in paperback 2021

Exclusive worldwide distribution by CRC Press, a member of Taylor & Francis Group
No claim to original U.S. Government works

ISBN 13: 978-1-77-463077-8 (pbk)
ISBN 13: 978-1-77-188637-6 (hbk)

Library and Archives Canada Cataloguing in Publication

Engineering technology and industrial chemistry with applications / edited by Reza K. Haghi, Francisco Torrens.

(Innovations in physical chemistry : monograph series)
Includes bibliographical references and index.
Issued in print and electronic formats.
ISBN 978-1-77188-637-6 (hardcover).--ISBN 978-1-315-10044-9 (PDF)

1. Chemistry, Technical. 2. Chemical engineering. 3. Materials. I. Haghi, Reza K., editor
II. Torrens, Francisco, editor III. Series: Innovations in physical chemistry. Monograph series

---

| TP145.E54 2018 | 660 | C2018-904420-9 | C2018-904421-7 |

CIP data on file with US Library of Congress

Apple Academic Press also publishes its books in a variety of electronic formats. Some content that appears in print may not be available in electronic format. For information about Apple Academic Press products, visit our website at **www.appleacademicpress.com** and the CRC Press website at **www.crcpress.com**

CONTENTS

ABOUT THE EDITORS

Reza K. Haghi, PhD, is an Infrared Analyst in the Environmental & Biochemical Sciences Group at the James Hutton Institute, Aberdeen, UK. He was formerly a research assistant at the Institute of Petroleum Engineering, Heriot-Watt University, Edinburgh, United Kingdom. Dr. Haghi is an expert in the development and application of spectroscopy techniques for monitoring hydrate and corrosion risks and has developed techniques for early detection of gas hydrate risks. He conducted an integrated experiment and modeling in his studies and has extended his research to monitoring system pH and risk of corrosion. He has published several papers in internationally well-known peer-reviewed scientific journals and has published several papers in conference proceedings as well as technical reports and lecture notes. He has extensive experience in spectroscopy, from both experimental and modeling aspects. During his PhD work, he developed various novel flow assurance techniques based on spectroscopy, as well as designed and operated test equipment. He received his MSc in advanced control system from the University of Salford, Manchester, United Kingdom.
E-mail: rk182@hw.ac.uk, rkhaghi@gmail.com

Francisco Torrens, PhD, is lecturer in physical chemistry at the Universitat de València in Spain. His scientific accomplishments include the first implementation at a Spanish university of a program for the elucidation of crystallographic structures and the construction of the first computational-chemistry program adapted to a vector-facility supercomputer.

He has written many articles published in professional journals and has acted as a reviewer as well. He has handled 26 research projects, has published two books and over 350 articles, and has made numerous presentations.
E-mail: Francisco.Torrens@uv.es

ABOUT THE INNOVATIONS IN PHYSICAL CHEMISTRY: MONOGRAPH SERIES

This book series aims to offer a comprehensive collection of books on physical principles and mathematical techniques for majors, non-majors, and chemical engineers. Because there are many exciting new areas of research involving computational chemistry, nanomaterials, smart materials, high-performance materials, and applications of the recently discovered graphene, there can be no doubt that physical chemistry is a vitally important field. Physical chemistry is considered a daunting branch of chemistry—it is grounded in physics and mathematics and draws on quantum mechanics, thermodynamics, and statistical thermodynamics.

Editors-in-Chief

A. K. Haghi, PhD
Editor-in-Chief, International Journal of Chemoinformatics and Chemical Engineering and Polymers Research Journal; Member, Canadian Research and Development Center of Sciences and Cultures (CRDCSC), Montreal, Quebec, Canada Email: AKHaghi@Yahoo.com

Lionello Pogliani, PhD
University of Valencia-Burjassot, Spain
Email: lionello.pogliani@uv.es

Ana Cristina Faria Ribeiro, PhD
Researcher, Department of Chemistry, University of Coimbra, Portugal
Email: anacfrib@ci.uc.pt

Books in the Series

- **Applied Physical Chemistry with Multidisciplinary Approaches**
  Editors: A. K. Haghi, PhD, Devrim Balköse, PhD, and Sabu Thomas, PhD
- **Chemical Technology and Informatics in Chemistry with Applications**
  Editors: Alexander V. Vakhrushev, DSc, Omari V. Mukbaniani, DSc, and Heru Susanto, PhD

- **Engineering Technologies for Renewable and Recyclable Materials: Physical-Chemical Properties and Functional Aspects**
  Editors: Jithin Joy, Maciej Jaroszewski, PhD, Praveen K. M.,
  and Sabu Thomas, PhD, and Reza Haghi, PhD

- **Engineering Technology and Industrial Chemistry with Applications**
  Editors: Reza Haghi, PhD, and Francisco Torrens, PhD

- **High-Performance Materials and Engineered Chemistry**
  Editors: Francisco Torrens, PhD, Devrim Balköse, PhD, and Sabu Thomas, PhD

- **Methodologies and Applications for Analytical and Physical Chemistry**
  Editors: A. K. Haghi, PhD, Sabu Thomas, PhD, Sukanchan Palit,
  and Priyanka Main

- **Modern Physical Chemistry: Engineering Models, Materials, and Methods with Applications**
  Editors: Reza Haghi, PhD, Emili Besalú, PhD, Maciej Jaroszewski, PhD,
  Sabu Thomas, PhD, and Praveen K. M.

- **Physical Chemistry for Chemists and Chemical Engineers: Multidisciplinary Research Perspectives**
  Editors: Alexander V. Vakhrushev, DSc, Reza Haghi, PhD,
  and J. V. de Julián-Ortiz, PhD

- **Physical Chemistry for Engineering and Applied Sciences: Theoretical and Methodological Implication**
  Editors: A. K. Haghi, PhD, Cristóbal Noé Aguilar, PhD, Sabu Thomas, PhD,
  and Praveen K. M.

- **Theoretical Models and Experimental Approaches in Physical Chemistry: Research Methodology and Practical Methods**
  Editors: A. K. Haghi, PhD, Sabu Thomas, PhD, Praveen K. M.,
  and Avinash R. Pai

LIST OF CONTRIBUTORS

Ana M. Amado
Química-Física Molecular, Departamento de Química, FCTUC, Universidade de Coimbra, P3004-535 Coimbra, Portugal
E-mail: amado.am@gmail.com, Tel/Fax: +351-239826541

Sevdiye Atakul
Department of Chemical Engineering, İzmir Institute of Technology, Gulbahce Urla Izmir, Turkey; Akzo Nobel Boya AŞ İzmir Turkey
E-mail: sevdiyeatakul@gmail.com

Eduarda F. G. Azevedo
Coimbra Chemistry Centre, Department of Chemistry, University of Coimbra, 3004-535 Coimbra, Portugal
E-mail: edy.gil.azevedo@gmail.com, Tel:+351-239-854460; Fax:+351-239-827703

Maria L. G. Azevedo
Otorhinolaryngology Service, Centro Hospitalar Baixo Vouga, Aveiro, Portugal
E-mail: luisaazevedo.md@gmail.com

Devrim Balköse
Department of Chemical Engineering, İzmir Institute of Technology, Gulbahce Urla Izmir, Turkey
E-mail: devrimbalkose@gmail.com

Vladimir I. Binyukov
Senior Scientist, Emanuel Institute of Biochemical Physics, Russian Academy of Sciences, 4 Kosygin str., Moscow 119334, Russia
E-mail: bin707@mail.ru

Hugh D. Burrows
Coimbra Chemistry Centre and Department of Chemistry, University of Coimbra, 3004-535 Coimbra, Portugal

E. Carvajal-Millan
CTAOA, Laboratory of Biopolymers, Research Center for Food and Development, CIAD, A. C., Hermosillo, Sonora 83000, Mexico
E-mail: ecarvajal@ciad.mx

Gloria Castellano
Departamento de Ciencias Experimentales y Matemáticas, Facultad de Veterinaria y Ciencias Experimentales, Universidad Católica de Valencia San Vicente Mártir, Guillem de Castro-94, E-46001 València, Spain

Pedro F. Cruz
Coimbra Chemistry Centre and Department of Chemistry, University of Coimbra, 3004-535 Coimbra, Portugal

Adriana P. Gerola
Chemistry Department, Federal University of Santa Catarina, 88040-900 Florianópolis, Santa Catarina, Brazil

Lenka Gřundělová
Centre of Polymer Systems, Tomas Bata University in Zlin, Czech Republic

Licínia L. G. Justino
Coimbra Chemistry Centre and Department of Chemistry, University of Coimbra, 3004-535 Coimbra, Portugal

M. Luísa Ramos
Coimbra Chemistry Centre and Department of Chemistry, University of Coimbra, 3004-535 Coimbra, Portugal
Tel.:+351-239-854453, Fax:+351-239-827703, E-mail: mlramos@ci.uc.pt

Shima Maghsoodlou
University of Guilan, Rasht, Iran

Ludmila I. Matienko
Head of Department, Emanuel Institute of Biochemical Physics, Russian Academy of Sciences, 4 Kosygin str., Moscow 119334, Russia
E-mail: matienko@sky.chph.ras.ru

Elena M. Mil
Leading Scientist, Emanuel Institute of Biochemical Physics, Russian Academy of Sciences, 4 Kosygin str., Moscow 119334, Russia

Antonín Minařík
Department of Physics and Materials Engineering, Faculty of Technology, Tomas Bata University in Zlin, Czech Republic; Centre of Polymer Systems, Tomas Bata University in Zlin, Czech Republic

Larisa A. Mosolova
Senior Scientist, Emanuel Institute of Biochemical Physics, Russian Academy of Sciences, 4 Kosygin str., Moscow 119334, Russia
E-mail: elenamil2004@mail.ru

Aleš Mráček
Department of Physics and Materials Engineering, Faculty of Technology, Tomas Bata University in Zlin, Czech Republic; Centre of Polymer Systems, Tomas Bata University in Zlin, Czech Republic
E-mail: mracek@ft.utb.cz

Edvani C. Muniz
Chemistry Department, Maringá State University, 87020-900 Maringá, Paraná, Brazil; Post-graduate Program on Materials Science & Engineering, Federal University of Technology, 86036-370, Londrina, Paraná, Brazil

Ahmed Mohamed Omer
Polymer Materials Research Department, Advanced Technologies and New Materials Research Institute (ATNMRI), City of Scientific Research and Technological Applications (SRTA-City), New Borg El-Arab City, Alexandria, Egypt

Sukanchan Palit
Assistant Professor (Senior Scale), Department of Chemical Engineering, University of Petroleum and Energy Studies, Post-Office Bidholi via Premnagar, Dehradun 248007, India; 43, Judges Bagan, Post-Office Haridevpur, Kolkata 700082, India
Tel.: 0091-8958728093, E-mail: sukanchan68@gmail.com, sukanchan92@gmail.com.

R. Pérez Leal
Faculty of Agro-Technological Sciences, Autonomous University of Chihuahua, Chihuahua 31125, Mexico

S. Rafiei
University of Guilan, Rasht, Iran
E-mail: saeedeh.rafieii@gmail.com

N. Ramírez-Chávez
Faculty of Agro-Technological Sciences, Autonomous University of Chihuahua, Chihuahua 31125, Mexico

Ana C. F. Ribeiro
Coimbra Chemistry Centre, Department of Chemistry, University of Coimbra, 3004-535 Coimbra, Portugal

Adley F. Rubira
Chemistry Department, Maringá State University, 87020-900 Maringá, Paraná, Brazil

Maysa Mohamed Sabet
Polymer Materials Research Department, Advanced Technologies and New Materials Research Institute (ATNMRI), City of Scientific Research and Technological Applications (SRTA-City), New Borg El-Arab City, Alexandria, Egypt

J. Salmeron-Zamora
Faculty of Agro-Technological Sciences, Autonomous University of Chihuahua, Chihuahua 31125, Mexico

Sajjad Sedaghat
Department of chemistry, College of Science Shahr-e-Qods Branch, Islamic Azad University, Tehran, Iran
E-mail: sajjadsedaghat@yahoo.com

Tamer Mahmoud Tamer
Polymer Materials Research Department, Advanced Technologies and New Materials Research Institute (ATNMRI), City of Scientific Research and Technological Applications (SRTA-City), New Borg El-Arab City, Alexandria, Egypt

Francisco Torrens
Institut Universitari de Ciència Molecular, Universitat de València, Edifici d'Instituts de Paterna, P. O. Box 22085, E-46071 València, Spain

Artur J. M. Valente
CQC, Department of Chemistry, University of Coimbra, 3004-535 Coimbra, Portugal
E-mail: avalente@ci.uc.pt

Gennady E. Zaikov
Professor, Emanuel Institute of Biochemical Physics, Russian Academy of Sciences, 4 Kosygin str., Moscow 119334, Russia
E-mail: chembio@sky.chph.ras.ru

PREFACE

This book provides an overview of the field of engineering technology and industrial chemistry, explains the basic underlying theory at a meaningful level that is not beyond beginners, and gives numerous comparisons of different methods with one another and with experiments. The large number of references, relating to all significant topics mentioned, should make this book useful to graduate students and academic and industrial researchers.

This book is essential for chemists, engineers, and researchers in providing mutual communication between academics and industry professionals around the world.

Not only does this book summarize the classical theories, but it also exhibits their engineering applications in response to the current key issues. Recent trends in several areas of chemistry and chemical engineering science, which have important application to practice, are discussed.

Some of the main highlights of this volume are:

- Information on the important problems of chemical engineering modeling and nanotechnology. These investigations are accompanied by real-life applications in practice.
- New theoretical ideas in calculating experiments and experimental practice.
- Highlights of applications of chemical physics to subjects that chemical engineering students will see in graduate courses.
- Introduction to the types of challenges and real problems that are encountered in industry and graduate research.
- Presentation computational chemistry examples and applications.
- A focus on concepts more than formal experimental techniques and theoretical methods.

PART I
The Petroleum Industry

CHAPTER 1

PETROLEUM ENGINEERING, PETROCHEMICALS, ENVIRONMENTAL AND ENERGY SUSTAINABILITY: A VISION FOR THE FUTURE

SUKANCHAN PALIT*

Department of Chemical Engineering, University of Petroleum and Energy Studies, Bidholi via Premnagar, Dehradun 248007, India

43, Judges Bagan, Haridevpur, Kolkata 700082, India

*Corresponding author. E-mail: sukanchan92@gmail.com, sukanchan68@gmail.com

CONTENTS

ABSTRACT

Human civilization is witnessing drastic challenges today as regards to energy and environment. Nowadays, energy and environmental sustainability stand in the midst of scientific vision and deep scientific discernment. Petroleum engineering is ushering in a new era in the field of scientific regeneration and deep scientific rejuvenation. Science and engineering of petroleum refining is in a state of deepening crisis as depletion of fossil fuel resources stands as a truly vexing issue in the scientific horizon. In this chapter, the author pointedly focuses on the immense scientific potential, scientific success and wide horizon in research pursuit in the field of petroleum refining. Petroleum refining is an intensely complex research endeavor. The world of technology and engineering science stands baffled as energy sustainability is in a state of distress with the ever-growing concern of depletion of petroleum resources. The chapter also deeply reviews the area of petrochemicals and energy sustainability with the sole vision of furtherance of science and engineering. Water science is another facet of environmental sustainability today. This treatise with deep and cogent insight relates the scientific success and the scientific profundity in the field of water technology, drinking water treatment, and desalination science. The scientific intricacies, the vast challenges, and the vision for the future are well researched and presented in deep details in this chapter.

1.1 INTRODUCTION

Science and technology in today's human civilization are moving at a drastic pace surpassing vast and wide scientific frontiers. Mankind's immense scientific prowess, the wide world of scientific adjudication, and the deep scientific rigor in petroleum engineering science will all lead a long and visionary way in the true emancipation of sustainability these days. Vision of science and mankind are ushering in a new era in the field of energy and environmental sustainability. Technology has few answers to the intricate problems of global water shortages. This treatise also gives wide glimpses on the success of environmental as well as energy sustainability in the true context of furtherance of science and engineering in our present day human civilization. The challenge and vision of petroleum refining are immense and groundbreaking. The technological vision of furtherance of petroleum refining and petrochemicals science is presented with lucid details in this well researched treatise. Science and mankind's vision are the two opposite sides of the coin today. In the context of depletion of fossil fuel resources, technology and engineering science of petroleum refining needs to be reshaped and restructured with the progress of scientific and academic rigor in research pursuit. Human's immense scientific prowess, the wide world of scientific girth and scientific determination, and the futuristic vision of nonrenewable energy will all lead a long and visionary way in the true emancipation and true realization of energy sustainability. The chapter widely focuses on the immense success of sustainable development as regards to energy and environment today. Nuclear science and space technology are the cornerstones of scientific research. Human civilization and its scientific endeavor need to be redefined and reenvisioned at this state of fossil fuel crisis today. Technology and petroleum engineering science at this crucial juxtaposition needs to be widely reshaped as human civilization treads through the lanes and by-lanes of scientific history and scientific vision.[12,13,14]

1.2 THE AIM AND OBJECTIVE OF THIS STUDY

Human civilization is witnessing drastic challenges and world of vision today. Scientific vision, scientific contemplation, and deep scientific discerning are the forerunners toward a greater visionary era in the field of energy and environmental sustainability. The sole aim and objective of this study is to present the success of petroleum refining and other energy technologies before the scientific audience in wider realization of energy and environmental sustainability. Science is today a colossus with a wider vision of its own. The author in this treatise pointedly focuses on the huge scientific potential of petroleum refining and petrochemicals in true emancipation of energy sustainability. The vision and challenge of petroleum refining are immense and far-reaching. The chapter researches on the present advances in petroleum refining and petrochemicals, and its applications in the true realization of energy sustainability. Sustainable development is the organizing principle for meeting human development goals while at the same time sustaining the ability of natural systems to provide natural resources and ecosystem services upon which the human economy and society wholly depends. While the modern concept of sustainable development is derived mostly from the 1987 Brundtland Report, it is also deeply rooted in earlier ideas about sustainable forest management and 21st century environmental concerns.[12,13,14] As the concept of sustainability developed, it has vastly shifted to focus on economic development, social development, and environmental protection for future generations. This treatise widely researches on the startling facts of human scientific research pursuit in energy and environmental sustainability with the vision of petroleum engineering science in mind.

1.3   THE NEED AND RATIONALE BEHIND THIS STUDY

Today, fossil fuel resources are slowly on the verge of depletion and have become a grave concern for scientists, engineers, and environmental scientists. The vision of science needs to be restructured and reframed with the passage of scientific history, scientific vision, and the civilization's timeframe. Renewable energy is the utmost need of this present day human civilization. Also, environmental sustainability should be the cornerstone of all major policy decisions of our planet today. The immense need and the rationale of this study go beyond scientific imagination and scientific profundity. There is an utter need for this area of study since this domain of science is still unexplored today. Human scientific vision and scientific forbearance are the true pillars of human civilization. Depletion of fossil fuel resources and the grave concerns of nonrenewable energy technology are challenging the scientific landscape. Technology has few answers to vast and varied questions of energy sustainability and nonrenewable technology, thus, the utmost need of this study.[12,13,14]

1.4   ENERGY AND ENVIRONMENTAL SUSTAINABILITY AND THE VISION FOR THE FUTURE

Energy and environmental sustainability are the pillars of scientific development and the scientific rigor today. Scientific advancements in petroleum engineering, petrochemicals, and sustainability wholly depend upon scientific and academic rigor. The vision for the future in sustainability research is wide and bright. The grave concerns of degradation of environment and energy and the deep scientific vision of both petroleum engineering and environmental engineering are all leading a long and visionary way in the true emancipation of science and engineering. Renewable energy is the scientific success of today's scientific endeavor. Presently, nuclear power, wind energy, and solar energy are the sole supports and pillars of success in renewable research. Technological vision and scientific motivation are the wide forerunners toward a newer visionary era in renewable technology. Renewable energy is the ultimate necessity of today's scientific endeavor and research exposition. Developed and developing economies are treading toward a newer visionary era in the field of environmental and energy sustainability. Technological profundity is in a state of immense scientific introspection in our present day human civilization. In the similar vein, water science and water technology need to be reframed and re-envisaged with the progress of scientific and academic rigor.[12,13,14]

1.5   THE VISION AND STATUS OF PETROLEUM ENGINEERING SCIENCE TODAY

Technology of petroleum refining is advancing at a rapid pace these days. Scientific vision and scientific fortitude are necessary in the progress of human civilization. Depletion of fossil fuel resources is redefining the future status of petroleum refining and petroleum engineering science. The vision and status of petroleum engineering science is moving toward newer scientific regeneration and the wider paradigm of scientific rejuvenation. The state of human civilization is at a vicious stake with the progress of scientific and academic rigor. Research pursuit and scientific motivation should be toward greater emancipation of energy and environmental sustainability. Energy, environment, food, and water are the vital components toward progress and emancipation of human mankind. Water and desalination science are in the path of immense scientific revamping and wide vision.

1.6 PETROCHEMICALS AND PETROLEUM REFINING TODAY

Petrochemicals and petroleum refining are in the state of newer scientific vision and scientific profundity. Scientific truth and scientific sagacity are in the path of immense scientific rejuvenation. Today, the world is moving toward new directions and vision. Petrochemicals, advanced petroleum refining, and renewable energy technology are the vital need of the hour. Scientific success, scientific fortitude, and the vast domain of scientific profundity will all lead a long way in the true realization and the true emancipation of petroleum refining. Technology and engineering science are in a state of immense scientific quagmire. Petrochemicals are the need of the hour as human civilization moves forward.

1.7 CRUDE PETROLEUM OIL

The compounds in crude petroleum oil are essentially hydrocarbons or substituted hydrocarbons in which the major elements are carbon at 85–95% and hydrogen at 10–14% and the rest with non-hydrocarbon elements—sulfur (0.2–3%), nitrogen (<0.1–2%), oxygen (1–1.5%), and organometallic compounds of nickel, vanadium, arsenic, lead, and other metals in traces (in parts per million or parts per billion concentration).[1] Inorganic salts of magnesium chloride, sodium chlorides, and other mineral salts are also accompanied with crude oil from the well either because of water from formation or water and chemicals injected during drilling and production.[1] Technology of petrochemicals and petroleum refining are on the verge of newer scientific regeneration and deep scientific vision. Crude petroleum oil is the nerve center of human scientific endeavor. The challenges are immense yet success is less. In such a situation, petroleum refining and the avenues of petrochemicals are the hallmarks and torch-bearers toward a newer visionary future.[1]

1.7.1   PETROLEUM PRODUCT AND TEST METHODS

Crude oil analysis is the crucial component in the scientific success of petroleum refining. Technological vision, scientific candor, and deep scientific vision are the pillars and supports toward a greater emancipation of science and engineering today. Petroleum refining is the heart of human scientific endeavor.[1] The hydrocarbon gases: methane, ethane, propane, and butane are present in the crude oil in its dissolved state. Methane has a high vapor pressure and it escapes from crude oil unless pressure above the vapor pressure is maintained. Petroleum product and test methods are of intense interest in the furtherance of science and engineering.[1] Distillation, in fact, is done at atmospheric pressure and then followed under vacuum.[1] The separation from hydrocarbons can be improved if the vapor and its condensates (reflux) are in intimate contact for some time during distillation and a reproducible distillation analysis is possible.[1] Such a method of distillation is known as true boiling point (TBP) distillation. The scientific vision, profundity, and discerning are of utmost importance in the furtherance of petroleum engineering science. American Petroleum Institute (API) gravity is a gravity measuring parameter for all petroleum oil. This is related as $API = 141.5/s - 131.5$ where 's' is the specific gravity of oil at 15.5°C with respect to water at the same temperature. The greater is the specific gravity the lower the API gravity will be. For water, API gravity becomes 10 and for oil it is greater than 10.[1] Crude oil having an API gravity as low as 9 has been found, which is heavier than water but most commonly it is always greater than 10.[1] Scientific vision, validation, and the targets of research endeavor are the forerunners toward a newer visionary eon in the field of petroleum testing.[1]

The next entity is the "characterization factor," which is commonly used with API gravity to judge the quality and many physical properties of crude oil.[1] This is defined as the ratio of the cubic root of the molal average boiling point (in Rankine) of oil to its specific gravity at 15.5°C. Bottom sediment and water (BSW) is a measure of the quantity of residual sediment mostly settleable from the crude oil and water.[1] Science and engineering of characterization methods are moving at a rapid pace with scientific progress and profundity. For liquid petroleum gas, the petroleum testing methods include smoke point, flash point, char point, distillation test, sulfur point, and corrosion.[1] For automotive fuels, the testing methods are American Standard for Testing Material Distillation, Octane number, corrosion, Reid vapor pressure, oxidation stability, and the application of additives. For high speed diesel, the testing methods include cetane number, diesel index, sulfur, corrosion, flash point, flame length, pour point, and viscosity measurement.[1] For lubricating oils, the testing methods are viscosity, Saybolt method, Redwood method, Brookfield method, Viscosity index method, cloud point, and pour point.[1] Vision of technology and engineering science are in the path of newer scientific regeneration and an enriched scientific evolution. Thus, petroleum product testing methods need to be revamped and reorganized with the passage of deep scientific vision and timeframe.

1.7.2   HEAT EXCHANGERS AND PIPE STILL FURNACES

Heat exchanger equipment exchanges heat between a hot and cold fluid, that is, cold fluid is heated by hot fluid without using any fire or combustion.[1] Heat exchanger equipments in a petroleum refinery are the vital components in petroleum refining procedure. The most widely used heat exchangers transfer heat from hot to cold fluid separated by a metallic surface. These types of heat transfer are called recuperators, where the fluids do not mix physically but only the heat flows through the metallic wall surface.[1] Today, technology revamping is of utmost importance in the furtherance of science and engineering. Double pipe and shell-and-tube exchangers are the common recuperators. In a double pipe exchanger, one of the fluids passes through an inner tube and the other fluid passes through the annular space between the outer and inner tubes. A shell-and-tube heat exchanger consists of a bundle of tubes inside a larger diameter outer tube or shell. In both exchangers, fluids may flow either in opposite (countercurrent) or parallel direction (cocurrent). The rate of heat transfer is dependent on the value of overall heat transfer coefficient (U) which is also a function of the film heat transfer coefficients (h_i and h_o) and the thermal conductivity of the metal (k_w)of the tube.[1] The greater the value of U the higher the rate of heat transfer. Pipe still furnaces are the other type of heat-exchange equipment. Validation of heat transfer science and deep scientific profundity behind the fundamentals of petroleum engineering science will all lead a long and visionary way in the true realization of fossil fuel technology in the near future.[1]

1.7.3   DISTILLATION AND STRIPPING

Distillation is a process in which more volatile substances in a mixture are separated from less volatile substances.[1] This separation can be carried out either by vaporizing a liquid mixture or by condensing a vapor mixture. In the actual practice of distillation, liquid containing the volatile substances is heated to generate vapor which is separated from the mixture followed by condensation of the vapors as distillate.[1] Technology of distillation and stripping are today ushering in a new age of scientific rejuvenation and scientific regeneration. Fundamentals of mass transfer in chemical process engineering and petroleum engineering need to be revamped and reenvisioned with the deep passage of scientific history and time. The yardsticks of technological validation are the necessities of scientific progress today. The science of distillation needs to be similarly reorganized with the progress of scientific and academic rigor.[1]

1.7.4   INSTRUMENTATION AND CONTROL IN A REFINERY

Modern chemical plants are automatically controlled with the objective of achieving highest productivity with minimum human intervention.[1] Process instrumentation and control stand as huge parameters in the scientific progress of petroleum refining today. Both the process control logic and strategies of control have undergone a massive change owing to the transformation of the analogue environment to a successful digital one. The automatic process control in a plant consists of three groups of elements: hardware, software, and the transmission lines.[1] Process control and integration are the heart of chemical engineering research endeavor. Hardware includes sensors/transducers, transmitters/signal conditioners, controllers, and final control elements (control valves, switches, solenoids, motors, etc.). A refinery process control and instrumentation is the vital backbone toward the furtherance of petroleum engineering science. In this chapter, the author deeply comprehends the scientific success, fortitude, and profundity behind the intricacies of petroleum refining today.[1]

1.7.5   PLANT MANAGEMENT AND ECONOMICS

Plant management and economics stand as important components in the scientific success and the furtherance of petroleum engineering science today. The science of economics is slowly evolving into a newer knowledge dimension and a newer technological profundity. Petroleum refining and petroleum engineering science are the vital components in the economic and scientific progress of a developed as well as developing nation today.[1]

The cost of an equipment may be obtained in two ways (1) by estimation and (2) the purchased price. Cost estimation is required when the equipment is fabricated. Estimated cost will include the material, labor, and the overhead components involved in manufacturing.[1] Scientific vision is in the course of immense restructuring as science of petroleum refining moves forward.[1]

Estimated cost will include the material, labor, and the overhead components in manufacturing. Material cost will include the direct and indirect components. Direct cost will include the cost or price of the materials, for example, plates, tubes, rods, beams, and so forth which can be identified with the equipment.[1] There are indirect methods also that include the cost of nuts, bolts, screws, welding, threading materials, and so forth. Energy audit is another avenue of emancipation of plant management and economics today.[1] Energy is consumed in any plant in the form of electricity and heat which are required to drive the machineries and to maintain the necessary operating conditions in the processing equipment. This is the challenge of science and the deep vision of technology today. The wide scientific vision and profundity of engineering science are the torchbearers toward a greater visionary eon in the field of petroleum refining today.[1] Energy economics in a similar vision has today gone beyond scientific vision and deep scientific imagination. The energy economics of petroleum refining and petroleum engineering science are today unfolding a newer visionary era in the field of scientific emancipation.

1.8   RECENT ADVANCES IN ENERGY AND ENVIRONMENTAL SUSTAINABILITY

Energy and environmental sustainability are the definite challenges and vision of our planet. Nowadays, technology and engineering science are in the threshold of a newer scientific rejuvenation. Science has no answers to the future of petroleum engineering, petroleum refining, and nonrenewable energy technology. Today, sustainable development whether it is infrastructural, energy, or environmental, are poised toward a newer innovation and vision. The success of human civilization lies in the hands of scientists and engineers. Various reports by international bodies targets sustainable issues and global energy research and development initiatives.

World Energy Council Report[2] elucidated on the 2013 Energy Sustainability Index. Energy issues needs to be readdressed and reenvisioned with the passage of scientific history, scientific vision, and time. Energy sustainability and progress of engineering science are two opposite sides of the visionary coin.[2] Technology and engineering science are in the path of immense scientific regeneration today. This report provides country-level details on the results of the 2013 Energy Sustainability Index prepared by the World Energy Council (WEC)in partnership with a private sector management consulting firm.[2] These profiles and the visionary index provide a comparative ranking of countries' ability to provide a stable, affordable, and environmentally-sensitive energy system and deeply highlight current challenges.[2] This is the upshot of 2013 World Energy Council Report.[2] The 2012 and 2013 reports' methodology is based on the guiding principle that energy sustainability involves the participation of both industry and policymakers. Technology, scientific vision, and academic rigor are in the path of immense restructuring and scientific discernment. Energy performance dimensions involve energy security, equity, and environmental sustainability. By 2030, the United Nations hopes that there will be universal access to modern energy services, a doubling of the share of renewable energy resources in the global energy mix and a doubling of the global rate of improvement in energy efficiency.[2] Validation of science and technology is the true necessity of human scientific endeavor today. Sustainability whether it is social or economic is the new avenue of research pursuit. Global demand for primary energy is expected to increase by 27 and 61% by 2050.[2] Scientific vision, the challenges of engineering science, and the visionary world of sustainable development will lead a long and visionary way in the true emancipation of science and technology.On our planet, 1.2 billion people still do not have access to electricity and 2.8 billion lack access to clean cooking facilities.[2] In such a crucial situation, the report presents a comprehensive view of energy sustainability and its opportunities and challenges before the readers. According to this report, energy security implies the effective management of primary energy supply from domestic and external sources, the reliability of energy infrastructure and the ability of energy facilitators to meet current and future demand. This report effectively elucidates the tremendous success of energy technologies in meeting sustainability needs with the sole aim in enriching the future of sustainability science.[2]

McKinsey and Company Report[3] deeply investigated environmental and energy sustainability with a visionary approach toward a developing country like India. In this treatise, the authors deeply tread through the success of application of energy and environmental sustainability in the developing country scenario. The world of sustainability is witnessing a renewed resurgence and a deep scientific vision. India is already one of the largest economies in the world poised for greater urbanization and a greater realization of environmental and energy sustainability. Rapid urbanization and rapid economic growth are the pillars of success in India today.[3] In 2008–2009, McKinsey and Company undertook a study deeply to identify and prioritize opportunities for India to meet the closely linked challenges of energy security and environmental sustainability that progresses with economic growth.[3] Technological vision, scientific contemplation, and deep scientific innovations are the hallmarks of present day in the developing country like India. This report is a result of that effort and ongoing scientific challenges. Mankind today stands in the juxtaposition of deep scientific vision and technological introspection. Its core purpose is to illustrate which measures have the greatest potential to reduce emissions and correspondingly energy use, and which are the most feasible given the substantial challenges in funding, regulation, technology, capacity, and market imperfections.[3] For India as a developing country, challenges are immense and ever-growing. The success of science and its implementation are the cornerstones of scientific research pursuit today. Technology has few answers to the surmounting difficulties and immense hindrances of sustainability implementation in India. Poverty, lack of economic growth, and lack of provision of basic human needs still stands as vexing issues 70 years after independence in India. Economically feasible solutions to the challenges of energy security and environmental sustainability continue to trouble the economic landscape of India.[3] The methodology is built upon McKinsey's research into greenhouse gas (GHG) abatement over the past 3 years in 18 countries.[3] The challenges of Indian economy are vast and varied. This report deeply highlights that there are considerable benefits from reducing emissions—they include improving energy security, promoting inclusive growth, improving quality of life, and the vast leadership emancipation in sustainable development.[3]

TWAS Report[4] deeply comprehended on the topic of sustainable energy for developing countries. The authors of this report elucidates upon the historic energy trends, rising consumption, and the transition to commercial forms of energy, increasing power and improving efficiency, decarbonization and diversification, sustainable energy challenges for developing countries, sustainable energy technologies, and the world of policies and actions.[4] Developing and emerging economies face a two-fold energy challenge in the 21st century. Meeting the needs of billions of people who still lack access to basic; modern energy services while simultaneously participating in a global transition to clean, low energy systems.[4] Technology is highly challenged today in the developing country scenario. Science and engineering have few answers toward the implementations and innovations of human civilization.[4] Vexing issues are immense and vital. Historically, humanity's use of energy has been marked by four difficult trends: (1) rising consumption and a transition from traditional sources of energy to commercial forms of energy; (2) steady improvement in the power and efficiency of energy technologies; (3) fuel diversification and decarbonization; and (4) improved pollution control measures.[4] These highly diversified issues are troubling the deep scientific scenario of developing world. Technology has few answers to the stunted economic growth of a developed nation. The authors deeply comprehended the challenges of science and sustainability in developing nation.[4]

United States Department of Energy Report[5] elucidated energy and environmental profile of the US petroleum refining industry in details. The authors gave a detailed overview of petroleum refining industry.[5] As a highly developed nation, the United States today stands in the midst of deep scientific vision and contemplation. The authors of this report touch upon the deep technical details of petroleum refining such as atmospheric and vacuum distillation, cracking and coking processes, catalytic reforming, alkylation, hydrotreatment, and the world of supporting processes such as sulfur treatment, chemical treatment, water treatment, and process heating.[5] Petroleum is the single largest source of energy for the United States of America. When measured in British Thermal Units, the nation relies on petroleum two times more than coal or natural gas and four times more than nuclear power, hydroelectricity, and other renewable energy sources. Technology and science are its pinnacles in scientific pursuit for highly advanced nation like the United States of America.[5]

Technology and science of petroleum engineering are in the path of newer rejuvenation and deep scientific regeneration. In this treatise, the author rigorously focuses on the advances made in the field of sustainability with the sole and vital aim of furtherance of science and engineering. Depletion of fossil fuel resources is a major concern toward scientific endeavor today. Scientific vision, forbearance, and profundity are the hallmarks toward a newer visionary era in petroleum refining.

United States Department of Energy Report[6] elucidated on the topic of energy bandwidth for petroleum refining processes. The Industrial Technologies Program (ITP)[6] is a research and development program within the U.S. Department of Energy, Office of Energy Efficiency, and Renewable Energy. The Industrial Technologies Program conducted a bandwidth study to analyze the most energy-intensive unit operations used in the U.S. refineries.[6] Technology and engineering science are highly advanced today and are encompassed by academic and scientific rigor. The authors of this report highlighted the following points:(1)petroleum refining process such as crude oil distillation, (2)fluid catalytic cracking, (3)catalytic hydrotreating, (4)catalytic reforming, and (5) alkylation.[6] The main hallmark of this treatise is the scientific exposition of energy bandwidth for five principal petroleum refining processes. Technological advancements, the vision for the future, and the vast scientific rigor in the field of petroleum engineering will veritably lead a long and visionary way in the true realization and true enrichment of scientific research pursuit in fossil fuel science today.[6]

The United States Environmental Protection Agency Report[7] deeply comprehended energy trends in selected manufacturing sectors and the vast opportunities and challenges for environmentally preferable energy outcomes. The petroleum refining industry includes establishments engaged in refining crude petroleum into refined petroleum products through multiple distinct processes including distillation, hydrotreating, alkylation, and reforming.[7] Academic and scientific rigor are gaining immense heights as science and engineering are ushering in a new era in petroleum refining.[7] Today depletion of fossil fuel resources is a practical blunder of human science and the progress of mankind. This report gives a veritable glimpse to the scientific vision and comprehension in the future of petroleum refining industry.[7]

The Energy Research Institute Report[8] deeply discussed with vast insight green growth strategies for oil and natural gas sector in India. The report gives an overview of India's oil and gas sector. The areas covered in this report are regulatory institutions and companies, taxes and subsidies, domestic oil and gas production, midstream and oil and gas pipelines, and gas imports infrastructure.[8] The report also highlights avenues of green growth of in the oil and gas sector. Growth strategies, especially green engineering and green technology are of immense importance in the scientific success of human civilization and the deep scientific vision of a nation's policies. The Indian scenario is exceedingly intricate. Technology has few answers to the development of a developing country like India. Resource constraints, price fluctuation, and the climate crisis have forced the developing nation like India a serious investigation of the existing resources and carbon-intensive growth strategies. Technology of petroleum refining needs to be reenvisioned and re-envisaged with the passage of scientific vision, profundity, and timeframe. Asia has been the driver of growth in the past decade, witnessing a compound annual growth rate of 7.5% from 1990 to 2012. From 1990 to 2012, Indian has veritably witnessed a compound annual growth rate (CAGR) of 6.5%, and its growth is expected to continue at more or less the same pace upto 2040.[8] There is a strong correlation between human development and per capita energy consumption in the case of developing nation like India. India, when compared to Brazil, Russia, China, and South Africa, ranks the lowest in Human Development Index (HDI).[8] Similarly, India's scores in the Energy Development Index (EDI) developed by the International Energy Agency (IEA) are also low. India ranks 41st among the 80 countries ranked by IEA in 2012.[8] This is the energy situation for a globally developing and challenged country like India. Mankind today is in a state of immense distress as regards energy utilization. The challenge of science and technology for India's case is ushering in a new eon in the field of economic progress. Petroleum refining and energy solutions are the success of economic progress of India. This report widely elucidates the example of India in the search for scientific truth and vision in green engineering and sustainability.[8]

Brandt[9] discussed with deep and cogent insight that oil depletion and the efficiency of oil production for the case of California, USA. This study widely explores the impact of oil depletion on the energetic efficiency of oil extraction and refining in California. Technology revamping and deep scientific vision are of immense importance in the true emancipation of petroleum refining these days. This study quantifies the energy efficiency impacts of oil depletion in California using methods of life cycle analysis (LCA) and net energy analysis (NEA). Scientific validation, wide and broad scientific vision, and scientific candor are the pillars of regeneration in petroleum industry today.[9]

The situation is not that grave in petroleum refining industry today. The challenge of science needs to be reenvisioned with the passage of scientific cognizance and vision. Hydrocarbons have played one of the critical roles in economic history by fueling globalization and industrialization. In spite of the global thrust in renewable energy, IEA's World Economic Outlook for 2040 projects that oil and gas will stand the single largest energy source throughout the projection period. Oil and gas resources are the nerves of a nation whether it is developed or developing. Developed as well as developing countries' oil and gas strategies are ushering in a new era of scientific regeneration and rejuvenation.

1.9   RECENT RESEARCH PURSUIT IN THE FIELD OF PETROLEUM REFINING

Petroleum refining advancements are the cornerstones of human research endeavor today. International Council on Clean Transportation report[10] deeply elucidated on the intricacies of petroleum refining. The authors deeply comprehended on petroleum refining at a glance. Science and technological cognizance of petroleum industry are moving at a drastic pace. This tutorial addresses the basic principles of petroleum refining, as they relate to the production of ultra-low-sulfur-fuels (ULSF), in particular gasoline (ULSG) and diesel fuel (ULSD).[10] The sole mission of the study targets (1) an organizing framework for the overall analysis,[10] (2) identify the technical factors that determine the refining cost of ULSG and ULSD production,[10] and (3) facilitate the interpretation of the results of this analysis. The success of petroleum refining industry today lies in the hands of scientists, geologists, chemical engineers, and petroleum engineers. This report gives a visionary approach toward the success of science and engineering in the wider domain of petroleum industry.[10]

Office of Industrial Technologies, the U.S. Department of Energy Report (2001)[11] vastly comprehended recent trends in research in petroleum refining. In February 2000, petroleum industry leaders signed a compact with the U.S. Department of Energy's Office of Industrial Technologies to work together though the Industries of the Future initiative. High priority research needs are:

- Energy and process efficiency
- Environmental performance

- Materials and inspection technology
- Distribution system and retail delivery services

The demonstrated success avenues are:[11]

- Waste heat process chiller
- Fouling minimization
- Robotics inspection system
- Radiation stabilized burner
- Low-profile fluid catalytic converter (FCC)
- Computational fluid dynamics Model of FCC
- Advanced process analysis for refining

The vast challenge and the vision of petroleum engineering research trends are extremely scientific, groundbreaking, and surpassing wide frontiers. Process integration, design, and modeling stand as major components of the success of petroleum refining today. Recent advances in petroleum refining are witnessing drastic challenges. The success of innovation in petroleum industry needs to be widely enriched with the passage of scientific history and timeframe.[11]

1.10   RECENT TECHNOLOGICAL VISION IN THE FIELD OF PETROCHEMICALS

Petrochemicals industry stands as a major industry in the furtherance of science and engineering in the paradigm of energy sustainability today. Ray Chaudhuri[1] deeply comprehended on the success and scientific potential of petrochemicals and petroleum engineering today. The author widely researched on the field of petrochemicals with the sole objective of furtherance of science and engineering. The author started with deep scientific vision on the various components of crude oil, petroleum products, and test methods and finally processing operations in a petroleum refinery. Petrochemicals stand as a major component and toward the scientific success of petroleum engineering science and petroleum refining as a whole. Mankind today stands in the midst of a difficult and ever-growing crisis of depletion of fossil fuel resources. Thus, comes the utmost need of energy sustainability. Energy, food, and water are the important components in a nation's economic growth today. Technological vision, the intricacies of petroleum technology, and future challenges of science will all lead a long and visionary way in the true emancipation of petrochemical technology today. Energy sustainability and petroleum refining/petroleum technology are the two opposite sides of the visionary coin. Progress of human civilization depends on the scientific success and scientific profundity of human research pursuit. Innovations in petroleum refining and petrochemicals, and the deep scientific vision behind it are the hallmarks of human scientific endeavor today. Petroleum is a fossilized mass that is accumulated below the earth's surface from time immemorial. Raw petroleum is known as crude (petroleum) oil or mineral oil. It is a mixture of various organic substances and is the source of hydrocarbons such as methane, ethane, propane, butane, pentane and various other paraffinic, napthenic, and aromatic hydrocarbons, the building blocks of today's organic industry. Various petroleum products, such as gaseous and liquid fuels, lubricating oil, solvents, asphalts, waxes, and coke are derived from refining crude oil. Many lighter hydrocarbons and other organic chemicals are synthesized by thermal and catalytic treatments of these hydrocarbons. Technology and engineering science are moving at a rapid pace in today's scientific regeneration. Technological and scientific validations are the pillars of success and scientific truth today. The challenge of human civilization needs to be reenvisioned and re-envisaged. The hydrocarbon processing industry is basically divided into three distinct activities: petroleum production, refining, and petrochemical manufacturing. Refineries produce cooking gas (liquefied petroleum gas, LPG), motor spirit (also known as petrol or gasoline), naptha, kerosene, aviation turbine fuel (ATF), high speed diesel (HSD), lubricating base oils, wax, coke, bitumen, and so forth, which are mostly a mixture of various hydrocarbons.[12,13,14] In a petrochemical plant, or in a petrochemical complex, pure hydrocarbons or other organic chemicals with a definite number and type of constituent element or compound are produced from the products in refineries. Technological advancements in petroleum refining are in a state of immense scientific contemplation. Today, technology is veritably linked with validation of science. Petrochemical engineering and petroleum technology are in the path of immense scientific regeneration and vast rejuvenation. Products from a petrochemical complex are plastics, rubbers, synthetic fibers, raw materials for soap and detergents, alcohols, paints, pharmaceuticals, and so forth. This treatise is a vastly researched and well observed piece of scientific work. Since petroleum is a mixture of hundreds and thousands of hydrocarbon compounds, there is a high possibility of the formation of new compounds. Technological dimensions are today geared up for newer challenges. The success of petroleum chemistry and refining are opening up new avenues of research in decades to come.

Sustainability whether it is environment or energy are ushering in a new era in the domain of furtherance of science and engineering. Sustainability encompasses the factors of energy, environment, empathy, and equity in the economic progress of a developed as well as a developing nation. The challenge of science is slowly evolving in this century. Research pursuit in the field of petrochemicals is the challenge of energy science today. Success of science and the vision of engineering science are the hallmarks of research pursuit today. The author in this treatise opens up a new chapter in the field of petroleum engineering science.

1.11   SUSTAINABLE DEVELOPMENT AND PETROLEUM REFINING IN TODAY'S WORLD

Sustainable development and the vast challenges behind it are slowly changing the scientific scenario today. Technological vision of water science and technology is immense and is today targeted toward drinking water treatment, industrial wastewater treatment, and provision of pure drinking water. The challenge and vision of science and engineering need to be reenvisioned with the progress of scientific vision, scientific rigor, and the world of timeframe. Energy and environmental sustainability are the scientific targets of human civilization today. Mankind is plunged into the deep crisis as environmental disasters and frequent catastrophes are challenging the wide scientific landscape these days. Environmental regulations and stringent environmental restrictions have changed the scientific fabric of human scientific endeavor today. The success, vision, and challenge need to be reenvisioned and re-enshrined with the passage of scientific candor and scientific contemplation. Global water research and development initiatives are linked to the visionary domain of environmental sustainability by an unsevered umbilical cord. Arsenic and heavy metal groundwater contamination are challenging the entire gamut of scientific vision and scientific research pursuit today. Mankind's immense scientific prowess, the vast scientific potential behind groundwater remediation, and the challenge of water research and development initiatives are all leading a long and visionary way in the true emancipation of water research and environmental sustainability today.

Energy sustainability and petroleum refining are the two opposite sides of the visionary coin today. Renewable energy is the need of the hour. Scientific splendor, the technological profundity, and the wide vision of energy sustainability are the forerunners toward a newer visionary era in the field of energy emancipation. Technology has few answers toward the immense intricate issues of energy technology today. Nonrenewable technology today needs to be phased out as science, engineering as mankind moves forward. Nuclear science, wind energy, biomass utilization, and solar energy are the utmost need of the hour. The challenge of renewable energy technology, thus, needs to be reenvisioned with the progress of science and technology.

1.12 OPPORTUNITIES, CHALLENGES, AND THE SCIENTIFIC SUCCESS BEHIND PETROLEUM REFINING

Opportunities and challenges in the field of petroleum refining are crossing vast visionary boundaries today. Technology has few answers toward the remarkable and the extremely promising area of renewable energy today. Technology and engineering science of petroleum refining are witnessing immense scientific rejuvenation nowadays. Energy science and engineering are the scientific vision of today. The immense challenges of depletion of fossil fuel resources have veritably challenged the scientific domain of chemical process engineering and petroleum engineering science. Petrochemicals and the vast area of renewable energy technology are the futuristic vision of today. Wind energy, biomass energy, wave energy, and solar energy are the challenges of today. Mankind is in a state of immense scientific conundrum and also in a state of deep division. Technology has been faltering because of the intricate issues of scientific profundity of renewable energy technology. The challenge lies in the hands of petroleum engineers, chemical engineers, and geologists on the important issue of depletion of fossil fuel energy. Our planet is today in dire straits as technology falters and stands baffled. Successive human generations and scientific rejuvenation needs to be streamlined as regards sustainable development and effective implementation of renewable technology.[12,13,14]

1.13   FUTURE RESEARCH TRENDS AND FUTURE FRONTIERS

The world of chemical process engineering and petroleum engineering science are ushering in a new era of scientific vision and rejuvenation. Technological validation, the challenges of science, and the immense needs of human society are all today leading a long way in the true redefinition of sustainability science. Technology is moving at a rapid pace surpassing visionary barriers today. Future research trends in sustainability need to target the provision of basic human needs such as energy, food, water, and electricity. Scientific vision today is in the era of deep revival. Renewable energy is reviving the causes and vision of energy science and engineering. Environmental sustainability is redefining the vast scope of water science, technology, and industrial wastewater treatment. The world of desalination science is rejuvenating the very visionary world of science and technology. Petroleum refining is also redefining the entire energy scenario. Future research trends are targeted toward newer refining technologies and newer innovations. Energy and environment are the two vast pillars of technological innovation and scientific validation. Research focus and ideals should be linked toward greater innovation in energy and environmental sustainability.

1.14   CONCLUSION, SUMMARY, AND FUTURE PERSPECTIVES

Petroleum refining and petroleum engineering science are in a state of immense scientific conundrum and at the same time scientific vision today. Technology revamping and scientific evolution are the key needs of civilization's progress. Future perspectives of science are veritably linked to energy and environmental sustainability, the greater realization of challenges and vision of engineering science. In the similar manner, petroleum industry and the evolution of energy engineering are ushering in a new eon in the field of furtherance of science in our human civilization today. The achievement of success in scientific innovation in fossil fuel science is slow and steady. Alternate sources of energy and renewable energy are the future thoughts and future research trends of tomorrow. Future of human civilization is in a state of immense distress and unending catastrophe due to the depletion of fossil fuel resources. Technology revamping and scientific regeneration are the only solutions of the future of human civilization. In this treatise the author pointedly focuses on the deep scientific vision, the immense scientific potential and the scientific enigma in the implementation of scientific innovations, and the successful targets toward sustainability. Human race needs to be veritably revamped and regenerated as science and engineering progresses forward. Renewable energy paradigm such as wind, solar, and biomass are the effective and feasible solutions toward a successful future. This paper widely researches on the success of sustainability and the true realization of science today. The path toward energy and environmental sustainability are vast and varied. This treatise highlights and investigates with immense foresight the futuristic vision of sustainability science particularly sustainability of energy today. The answers are few yet the challenge and the vision are many. This chapter unfolds the intricacies of petroleum refining and opens new windows of scientific innovation in decades to come.

1.15   ACKNOWLEDGMENT

The authors with deep respect acknowledges the contribution of his late father Shri Subimal Palit, an eminent textile engineer from Kolkata, India who taught the author the rudiments of chemical engineering.

KEYWORDS

- energy
- petroleum
- vision
- sustainability
- petrochemicals

REFERENCES

1.   Chaudhuri, U. R. *Fundamentals of Petroleum and Petrochemical Engineering;* CRC Press/Taylor and Francis Group: United States of America, 2011.
2.   World Energy Trilemma. Energy Sustainability Index Report, World Energy Council and Oliver Wyman, 2013.

3. McKinsey and Company. Environmental and Energy Sustainability: An Approach for India, August, 2009.
4. TWAS The Academy of Sciences for the Developing World Report. Sustainable Energy for Developing Countries, 2008.
5. United States Department of Energy Report. Energy and Environmental profile of the U.S.A. Petroleum Refining industry, November 2007.
6. U.S. Department of Energy Report. Office of Energy Efficiency and Renewable Energy Industrial Technologies Program, 2006.
7. U.S. Environmental Protection Agency Report. Energy Trends in Selected Manufacturing Sectors: Opportunities and Challenges for environmentally preferable energy outcomes, March 2007.
8. The Energy Research Institute Report. Green Growth Strategies for Oil and Natural Gas Sector in India, Supported by Global Green Growth Institute, 2015.
9. Brandt, A. R. Oil Depletion and the Energy Efficiency of Oil Production: The Case of California. *Sustainability* 2011, 3(10), 1833–1854.
10. International Council on Clean Transportation. An Introduction to Petroleum Refining and the Production of Ultra Low Sulphur Gasoline and Diesel Fuel, October 24, 2011.
11. Office of Energy Efficiency and Renewable Energy. U.S. Department of Energy Report, Office of Industrial Technologies, 1998.
12. Occelli, M. *Advances in Fluid Catalytic Cracking: Testing, Characterization and Environmental Regulations;* CRC Press/Taylor and Francis Group, 2010.
13. www.google.com
14. www.wikipedia.com

MODELING, SIMULATION, OPTIMIZATION, AND CONTROL OF A FLUID CATALYTIC CRACKING UNIT IN A PETROLEUM REFINERY: A FAR-REACHING AND COMPREHENSIVE REVIEW

SUKANCHAN PALIT*

Department of Chemical Engineering, University of Petroleum and Energy Studies, Bidholi via Premnagar, Dehradun 248007, India

43, Judges Bagan, Haridevpur, Kolkata 700082, India

*Corresponding author. E-mail: sukanchan68@gmail.com, sukanchan92@gmail.com

CONTENTS

ABSTRACT

Scientific vision and scientific research pursuit in today's human civilization are witnessing drastic changes. Human civilization and scientific forbearance needs to be reenvisioned and re-envisaged with the passage of scientific history, scientific vision, and time. Energy sustainability and petroleum engineering science need to be revamped with the growing concern for depletion of fossil fuel resources. Fluid catalytic cracking (FCC) unit is a vital and prominent procedure in petroleum refining. Technological vision, scientific advancement, and the vast validation of science are the forerunners of greater emancipation of petroleum engineering science today. The author in this treatise pointedly focuses on the immense scientific potential, the deep scientific vision, and the scientific profundity behind the modeling, simulation, optimization, and control of the fluid catalytic cracking unit. The author trudges with immense thoughtfulness and clarity, the vast domain of mathematical modeling, cracking kinetics, riser models, stripper/disengage models, regenerator models, and the scientific vision of unsteady state and dynamic modeling of FCC unit. This is a comprehensive review of deep scientific understanding and cognizance in the field of optimization and control of the FCC unit.

2.1 INTRODUCTION

Science and engineering are today moving from one scientific paradigm toward another. Today, the technology of petroleum refining is ushering in a new era of scientific regeneration and scientific rejuvenation. Fluid catalytic cracking is one of the major areas of scientific research pursuit in petroleum refining today. Mankind's immense scientific prowess and determination are witnessing extensive scientific revamping as science and engineering moves toward a newer eon of scientific endeavor. Futuristic vision, scientific girth, and scientific profundity will all lead a long and visionary way to the true realization and true emancipation of petroleum engineering science today. In this treatise, the author rigorously points out toward the immense scientific research pursuit in the field of modeling, simulation, optimization, and control of fluid catalytic cracking (FCC) unit with the sole vision of furtherance of science and engineering. Optimization and control of FCC unit is an uncovered area in the furtherance of science and engineering today. Petroleum engineering science is witnessing immense scientific revamping today. In this well-researched paper, the author delves deep into the ever-growing crisis of depletion of fossil fuels and the vast solution areas in terms of the design of petroleum refining units such as FCC unit. Design of a FCC unit stands as a major component in the modeling, simulation, optimization, and control of a petroleum refining unit. Technological vision, scientific clarity, and thoughtfulness are the pallbearers toward a greater vision and a greater emancipation in the design, modeling, and simulation of a petroleum refining unit. Refiners are being forced to make significant investments because of growing demand for refinery products combined with the decreasing quality of crude oils and tighter product specifications due to environmental constraints, and regulations. In this vital context, the use of advanced process engineering tools and methodologies has become immensely essential for refiners, not only for design but also in the tasks of process control, optimization, scheduling, and planning. Today optimization and control are the two areas that need to be restructured and reenvisioned as the science of modeling and simulation moves from one visionary paradigm toward another. In this treatise, the grave concerns of fossil fuel depletion are taken into account and science of modeling and simulation are reenvisioned and deeply envisaged with the passage of scientific history and time.[19,20]

2.2   THE AIM AND OBJECTIVE OF THIS STUDY

Petroleum engineering science is moving from one visionary paradigm toward another. Petroleum refining and the design of their units are today surpassing visionary scientific frontiers. Today, scientific vision, deep scientific introspection, and wide technological validation are the forerunners of the greater scientific emancipation of petroleum refining.[19,20] Depletion of fossil fuel resources is an area of immense concern to the scientific world. The aim and objective of this study truly surpass broad scientific imagination and open new windows of scientific innovation in petroleum engineering in the decades to come. The world of modeling, simulation, optimization, and control of petroleum refining units is an area of immense scientific introspection and scientific comprehension. Thus, there is a vital and utmost need of application of multi-objective optimization and genetic algorithm. Technology is highly challenged today with the passage of scientific history, scientific vision, and time.[19,20] This treatise not only gives a glimpse of the modeling techniques of all the parts of a FCC unit but also goes deeper into the unknown world of optimization and control of petroleum refining units.[19,20] The author of this treatise rigorously points out toward the vast scientific potential and the deep scientific understanding in the application of genetic algorithm and multi-objective optimization in the design of FCC unit.[19,20] These mathematical tools will open a newer world in the field of applied mathematics, chemical process engineering, and petroleum engineering science.[19,20]

2.3   THE NEED AND THE RATIONALE OF THIS STUDY

Today, petroleum engineering stands in the midst of deep scientific vision and forbearance. Technology needs to be revamped with respect to the future of petroleum refining and the avenues of scientific challenges.[19,20] The need and the rationale behind this study is immense and needs to be pondered with vision and challenge. Fossil fuel is slowly on the wane and is dwindling. At such a crucial juncture of science and engineering, new innovations and visionary challenges in the design of petroleum refining unit are of utmost importance. Mankind's immense scientific girth and determination, the success of science, and the academic rigor will all lead a long and visionary way in the true realization of petroleum engineering science today.[19,20] The rationale behind this study goes beyond scientific judgment and scientific sagacity. The technology of petroleum engineering science is vast and groundbreaking with the visionary passage of scientific history.[19,20] The needs of petroleum engineering are the immense rationalization toward scientific design, application of mathematical tools, and the academic rigor behind optimization and control. The scientific clarity, the scientific lucidity, and the immense scientific optimism are the pallbearers toward a greater visionary future in modeling and simulation.[19,20,16,5]

2.4   LITERATURE REVIEW

Scientific vision and scientific profundity are the pillars of today's scientific endeavor. Immense scientific and academic rigor are the challenges of petroleum engineering science today. Scientific literature is abundant and far-reaching in the field of petroleum engineering science. Design of a FCC unit is an important component in the furtherance of science and technology. Similarly, chemical process engineering and applied mathematics are witnessing drastic challenges in the path toward scientific rejuvenation. Pinheiro et al.[15] deeply discussed with lucid insight, process modeling, simulation, and the control of FCC.[15] Scientific vision, scientific fortitude, and deep scientific discerning are the pillars of research pursuit in modeling and simulation of chemical engineering and petroleum engineering systems today.[15] This is a phenomenal piece of review work.[15] This treatise focuses on the FCC process and widely reviews recent developments in its modeling, monitoring, control, and optimization.[15] This vastly challenging endeavor exhibits complex behavior, requiring detailed models to express the nonlinear effects and wide interactions between input and control variables that are vastly observed in industrial practice.[15] The currently available FCC models differ enormously in terms of their scope, level of detail, modeling hypothesis, and solution approaches used.[15] The scientific outcome, deep scientific knowledge, and discernment are the truths of research pursuit today.[15] To widely improve the existing modeling approaches, this comprehensive review describes and compares the different mathematical targets that have been vastly applied in the modeling, simulation, control, and optimization of this important downstream unit.[15] Han et al.[10] described with cogent insight dynamic modeling and simulation of a FCC process with much emphasis on process modeling.[10] The vast technological vision, the wide foresight, and scientific fortitude will veritably lead a long and effective way in the true realization of modeling and simulation of FCC unit.[10] The purpose of this study is to develop a dynamic model of a typical FCC unit that consists of the reactor, regenerator, and the catalyst transfer lines. A distributed parameter model is presented for the reactor riser to predict the distributions of the catalyst and gas-phase velocities, the molar concentrations of four-lump species, and the temperature.[10] Jimenez-Garcia et al.[12] with cogent insight described the buildup of the FCC unit and the deep scientific vision of the catalytic cracking environment.[12] This work is devoted to briefly review some propositions about the use of FCC converters and how these aspects demand knowledge of riser reactors, change in the methodology to estimate effective reaction rates, and catalyst activity, as well as process control actions and its effect on the energy balance of the converter and its operational difficulties.[12,16,5]

2.5   ENERGY SUSTAINABILITY AND THE VISION OF SCIENCE

Energy sustainability and holistic sustainable development are the pallbearers toward a greater visionary era in the field of petroleum engineering and energy engineering. Technological profundity and scientific forbearance are the pillars of scientific research pursuit today. Science today is a huge colossus with a definite vision and willpower of its own. This century is a revolutionary century with regards to advancements in nuclear engineering and space technology.[19,20] Sustainability science needs to be pronounced and reenvisioned with the passage of human history and human scientific endeavor. The vision and the challenge of science need to be enriched and restructured as human civilization trudges into a visionary century of engineering science and petroleum engineering. Technological innovation and vast scientific motives are of utmost need in the true realization of energy and environmental sustainability today.[19,20] Human civilization is today gearing toward immense challenges this century. Electricity, water, and infrastructure are the pillars of progress.[19,20] The question of the provision of pure drinking water encompasses the very scientific truth of environmental sustainability. Scientific divinity, the greatness of scientific profundity, and the success of scientific progress will all lead a long and visionary way in the true realization of energy and environmental sustainability today. The vision of the future for a deep scientific understanding of environmental and energy sustainability lies in the hands of future scientists and engineers.[19,20] Mankind today stands in the midst of deep crisis and introspection. In the similar vein, petroleum engineering science is in the midst of intricate scientific arguments and scientific comprehension.[19,20] This century will surely and veritably open up new windows of scientific innovation and scientific instincts in decades to come.[19,20,16,5]

2.6 PETROLEUM REFINING AND NEXT-GENERATION TECHNOLOGIES

The status of energy sustainability in today's human civilization is vast and versatile. The visionary ideas and the intricate words of Dr. Gro Harlem Brundtland, former Prime Minister of Norway need to be redefined and reemphasized.[19,20] Petroleum refining in the similar manner is the path of immense scientific regeneration. Petroleum refining is a unique and critical link in the petroleum supply chain, from the wellhead to the pump. The other links add value to petroleum by moving and storing it (e. g., lifting crude oil to the surface, moving crude oil from oil fields to storage facilities and then to refineries, moving refined products from the refinery to terminals, and end-use locations, etc.). Refining adds value by converting crude oil (which in itself has little end-use value) into a range of refined products, including transportation fuels. Technology and knowledge dimensions are of vital need as science and engineering of petroleum refining move forward. The scientific truth, scientific forbearance, and deep scientific discernment are the pillars of research pursuit today. The technology of petroleum refining needs to be reframed as human civilization and human scientific research pursuit gains new and visionary heights today. Petroleum refineries are large, capital-intensive manufacturing facilities with extremely complicated processing schemes.[19,20] They convert crude oils and other input streams into dozens of refined co-products which include the followings:

Liquefied petroleum gases	Petrochemical feedstocks
Gasoline	Lubricating oils and waxes
Jet fuel	Home heating oils
Kerosene (for lighting and heating)	Fuel oil (for power generation, marine fuel, industrial and district heating)
Diesel fuel	Asphalt (for paving and roofing uses)

Of these, the transportation fuels have the highest value while fuel oils and asphalt have the lowest value. Many refined products, such as gasoline, are produced in multiple grades to meet different specifications and standards. Scientific vision in petroleum refining is wide and far-reaching. Refining technology stands today in the midst of deep introspection and reenvisioning.

Oil refining is a key activity in chemical process industries. The goal of oil refining is two-fold: (1) production of fuels for transportation, power generation, and heating; (2) production of raw materials for chemical process industries.[19,20]

More than 660 refineries in 116 countries are currently in operation, producing more than 85 million barrels of refined products per day.[19,20] Each refinery has a unique physical configuration, as well as unique operating characteristics and economics. Scientific vision, profundity, and truth will all lead a long and visionary way in the true realization of petroleum engineering science today.[19,20] A refinery's configuration and performance characteristics are redefined by the refinery's location, duration, availability of funds for capital investment, available crude oils, product demand, product quality requirements, environmental restrictions, market specifications, and requirements for refined products.[19,20] Validation of science of petroleum engineering is gaining new heights as human civilization moves forward toward a newer scientific era.[19,20]

2.7   THE CHEMICAL COMPOSITION OF CRUDE OIL

Hundreds of different crude oils (usually identified by geographic origin) are vastly processed, in greater or lesser volumes in the world's refineries. Each crude oil is unique and is a complex mixture of thousands of compounds. Most of the compounds in crude oils are hydrocarbons (organic compounds composed of carbon and hydrogen atoms).[19,20] Other compounds in crude oil not only contain carbon and hydrogen but also small (but important) amounts of other ("hetero") elements—most notably sulfur, as well as nitrogen and certain metals (e.g. nickel, vanadium, etc.).[19,20] The compounds that make up crude oil range from the smallest and simplest hydrocarbon molecule—CH_4 (methane)—to large, complex molecules containing up to 50 or more carbon atoms (as well as hydrogen and hetero-elements).[19,20] Technological vision and deep scientific understanding are of utmost importance in the avenues of scientific and academic rigor in the field of petroleum engineering.[19,20]

Paraffins, aromatics, and naphthenes are natural constituents of crude oil, and are produced in various refining operations as well. Olefins usually are not present in crude oil; they are produced in certain refining operations that are dedicated mainly to gasoline production. The heavier (denser) the crude oil, the higher its C/H ratio.[19,20] Due to the chemistry of oil refining, the higher the C/H ratio of a crude oil, the more intense and costly the refinery processing required to produce given volumes of gasoline and distillate fuels.[19,20] Thus, the chemical composition of a crude oil and its various boiling range fractions influence refinery investment requirements. The technology of petroleum refining is highly advanced today and surpassing wide and vast visionary boundaries. In this treatise, the author rigorously points out toward the scientific success and the deep scientific vision in the true emancipation of petroleum engineering science.[19,20]

2.8   VISION OF ATOMIC STRUCTURE, SPECTROSCOPY, AND PETROLEUM ENGINEERING SCIENCE

Atomic structure and spectroscopy stands as a major research avenue in the field of petroleum engineering and applied chemistry.[11] Scientific vision, scientific profundity, and deep scientific discerning are of utmost need in the path of science and engineering today. Depletion of fossil fuel sources is a vexing issue of immense concern in the path toward scientific cognizance in petroleum engineering. The challenge of spectroscopy science needs to be reenvisioned with the passage of scientific history, scientific vision, and the timeframe of human civilization. Technology, science, and engineering are witnessing drastic challenges with the furtherance of scientific research pursuit.[11] Research, teaching, learning, and practice of spectroscopy are ushering in a new era in the field of engineering science today. Spectroscopy is defined as the science that deals with interactions between electromagnetic radiation and matter.[11] This research avenue has recently experienced a promising and explosive growth as a result of varied and visionary innovations in methodologies and instrumentation, which offer the possibilities for new applications and novel methods of analysis to solve intricate analytical problems and difficult scientific questions.[11] Today, technology has few answers to the immense crisis of fossil fuel sources. Petroleum technology needs to be once more redefined with the progress of scientific innovation. Spectroscopic techniques are the challenge of the day.[11] Research scientists and analytical scientists are deeply confronted with the increasing complexity of real-life sample analysis.[11] Thus, the need of an in-depth scientific analysis. In order to obtain high-quality analytical data, the primary objective of the analytical scientist must be ideally to obtain an artifact-free sample of the analysis. This is an ever-growing and challenging area of scientific research and needs to be restructured and redefined with the progress of scientific and academic rigor.[11] It is often the case that many sampling programs frequently select the sampling methods, based on what equipment is available rather than what question is to be answered or what problem is to be addressed. Technological vision, scientific thoughts, and vast scientific rigor are the pallbearers toward a greater challenge of spectroscopy science today.[11] The vision of petroleum engineering in such a situation is wide and bright today. Sampling methodology differs greatly upon whether the sample is in gaseous, liquid, or solid phase.[11] If the sample is in the liquid or solid phase, is the sample an aerosol or particle that remains in a particular gaseous phase? In or on what medium shall the sample be collected and retained?.[11] These questions need to be veritably addressed as the science of spectroscopy progresses.[11] Technological challenges and scientific hindrances are of utmost importance as academic rigor reaches immense heights.[11]

An early example of colorimetric analysis is Nessler's method for ammonia, which was introduced in 1856.[11] Nessler found that adding an alkaline solution of HgI_2 and KI to a dilute solution of ammonia produces a yellow to reddish brown colloid, with the colloid's color depending on the concentration of ammonia.[11] By visually comparing the color of a sample to the colors of series of standards, Nessler was able to determine the concentration of ammonia. Colorimetry, in which a sample absorbs visible light, is one big example of a spectroscopic method of analysis.[11] At the end of the 19th century, spectroscopy was limited to the absorption, emission, and scattering of visible, ultraviolet, and infrared electromagnetic radiation.[11] Since its introduction, spectroscopy has widened its research rigor to include other forms of electromagnetic radiation—such as x-rays, microwaves, and radio waves—and other energetic particles—such as electrons and ions.[11] Technology and engineering science of spectroscopy and their applications are being redefined and reenvisioned today with the progress of research pursuit.[11] The overview on the subject of spectroscopy involves the interaction of ultraviolet, visible, and infrared radiation with matter.[11] The scientific potential, the scientific success, and the vast and versatile vision of spectroscopy science will all lead a long and visionary way in the true realization of petroleum refining technology today. Because these techniques use optical materials to disperse and focus the radiation, they are often identified as optical spectroscopies.[11] The author has use the simpler term spectroscopy for scientific and research convenience, in place of optical spectroscopy with a deeper vision toward scientific pursuit in the area of analytical techniques.[11] Infrared spectroscopy is certainly one of the most analytical procedures available to today's scientists.[11] One of the great advantages of infrared spectroscopy is that virtually any state can be deeply pondered and reconciled. Liquids, solutions, pastes, powders, films, fibers, gases, and surfaces can all be deeply investigated with a judicious choice of the sampling technique.[11] The wide learning objectives in the field of infrared spectroscopy are the understanding of the origin of electromagnetic radiation, determination of frequency, wavelength, wavenumber, and energy change associated with infrared radiation.[11]

2.9 CHEMICAL KINETICS, THERMODYNAMICS, AND PETROLEUM REFINING

Petroleum refining today stands in the midst of deep scientific vision and scientific forbearance. Technology is highly challenged today as scientific and academic rigor needs to be redefined with the passage of scientific history, scientific profundity, and the timeframe.[15,19,20] Chemical kinetics and thermodynamics play crucial role in the success of scientific research pursuit in petroleum refining. Research endeavor in chemical kinetics and thermodynamics is today changing the scientific landscape of petroleum refining.[19,20] The thermodynamic study of technological processes has two definite objectives: (1) determination of the overall thermal effect of chemical transformations that take place in the industrial processes, (2) determination of the equilibrium composition for a broad range of temperatures and pressures in order to deduce optimum working conditions and performances.[19,20] Petroleum refining and modeling as well as simulation of FCC unit stands as a major component in the success and furtherance of science and engineering. Technological hindrances and scientific barriers need to be reorganized as petroleum engineering stands in the midst of deep scientific vision and deep introspection.

2.10   SCIENTIFIC COGNIZANCE OF FLUID CATALYTIC CRACKING (FCC)

Fluid catalytic cracking today stands in the midst of deep scientific cognizance and broad scientific understanding. FCC and its wide vision are opening up new scientific ventures and newer scientific innovations. The scientific cognizance of FCC is reaching immense heights as science and engineering of petroleum engineering moves from one visionary paradigm toward another. Today, technological challenges of design, modeling, and simulation of FCC unit are surpassing wide and vast scientific frontiers. The vision of optimization and control of FCC unit needs to be readdressed and re-envisaged with the passage of scientific vision, scientific forbearance, and the time of human innovation. Today science and engineering are a huge colossus with a definite vision of its own.[19,20]

2.11   MODELING APPROACHES IN DESIGN OF FCC UNITS IN PETROLEUM REFINERY

Modeling approaches in design of a FCC unit in petroleum refinery is today crossing wide and vast visionary boundaries. Mankind's vast scientific girth and determination, the greatness of science and technology, and the vast scientific rigor of petroleum engineering are the forerunners of a newer century today. Modeling and simulation of petroleum refining units are gaining new heights as applied mathematics and chemical process engineering usher in an era of deep scientific understanding. Technological challenges and scientific motivation are immense in this century.[19,20] Catalysis and chemical reaction engineering are witnessing a new light at the end of the scientific tunnel. Fluid Catalytic Cracking is as exceedingly complex chemistry as petroleum refining. Modeling and simulation of cracking units and other units in petroleum refining are the cornerstones of scientific research today. Optimization and control are the other major avenues of scientific research pursuit today. Technology is replete with immense profundity and scientific divinity. In this treatise, the author pointedly focuses on the success of modeling, simulation, optimization, and control of petroleum refining unit and its further research endeavor and vision.[19,20]

2.12   RISER MODELING AND THE VISION FOR THE FUTURE

Today, riser modeling stands as a major component in the furtherance of science and engineering of petroleum technology.[15] The challenge and vision of science needs to be revisited and re-envisaged with the course of scientific history and time.[15] The dramatic changes in petroleum engineering science are transforming the wide scientific scenario and the success and targets of science.

The feed vaporization and the cracking reactions occur inside the riser, while catalyst, hydrocarbon droplets, and hydrocarbon vapors travel upwards. Therefore, there are several chemical and physical complex phenomena occurring simultaneously inside the riser. The rate of feed vaporization in the riser affects its performance to a large extent. Vaporization modeling involves three phases: catalyst particles, hydrocarbon liquid droplets, and hydrocarbon vapors.[15] The vision of process modeling and simulation is entering a new era in the field of petroleum engineering today.[15] Validation of science and technological vision are of immense importance in the avenues of scientific research pursuit today. Immensely complex chemical and physical phenomena occur inside the riser.[15] Vaporization modeling also requires different balances to account for the sensible heat gain of the liquid droplets (while the temperature in the droplet is below vaporization temperature), the vaporization of the liquid droplets with mass transfer from the droplet to the gas phase, and the heat transfer between the catalyst and gas phases after vaporization is complete.[15]

Although the various modeling approaches for the riser can be widely found in literature, a one-dimensional (1D) model is the most commonly used method to describe the FCC units.[15] With the more complex three-dimensional (3D) models, two kinds of approaches are applied to elucidate the riser hydrodynamics: the Eulerian–Eulerian and the Eulerian–Lagrangian approaches, both implemented by computational fluid dynamics (CFD) modeling procedures.[15] The Eulerian–Eulerian considers both gas and solid continuum phases.[15]

Scientific vision in the field of petroleum engineering science is today entering into a new phase. Technological profundity, scientific fortitude, and the world of challenges are briskly opening up new dimensions of research endeavor in decades to come.[15] The science of modeling and simulation is being challenged today and the technological validation of modeling and simulation needs to be reenvisioned with the progress of scientific and academic rigor.[15]

2.13   REGENERATOR MODELING APPROACHES

Regenerator modeling approaches are changing the scientific scenario. Technology and science of process design are ushering in a new era of process modeling and simulation. Mankind's wide scientific prowess and profundity are challenging the scientific fabric today. The technology of petroleum refining is today opening new avenues in research and scientific innovation. Restructuring petroleum refining science is veritably revolutionizing the grave cause of fossil fuel depletion. Today, science of mathematical tools does not lag behind. Technological vision, scientific motivation, and the challenge of chemical process design are virtually answering the intricate questions of modeling, simulation, and optimization of FCC unit.

In FCC modeling, investigations of the regenerator hydrodynamics and steady-state behavior have been conducted by numerous researchers. Moreover, the combustion phenomena in the regenerator, such as the CO post-combustion reaction and the prediction of the CO_2/CO ratio in the flue gas, have also been a matter of deep introspection.[15,16,5] FCC regenerators usually consist of large fluidized bed reactors with complex hydrodynamics, where a strongly exothermic reaction of the combustion of coke on the catalyst takes place. Inside the regenerator, two distinct regions can be envisioned: the dense region and the dilute region (frequently referred to as the freeboard).[15,16,5]

2.14 SCIENTIFIC RESEARCH PURSUIT IN THE FIELD OF MODELING AND SIMULATION OF FCC UNIT

Mathematical tools and applied mathematics are veritably changing the face of scientific research pursuit in designing of an FCC unit today. Scientific vision, scientific fortitude, and scientific sagacity are today the forerunners of a newer era of research pursuit in process design. Chemical process design, the world of chemical process engineering, and the scientific success are veritably changing drastically. Petroleum engineering science needs to be readdressed and reenvisioned with resounding success and with deep scientific understanding. Research pursuit in the field of modeling and simulation of FCC unit is both far-reaching and widely proven. Technological profundity and deep scientific cognizance are the pillars of modern civilization. This century is a period of nuclear technology and space technology advancements. The wide world of applied mathematics and mathematical tools will open new scientific ventures in design and modeling of chemical engineering systems in decades to come.[16,5]

Dagde et al.[6] discussed with deep and cogent insight, modeling and simulation of industrial FCC unit, and made a detailed analysis based on a five-lump kinetic scheme for gas-oil cracking.[6] Models which describe the performance of the riser and regenerator reactors of FCC unit are presented in detail.[6] The riser-reactor is effectively modeled as a plug flow reactor operating adiabatically, using five-lump kinetics of the cracking reactions.[6] The regenerator-reactor is divided into a dilute region and a dense region, with the dense region divided into a bubble phase and an emulsion phase.[6] Technological vision and scientific candor are the pallbearers toward a greater emancipation of petroleum engineering science and petroleum refining today. Modeling and simulation of FCC unit stand in the midst of immense scientific introspection and technological profundity.[6] The bubble phase of the regenerator is modeled as a plug flow reactor, while the emulsion phase is modeled as a continuous stirred tank reactor.[6] The models are then validated using plant data obtained from a functional industrial FCC reactor.[6] Scientific regeneration and scientific adjudication are of utmost need in this hour of history and time. Simulation results indicate that catalyst to gas oil ratio and inlet air velocity have significant scientific results on the performance of the riser and regenerator reactors, respectively.[6] The yield of gasoline and other products of the catalytic cracking process increases as the height of riser-reactor increases, with a maximum yield of gasoline (of about 0.45 mass fraction) occurring about halfway up the riser-height.[6] FCC unit consists of a reaction section and a fractionating section that operate together as an integrated process unit.[6] The reaction section has two reactors: (i) the riser-reactor, where almost all the endothermic cracking reactions and coke deposition on the catalyst occur, and (ii) the regenerator-reactor, where the air is used to burn off the accumulated coke on the catalyst.[6] The catalyst-regeneration process also provides the heat required for the endothermic cracking reactions in the riser-reactor.[6] In the FCC unit, the catalyst enters the riser-reactor as a dense bed and is pneumatically conveyed upwards by the dispersing steam and vaporizing gas-oil feed.[6] It is during this period of conveying the catalyst that catalytic cracking of gas-oil takes place through the efficient catalyst and gas-oil contact. The catalyst later becomes deactivated due to coke deposition on it, and the deactivated catalyst then passes through the spent-catalyst slide-valve in the riser-reactor and enters the top of the regenerator.[6] Today, science and technology of petroleum refining are advancing at a rapid pace surpassing one visionary frontier over another.[6] The scientific vision of petroleum refining and modeling and simulation are witnessing immense scientific challenges, and are replete with vision and forbearance. This treatise opens new knowledge dimensions in the field of modeling and simulation of a vital petroleum refining unit. Dadge et al.[6] reenvision and re-envisage the immense scientific success and scientific fortitude behind the modeling and simulation of an FCC unit. The scientific profundity and the deep scientific adjudication behind modeling and simulation of FCC unit have opened up new windows of innovation and scientific instinct in years to come.[6] Modeling and simulation of FCC unit in petroleum refinery are redefining the science of applied mathematics, chemical process engineering, and petroleum engineering science.[6] This research work goes beyond scientific imagination and culminates in wide scientific vision and deep scientific introspection.[6] The major purpose of the regenerator is to oxidize the coke on the spent catalyst with oxygen to form CO, CO_2, and H_2O, thereby reactivating the catalyst.[6] Compressed combustion-air enters the regenerator from the bottom through a grid distribution pattern designed to provide efficient mixing of air with deactivated catalyst, resulting in a fluidized-bed catalyst-regeneration operation.[6]

This treatise evolves into a new direction of scientific discernment and scientific profundity with the sole objective of furtherance of engineering and technology.[6] Technological vision and scientific objectives are at its utmost best as petroleum refining research pursuit enters into a new era.[6] The FCC unit is quite complex and intricate from the point of view of process modeling/simulation and control.[6] The immense complexity of the FCC unit is attributed to the immense interaction between the riser and the regenerator reactors, and the vast uncertainty in the cracking reactions, coke deposition, and coke burning kinetics.[6]

Ahmed et al.[1] described the modeling and simulation of FCC unit in lucid details. This extensive study depicts the dynamic model for an industrial universal oil product (UOP) fluid catalytic cracking unit on the basis of mass and energy balance, and the transfer functions are found with deep scientific discerning.[1] Technological validation and scientific comprehension are of utmost need as science and engineering of petroleum refining ushers in a new era in the field of sustainability science.[1] Science and technology of energy sustainability today needs to be reframed and restructured as petroleum refining and fuel science move toward the impending crisis of depletion of fossil fuel sources.[1] In this study, the riser and the regenerator are simulated by using MATLAB/Simulink software.[1] The immense challenge and the wide vision of process modeling are opening a new era in the furtherance of science and engineering.[1] The vast dynamic behavior of the process is carried out by measuring the temperature response of the riser and regenerator to step change in gas oil flow rate, gas oil temperature, catalyst flow rate, air flow rate, and temperature. Optimization of process parameters are the vast challenges of science and engineering of modeling and simulation today.[1] The fluid catalytic cracking (FCC) unit is the essential transformation unit in most refineries and it stands as a complex process in the petroleum refinery.[1] There are many mathematical models for FCC in literature, some of them use a very simplified cracking process description, and few of them present a wider integration between regenerator and reactor.[1] Most of the mathematical models are based on models with a higher degree of empiricism and makes use of pseudo-components corresponding to different groups of species, usually called lumps.[1] Applied mathematics and chemical process engineering are two opposite sides of the visionary coin today. Lumping mechanism widely uncovers the scientific success and the scientific vision of mathematical modeling of petroleum refining and the holistic domain of petroleum engineering science.[1] From the present study, the following strong conclusions were drawn regarding the dynamic behavior of the FCC unit. A simple overall dynamic model for the riser and the regenerator was widely developed. The model simulates the entire FCC unit.[1] The steady-state results of the simulation were compared with industrial data and showed excellent results. The challenges and vision of petroleum refining and petroleum processing technology are slowly evolving and the scientific clarity is being reenvisioned. In this treatise, the author pointedly focuses on the technological vision, the scientific clarity, and the vast scientific profundity in the domain of modeling and simulation of FCC unit in a petroleum refinery.[1]

Bhende et al.[3] delineated lucidly the modeling and simulation of FCC unit for the estimation of gasoline production. Petroleum refining and petroleum engineering science are the visionary domains of science which need to be reenvisioned and restructured with the passage of history, human scientific vision, and time.[3] Nowadays, mathematical tools, applied mathematics, and chemical process engineering are in the path of newer scientific regeneration and deep scientific progeny. Fluid catalytic cracking is one of the pivotal processes in the petroleum refining industry for the conversion of heavy gas oil to gasoline and diesel. The technology of petroleum refining is highly advanced today with research pursuit surpassing wide frontiers.[3] This work describes the development of a mathematical model that can simulate and reenvision the behavior of the FCC unit, which consists of feed and preheat system, riser, stripper, reactor, regenerator, and the main fractionator.[3] Scientific destiny, scientific discernment, and deep scientific knowledge are the forerunners toward a greater visionary era in the field of modeling, simulation, and optimization of FCC as well as fixed bed reactors. This model describes the seven main sections of the entire FCC unit: (1) the feed and preheating system, (2) riser, (3) stripper, (4) reactor, (5) regenerator, and (6) main fractionators. This model is able to predict veritably and describe the compositions of the final production rate and the distribution of the main components in the final product.[3] For the present study, a refinery process is simulated in Aspen HYSYS 7.3 environment.[3] Technological profundity and scientific vision are the pillars and supports toward a newer visionary eon in the field of petroleum engineering science today.[3] This paper widely supports the immense scientific potential and the deep scientific understanding in the scientific endeavor of the design of an FCC unit in a petroleum refinery.[3]

Elamurugan et al.[8] discussed with deep and cogent insight, the modeling and control of FCC unit in petroleum refinery.[8] For over 60 years, catalytic cracking has been one of the pivotal processes in petroleum refining, having undergone significant scientific development. Scientific divinity, scientific progeny, and the deep scientific discerning of FCC are ushering in a new era in the field of science and engineering today. The FCCU has become the veritable "test bench" of many advanced control methods.[8] Technological vision and scientific fervor are the cornerstones of scientific advancements today.[8] Today, both the industry and the academia are expressing immense concern in the development of new control algorithms, and in their efficient industrial FCC implementation.[8] Scientific contemplation and the wide world of petroleum refining are changing the face of scientific research pursuit today.[8] Analysis and control of FCC process have been known as challenging problems due to the following process characteristics: (1) very complicated and latent knowledge of hydrodynamics, (2) complex kinetics of both cracking and coke burning reactions, (3) strong interaction between reactor and regenerator, and (4) many versatile operating constraints.[8] The forays into scientific endeavor are today veritably far-reaching and surpassing visionary frontiers.[8] Scientific vision is today in a state of immense crisis as depletion of fossil fuel resources stands as a major issue in the furtherance of science and engineering.[8] The modeling of complex chemical systems for the simulation of process dynamics and control has been exceedingly motivated by the economic incentives for improvement of plant operation and design. Scientific and technological validation also stands in the midst of contemplation and vision. Fluidized catalytic cracking is an important process in petroleum refineries.[8] It upgrades heavy hydrocarbons to lighter and more valuable products by cracking, and is the major producer of gasoline in refineries. Fluidized catalytic cracking units present a vast challenge in multivariable control problems. The selection of good inputs (manipulated variables) and outputs (measured variables) is an important and vexing issue.[8,15]

2.15 OPTIMIZATION OF FCC UNITS

Optimization of an FCC unit stands as a major component in the visionary world of petroleum refining. Today, chemical process engineering and petroleum engineering science are the two opposite sides of the visionary engineering coin. Science and technology are in the path of immense scientific rejuvenation and scientific regeneration. Scientific cognizance, scientific forbearance, and validation of technology will all go a long way in the true emancipation of petroleum refining and petroleum engineering science today. Mathematical tools and applied mathematics are the supporting pillars toward greater scientific advancements in the field of petroleum engineering. Technology is rapidly advancing today and veritably changing the scientific landscape. Fluid catalytic cracking is replete with immense scientific vision and scientific intricacies. Scientific validation and technological vision are also changing the veritable scientific fabric of petroleum engineering science. The challenges and vision of engineering and science are immense and crossing wide and vast visionary boundaries.

2.16   THE WORLD OF VISION OF THE CONTROL OF FCC UNITS

Process control and process design of a FCCU stands as a vital pillar in the furtherance of science and engineering of chemical process engineering. Technological validation and scientific fortitude are of utmost need in the progress of scientific and academic rigor in petroleum engineering science. Process integration of petroleum refining units is another major facet of scientific research pursuit today. Chemical process engineering and petroleum engineering are today linked by an unsevered umbilical cord. Huge economic incentives are usually tied to operating FCC processes at optimal conditions, because these units are exceedingly responsible for approximating 40% of the gasoline pool. In general, this pertains to physical and operating constraints of various kinds (blower capacity, limit material temperatures, etc.). Given the immense complexity of the FCC process, achieving this goal while simultaneously ensuring the safety of operation requires the use of wide-ranging supervision strategies resulting in real time optimization research avenues. A key aspect in most of these strategies is the presence of various layers, each addressing a specific set of operational objectives arranged in a wide range of hierarchy decisions. Pinheiro et al.[15] discussed with immense insight the scientific success and deep scientific profundity in the research avenue in the control of FCCU.[15] Besides the need to operate close to process constraints, the strong interaction between the individual control loops and the nonlinear behavior of the unit constitute major challenges and major research profundity in the design and implementation of the basic regulatory layer.[15] This well-researched paper deeply ponders the immense scientific potential and the wide research avenues in the process control of FCC unit in a petroleum refinery.[15]

2.17   RECENT SCIENTIFIC ENDEAVOR IN THE FIELD OF OPTIMIZATION OF FCC UNITS

Optimization of FCC units in petroleum refinery is the target and vision of scientific research pursuit today. Technological vision and scientific validation are the forerunners toward a newer era of scientific rejuvenation in the field of modeling, simulation, and optimization of FCC unit. Human scientific endeavor and human civilization are today in the path of immense scientific regeneration. Today, the optimization of FCCU stands in the midst of deep scientific introspection and wide scientific vision. Today, design of a petroleum refining unit needs lot of reenvisioning and re-discernment. Genetic algorithm, evolutionary computation, and multi-objective optimization are the wide avenues of science which have application areas in the design of FCCU or other petroleum refining units. The challenge of science, the vision of the efficiency of the process, and the world of technological and engineering vision will all lead a long and effective way in the true emancipation, and realization of petroleum engineering science today. Sankararao et al.[17] lucidly described multi-objective optimization of an industrial FCCU using two jumping gene adaptations of simulated annealing. Multi-objective optimization and multi-objective simulated annealing are two research avenues of evolutionary computation today.[17] The technology of optimization is highly advanced today. Multi-objective optimization of industrial chemical engineering operations using simulated annealing (SA), genetic algorithm (GA), and their adaptation often requires considerable computation time.[17] Science and technology of genetic algorithm, and evolutionary computation are ushering in a new era in the domain of scientific comprehension and scientific fortitude. Since the computational time is high, any adaptation to speed up the procedure is thus desirable.[17] In this paper, the author rigorously points out toward the application of jumping gene adaptations. The performance of the evolutionary technique, multi-objective simulated annealing (MOSA) is found to improve by the use of two adaptations, inspired by the concept of jumping genes in biology.[17] The scientific conundrum is the status of this century. Deep scientific vision is of utmost need as engineering science of petroleum engineering enters a newer era. Kasat et al.[14] deeply discussed with cogent insight multi-objective optimization of industrial FCC units using elitist non dominated sorting genetic algorithm (NSGA).[14] This is an improved version of genetic algorithm and goes beyond scientific imagination. This study provides deep and cogent insight into the optimal operation of the FCCU. A five-lump model is used to characterize the feed and the products.[14] The model is tuned using industrial data. The widely scientific vision of NSGA is pronounced and reenvisioned in this research work.[14] The elitist NSGA-II is used to solve a three objective function optimization problem.[14] Kasat et al.[13] discussed lucidly multi-objective optimization of an industrial FCCU using GA with the jumping genes operator.[13] The multi-objective optimization of industrial operations using the visionary technique of genetic algorithm and its variants often requires inordinately large amounts of computational time.[13] Thus there is need of a robust technique. Any adaptation to speed up the solution procedure is thus exceedingly desirable. An adaptation is, thus developed in this study that is inspired from natural genetics.[15,13]

Scientific forbearance and deep scientific profundity are the cornerstones of scientific research pursuit today. The validation of engineering science is the utmost need of the hour. This paper gives immense visionary angle to the success of optimization techniques in designing a FCC unit. Robust optimization techniques and the wide vision of science are of utmost necessity in the progress of scientific and academic rigor.

2.18   RECENT SCIENTIFIC RESEARCH ENDEAVOR IN THE FIELD OF PROCESS CONTROL OF FCC UNITS

Research endeavor in the field of process control of FCC unit needs to be restructured and rejustified with the passage of scientific history, scientific vision, and time. Today process design, process control, and process integration stands in the midst of deep scientific understanding and scientific introspection. Technology and engineering science are the pallbearers toward a newer eon of scientific regeneration and scientific rejuvenation. Depletion of fossil fuel resources, the issues of environmental protection, and the vicious crisis of maintenance of air quality norms have urged the scientists and engineers to pursue groundbreaking research. In this section, the author rigorously points out toward the scientific success, the deep scientific revelation, and the futuristic vision of design of FCC unit with specific inclination toward modeling and simulation. Process control is a vital component in the greater emancipation of process design science today.[15]

Brijet et al.[4] deeply comprehended the design of the model predictive controller (MPC) for FCCU. FCC stands as a major pillar toward the furtherance of science and engineering in petroleum technology and technological paradigm. FCCU plays the most vital role in modern refinery process because it has been widely used for producing more economic refinery products.[4] The implementation of control algorithms for such model predictive system is often complicated due to diverse variations in process dynamics and control.[4] Scientific vision is the forerunner toward the greater emancipation of process dynamics and control today. Petroleum refining is in a state of immense conundrum today. Model Predictive Controller can control highly interactive systems with many control variables and most vitally, MPC provides a systematic method of dealing with constraints on inputs and states.[4] FCC unit is widely recognized as an important component of scientific research pursuit today in petroleum refining. Various control strategies have been proposed in the past for FCC unit. In this scholarly treatise, MPC for FCCU has been designed using MATLAB/Simulink. Simulink investigations show the effectiveness of the proposed control scheme in reducing the error of the system process.[4,16,5]

Tootoonchy et al.[18] discussed lucidly the fuzzy logic modeling and controller design for a FCCU. This paper widely, with cogent foresight, examines the procedure for nonlinear modeling and Fuzzy controller design of FCCU, of Abadan Refinery in Iran.[18] In 2006 alone, FCCUs refined one-third of the crude oil worldwide.[18] Fuzzy logic is the perfect choice for uncertain, dynamic, and nonlinear processes where the mathematical model of the plant cannot be produced, or if realizable, a great deal of approximation is involved.[18] The heuristic approach of Fuzzy logic controllers is a visionary area of scientific forbearance and profundity and needs to be widely restructured.[18,15,16,5]

Process dynamics and control has a visionary status in science and engineering today. The wide world of chemical process engineering today is in a state of scientific resurgence. The question of efficiency of a chemical process always depends on the success of process control and process integration. This treatise widely elucidates the immense success of process dynamics and control in the course of research work in chemical engineering and petroleum engineering science.[15,2,7,16,5]

2.19 FUTURE WORK AND FUTURE PERSPECTIVES IN THE MODELING AND SIMULATION OF FCC UNITS

Modeling and simulation of FCC unit is a challenge of science and engineering today. Technological validation, scientific divinity, and the world of challenges and hindrances are today reframing the entire world of scientific research pursuit today. Mankind's immense scientific girth and determination are veritably changing the future work and perspectives of technology and science today. This treatise pointedly focuses on the vast world of process design of petroleum refining units with the sole vision of scientific proliferation and the deep scientific understanding. Futuristic vision and future perspectives in the field of petroleum refining are groundbreaking and crossing wide boundaries. Revamping of petroleum refining units and their effective designing is the utmost need of the hour. The challenges and the vision need to be reenvisioned and revamped with the passage of scientific vision, scientific forbearance, and the wide timeframe. The technology of petroleum refining is highly advanced today as science and engineering move toward a newer decade. As science and engineering move steadily with scientific regeneration and wider scientific vision, future perspectives in petroleum refining need to be revamped and restructured.

2.20 FUTURE RESEARCH TRENDS IN THE FIELD OF OPTIMIZATION AND CONTROL OF FCC UNITS

Future research trends are challenging the scientific scenario today. Efficiency, effectivity, and the need for air quality norms have drastically changed the holistic world of fuels and petroleum engineering science. Optimization and control are the two far-reaching domains which need to be pondered and effectively reenvisioned with the passage of scientific history, scientific vision, and time. Future of petroleum engineering science needs to be re-envisaged and reassessed as the depletion of fossil fuel resources changes the visionary scientific fabric. A futuristic vision of petroleum refining is today opening new windows of scientific contemplation and deep scientific forbearance. Today, petroleum refining and engineering science is the challenge and vision of science. Optimization and control are the newest rejoinder of scientific vision and scientific forbearance. Process control and deep scientific understanding are the forerunners of process design and process integration of a chemical plant today. Process optimization of a chemical and petroleum engineering system is today veritably opening new knowledge dimensions in decades to come. Process optimization, process modeling and simulation, process control, and deep scientific vision are the forerunners toward a newer world of chemical process engineering and petroleum engineering science today. The technology of applied mathematics and process modeling has no visible boundaries. Technological candor, scientific contemplation, and scientific vision are the feasible cornerstones of scientific research pursuit today. Petroleum engineering science today stands in the midst of a deep divide with a grave concern regarding energy sustainability issues. Future research trends are targeted toward the optimization issues concerning multi-objective optimization and application of the genetic algorithm. Technological vision today is in a state of immense stress and difficulties. Yet, science is moving at an extremely rapid pace. Human scientific endeavor is also in a state of scientific stress. The rationale should, thus, be toward the immense scientific vision and deep scientific comprehension.[15,2,7,16,5]

Futuristic vision in the field of process modeling and simulation of FCCU is wide and far-reaching. Scientific forbearance and deep scientific discerning are the pillars of scientific research pursuit today. Intricacies of process design and optimization are ushering in a new era in the field of chemical process engineering and petroleum engineering. FCC is a vital component in the furtherance of science and engineering today. Mankind's immense scientific prowess, scientific girth, and determination will all lead a long and visionary way in the true realization of chemical process design today. The candor of science, the vision of engineering science, and the futuristic challenges are the forerunners toward a better scientific order in the field of sustainability science.[2,7]

2.21 SCIENTIFIC VISION, SCIENTIFIC CHALLENGES, AND SCIENTIFIC BARRIERS IN DESIGN

Scientific vision and the vast challenges in process design are drastically changing the course of future scientific endeavor and the futuristic vision of petroleum engineering science. Mankind and human scientific endeavor today stands in the juxtaposition of scientific optimism and introspection. Design, modeling, and simulation are today replete with scientific vision and scientific determination. Scientific barriers and deep scientific and

technological struggles are veritably widening the futuristic vision of petroleum engineering science. Barriers and intricacies in modeling and simulation of petroleum refining units are far-reaching and groundbreaking. The
success of scientific research pursuit is opening up new windows of innovation and scientific determination in decades to come. Energy sustainability is the other area of grave concern today as depletion of fossil fuel
resources gains new scientific heights. Scientific candor and technological
profundity are the forerunners toward a newer visionary era in petroleum
refining. Today, energy crisis and environmental issues are destroying the
wide scientific fabric. Petroleum refining is the fruit of science in the present
day human civilization. Energy sustainability is the utmost need of the hour
as engineering science of petroleum refining surpasses visionary frontiers.
Sustainability issues and global water crisis are gaining immense heights in
the scientific panorama today. Technological forays, scientific challenges,
and deep scientific vision will all lead a long way in the true emancipation
and realization of holistic sustainable development today. The vision and
scientific challenges are absolutely awesome and far-reaching as science
moves toward a newer decade and the second half of this century. Scientific
landscape and scientific frontiers need to be readdressed and reframed as
petroleum engineering science is faced with difficulties and hindrances.
Process modeling and simulation of FCCU along with multi-objective optimization applications are the cornerstones of research today. The author in
this treatise repeatedly points out toward the ultimate aim of robust applications of the genetic algorithm and multi-objective optimization in process
modeling, simulation, optimization, and control of FCC unit.[15,2,7]

2.22 CONCLUSION, SUMMARY, AND FUTURE PERSPECTIVES

Petroleum engineering science and petroleum refining stand at the crucial
juncture of scientific optimism and deep scientific comprehension today.
Technology is highly challenged today. Future perspectives of science
need to be readdressed and re-envisaged with the success of scientific
history and scientific vision. In this treatise, the author rigorously points
out toward the immense scientific success and the scientific forbearance
behind the application of mathematical tools in the design of an FCC unit.
The science and technology of petroleum engineering is witnessing wide
challenges and hindrances as depletion of fossil fuel resources cautions
the path of science and engineering. Technology revamping in petroleum refining and scientific validation are the utmost need of the hour.

The world of challenges and the vast vision behind the design of petroleum refining units need to be readdressed and reenvisioned with each step of research pursuit. This paper goes beyond scientific vision and scientific imagination and opens a new chapter in energy sustainability and petroleum refining. Today 21st century stands in the midst of environmental crisis and sustainability challenges. The success of scientific endeavor needs to be streamlined with the passage of scientific history and visionary timeframe. The scientific discernment and the scientific progeny are ushering in a newer eon in the field of energy sustainability and renewable energy technology. Today, the wide vision of renewable energy stands at the crucial stage of scientific optimism and deep scientific introspection. This chapter crucially details the immense new research advances in the field of modeling, simulation, optimization, and control of FCCU. This comprehensive review is a well detailed uphill task which targets scientific forbearance and deep scientific understanding. Technology and engineering are in the path of immense scientific rejuvenation and deep scientific girth. The challenge of energy technology today lies in the hands of technological and scientific vision of renewable energy technology. The vision of petroleum refining needs to be once again restructured as science readdresses itself toward greater realization of sustainability.

2.23   ACKNOWLEDGMENT

The author deeply acknowledges the contribution of Shri Subimal Palit, the author's late father, and an eminent textile engineer from India who taught the author the rudiments of chemical engineering science.

KEYWORDS

- vision
- catalyst
- cracking
- modeling
- simulation
- optimization
- control

REFERENCES

1. Ahmed, D. F.; Ateya, S. K. Modelling and Simulation of Fluid Catalytic Cracking Unit. *J. Chem. Eng. Process Technol.* **2016,** *7*(4), 308. DOI: 10.4172/2157-7048.1000308.

2. Bandyopadhyay, S.; Saha, S.; Maulik, U.; Deb, K. A Simulated Annealing-Based Multiobjective Optimization Algorithm: AMOSA, (2008). *IEEE Trans. Evol. Comput.* **2008,** *12*(3), 269–283.

3. Bhende, S. G.; Patil, K. D. Modeling and Simulation for FCC Unit for the Estimation of Gasoline Production. *Int. J. Chem. Sci. Appl.* **2014,** *5*(2), 38–45.

4. Brijet, Z.; Bharathi, N.; Shanmugapriya, G. Design of Model Predictive Controller for Fluid Catalytic Cracking Unit. *Int. J. Control Theory Appl.* **2015,** *8*(5), 1917–1926.

5. Chang, J.; Meng, F.; Wang, L.; Zhang, K.; Chen, H.; Yang, Y. CFD Investigation of Hydrodynamics, Heat Transfer and Cracking Reaction in a Heavy Oil Riser with Bottom Airlift Loop Mixer. *Chem. Eng. Sci.* **2012,** *78,* 128–143.

6. Dagde, K. K.; Puyate, Y. T. Modelling and Simulation of Industrial FCC Unit: Analysis Based on Five-Lump Kinetic Scheme for Gas-Oil Cracking. *Int. J. Eng. Res. Appl.* **2012,** *2*(5), 698–714.

7. Deb, K. Multi-Objective Optimization Using Evolutionary Algorithms: An Introduction, Kanpur Genetic Algorithm Report No.-2011003, 2011.

8. Elamurugan, P.; Dinesh, D. K. Modeling and Control of Fluid Catalytic Cracking Unit in Petroleum Refinery. *Int. J. Comput. Commun. Inf. Technol.* **2010,** *2*(1), 55–59.

9. Elnashaie, S. S. E. H.; Elshishini, S. S. *Modeling, Simulation and Optimization of Industrial Fixed Bed Reactors;* Gordon and Breach Science Publishers: S.A., 1993.

10. Han, I.-S., Chung, C. -B. Dynamic Modeling and Simulation of a Fluidized Catalytic Cracking Process. Part I: Process modeling. *Chem. Eng. Sci.* **2001,** *56,* 1951–1971.

11. Harvey, D. *Modern Analytical Chemistry;* McGraw Hill Higher Education: USA, 2000.

12. Jimenez-Garcia, G.; Aguilar-Lopez, R.; Maya-Yescas, R. The Fluidized-Bed Catalytic Cracking Unit Building its Future Environment. *Fuel* **2011,** *90,* 3531–3541.

13. Kasat, R. B.; Gupta, S. K. Multiobjective Optimization of an Industrial Fluidized-bed Catalytic Cracking Unit (FCCU) Using Genetic Algorithm (GA) with the Jumping Genes Operator. *Comput. Chem. Eng.* **2003,** *27,* 1785–1800.

14. Kasat, R. B.; Kunzru, D.; Saraf, D. N.; Gupta, S. K. Multiobjective Optimization of Industrial FCC Units Using Elitist Nondominated Sorting Genetic Algorithm. *Ind. Eng. Chem. Res.* **2002,** *41,* 4765–4776.

15. Pinheiro, C. I. C.; Fernandes, J. L.; Domingues, L.; Chambel, A. J. S.; Graca, I.; Oliveira, N. M. C.; Cerqueira, H. S.; Ribeiro, F. R. Fluid Catalytic Cracking (FCC) Process Modeling, Simulation, and Control. *Ind. Eng. Chem. Res.* **2011,** *51,* 1–29.

16. Ramteke, M.; Gupta, S. K. Kinetic Modeling and Reactor Simulation and Optimization of Industrial Important Polymerization Processes: A Perspective, *Int. J. Chem. React. Eng.* **2011,** *9,* Review R1.

17. Sankararao, B.; Gupta, S. K.; Multi-Objective Optimization of an Industrial Fluidized-bed Catalytic Cracking Unit (FCCU) Using Two Jumping Gene Adaptations of Simulated Annealing. *Comput. Chem. Eng.* **2007,** *31,* 1496–1515.

18. Tootoonchy, H.; Hashemi, H. H. Fuzzy Logic Modeling and Controller Design for a Fluidized Catalytic Cracking Unit. *Proceedings of the World Congress on Engineering and Computer Science,* San Francisco, USA, Oct 23–25, 2013.

19. www.google.com (accessed Sept 1, 2017).

20. www.wikipedia.com (accessed Sept 1, 2017).

PART II
Nanotechnology

MODELING AND SIMULATION OF ELECTROSPINNING PROCESS BY SOLVING THE GOVERNING EQUATIONS OF ELECTRIFIED JETS AND USING FEniCS

S. RAFIEI*

University of Guilan, Rasht, Iran

Corresponding author. E-mail: saeedeh.rafieii@gmail.com

CONTENTS

ABSTRACT

Electrospinning is a sophisticated material process to manufacture well-tailored nanofibers for fiber reinforcement, tissue scaffolding, drug delivery, nanofiltration, and protective clothing. However, experimental results need to be controlled systematically through theoretical models for the optimization of single and bicomponent nanofiber diameter and alignment. This study is concerned with modeling and simulation of electrospinning process by solving the governing equations of an electrified jet using FEniCS software packages for finite element method. Jet diameter changes are focused in this model as the most important physical parameter. Electrospinning effective parameters on nanofiber structure including solvent evaporation, electrical field, viscosity, and flowing rate are simulated. Experimental nanofiber production in the same spinning condition with simulated spinning is applied to validate the model. Comparison between theoretical and experimental nanofiber diameter in different solution concentration presents good compatibility of the model. The relative error shows that for the higher nozzle diameter, the model can have an acceptable prediction for final diameter.

3.1 INTRODUCTION

Electrospinning is an established nanoindustrialized technology producing continuous nanofibers from polymer solutions in high electric fields.[1-4] Due to the very high aspect ratio, specific surface area, flexibility in surface functionalities, and superior mechanical properties of electrospun nanofibers, many investigations have been done in using them for a variety of applications including drug delivery, tissue engineering, conductive nanowires, nanosensors, biochemical protective clothing for the military, and wound dressing. Currently, nanofibers and nanostructures comprising single or blended materials have been widely produced. However, there is now an increasing need for producing coaxial compound nanofibers and core-shell types of nanostructures due to their unique properties.[2,5,6]

In each of their applications, the control of fiber diameter is essential to achieve specific functionalities to structural networks and increase their available surface areas. Moreover, mechanical properties of electrospun fibers in relation to their diameters have been well established.

Coaxial electrospinning for core-shell bicomponent nanofibers production was introduced in the last decade as a branch of nanotechnology, by which their applications are being explored and developed in biotechnology, drug delivery, and nanofluidics.[7-9] The novelty of coaxial electrospinning lies in coaxial two liquid nanojets which consist of a shell-forming liquid on the outer side and an inert immiscible liquid on the inner side, the latter constituting a very broad and general family of liquid nanotubular templates requiring no molecular-level assembly.[10,11]

In both simple and coaxial electrospinning, electrohydrodynamic (EHD) forces cause meniscus exiting from the tip of a capillary, when subject to an appropriate electric field, and deforming into a conical shape (Taylor cone).[3] The jet generally desires to break up into a spray of highly charged nanodroplets namely, "electrospray," unless the jet solidifies before such physical natural disruption occurs.[12,13] According to this fact, solvent evaporation plays a critical role in nanofiber and core-shell nanofiber formation. A rapid solvent evaporation accompanied by jet stretching due to electric forces and jet instabilities is ultimately responsible for the diameter, structure, and properties of the final solidified nanofibers and core-shell nanofibers.[14,15]

Systematic understanding of flow behavior in processes such as electrospinning can beextremely difficult because of the variation of different forces that may be involved, including capillarity, viscosity, inertia, gravity, as well as the additional stresses resulting from the extensional deformation of the microstructure within the fluid. Theoretical and numerical comprehensions of these phenomena can assist in overcoming the existing restrictions throughout the process. Several theoretical works about different aspects of electrospinning with the goal of process efficiency enhancement and well understanding of its underlying physics with focus on nanofiber diameter prediction are done.[16,17] However, there is still no entirely adequate mathematical and physical theory for electrically driven, viscoelastic core-shell jets similar to other technologies.[18] A novel model and simulation are introduced in this research to optimize these valuable techniques. Moreover, the results have been compared with experimental data to validate the model.

3.2   NUMERICAL METHOD

The electrospinning process is a fluid dynamics related problem which requires a numerical and mathematical vision for controlling the property and geometry of produced nanofibers. Although information on the effect of various processing parameters and constituent material properties can be obtained experimentally, theoretical models offer in-depth scientific understanding, which can be useful to clarify the affecting factors that cannot be exactly measured experimentally. Results from modeling also explained how processing parameters and fluid behavior lead to nanofibers with appropriate properties.

In this study, a new method, based on phase-field model was applied to model electrospinning process in order to superior control of electrospun jet diameter by applying FEniCS package software. Phase-field models are an increasingly popular choice for modeling the motion of multiphase fluids.

3.2.1 PHASE-FIELD MODEL

In this model, Ω is assumed sufficiently smooth. The unit outward normal to the boundary is denoted n. A binary mixture is contained in Ω and c denotes each components' concentration. The evolution of the mixture is assumed governed by the Cahn–Hilliard equation. In strong form, the problem can be stated as:

find $\varphi : \Omega \times (0,T) \rightarrow R$ such that:

$$\frac{\partial \varphi}{\partial t} = \nabla \cdot \left(M_\varphi \nabla \left(\mu_\varphi - \lambda \Delta \varphi \right) \right) \qquad \Omega \times (0,T) \tag{3.1}$$

$$\varphi = g \qquad \Gamma_g \times (0,T) \tag{3.2}$$

$$M_\varphi \nabla \left(\mu_\varphi - \lambda \Delta_\varphi \right) \cdot n = s \qquad \Gamma_s \times (0,T) \tag{3.3}$$

$$M_\varphi \left(\lambda \nabla_\varphi \right) \cdot n = 0 \qquad \Gamma \times (0,T) \tag{3.4}$$

$$\varphi(x,0) = \varphi_0(x) \qquad \Omega \tag{3.5}$$

Where, M_φ is the mobility, g is gravity, μ_φ represents the chemical potential of a regular solution in the absence of phase interfaces and λ is a positive constant (chemical potential).[19,20]

3.2.2 GOVERNING EQUATIONS

In this approach, a set of mass, momentum conservation, Maxwell and Fick's equations for fluids together with the advection-diffusion equation of the diffuse-interface (in the phase-field model) are used to solve as follows:

$$\nabla \cdot u = 0 \tag{3.6}$$

$$\rho\left(\frac{\partial u}{\partial t} + u \cdot \nabla u\right) = -\nabla \cdot p - \frac{1}{2} E \cdot E \nabla \varepsilon + \nabla \cdot (\varepsilon E) E + \nabla \cdot \mu\left(\nabla u + \nabla u^T\right) \quad (3.7)$$

$$\nabla \cdot (\varepsilon_r E) = \rho_e \quad (3.8)$$

$$\nabla \times E = 0 \quad (3.9)$$

$$\frac{\partial \varphi}{\partial t} + \nabla \cdot (\varphi u) = -\nabla \cdot \left[-\Gamma(\varphi)\nabla \eta\right] \quad (3.10)$$

Where, u, t, ρ, P, μ, φ, η, ε_r, E, and ρ_e denote velocity, time, density, pressure tensor, viscosity, order parameter (in Cahn–Hilliard method), chemical potential, relative dielectric constant, electric field, and volumetric charge density, respectively.

The scalar variable φ distinguishes between two phases (two components) with different values of φ. Furthermore, it indicates an interface as a finite volumetric zone between the phases across where φ varies continuously. Inside the interface, ρ is assumed to be a sine-curve function of φ that varies between the two given constants s for the polymeric (or liquid) phases. On the other hand, the viscosity μ varies between this two phases as a linear function of ρ.

In the case of having two fluids, in bicomponent nanofiber-relative dielectric constant ε can be obtained by:

$$\varepsilon_r = \frac{\varepsilon_{f1}}{\varepsilon_{f2}} \quad (3.11)$$

Where, ε_{f1} and ε_{f2} are dielectric constants of the first and second fluids, respectively. Solvent evaporation rate can be calculated according to Fick's law. Mass conservation of the mixture of polymer and solvent in the solution requires $c_1 V_1^- + c_2 V_2^- = 1$. The axisymmetric Fick's law of mass diffusion in the radial direction gives:

$$J_i = \frac{Dc_i}{RT} \nabla \eta_i \quad (3.12)$$

$$J_i = D\frac{\partial c_1}{\partial r} \quad (3.13)$$

Where, J is the "diffusion flux," of which the dimension is the amount of substance per unit area per unit time, so it is expressed in such units as $mol·m^{-2}·s^{-1}$. Diffusion flux measures the amount of substance that will flow through a unit area during a unit time interval. D is the diffusion coefficient or diffusivity. Its dimension is area per unit time, so typical units for expressing would be m^2/s. The index i denotes the ith species, c is the concentration, R is the universal gas constant (8.3144598 J/K/mol), T is the absolute temperature of spinning chamber (25°C=298.15 K), and η is the chemical potential (J/mol).

So, it can be written as:

$$\frac{\partial c_i}{\partial t} = \frac{\partial}{\partial r}(J_i) \tag{3.14}$$

Using initial and boundary condition, jet diameter and solvent evaporation parameters are related as below:

$$r = R(t): \frac{dR(t)}{dt} = -\frac{k_g M_S}{RT\rho_s}(p_s - p_\infty) \tag{3.15}$$

Where $R(t)$ is the jet radius in time t; k_g (cm/g) is the coefficient of solvent mass transfer at the jet surface; M_s is the solvent molar weight; T is the absolute temperature near the surface; R=8.3144 J/mol is the universal gas constant; p_∞ is the solvent vapor pressure in the atmosphere far from the jet surface (p in this study); p_s is the solvent saturation vapor pressure near the jet surface, and ρ_s is the solvent density. [14,21]

The constants and known parameters used in this simulation,—the same as experimental part are as follows:

The mass densities of solvent ($M_s = 18$) are $\rho_s = 1$ g/cm³. The polyvinyl alcohol (PVA) diffusion coefficient in infinite water is $D = 10^{-3}$ cm²/s. Solvent saturation vapor pressure at 25°C is 500 Pa. The solvent mass transfer coefficient at the jet surface is $k_g = 0.2 \times 10^{-2}$ m/s, by analogy with similar solvents.

So the Equation 3.15 can be written as:

$$\frac{dR(t)}{dt} = -\frac{k_g M_S}{RT\rho_s}(p_s - p) \tag{3.16}$$

According to this equation, the diameter of the jet can be calculated for each step.

3.3 EXPERIMENTAL

The PVA (Mw = 72000 Da) was obtained from Merck company and its solution was prepared by dissolving PVA in deionized water. The solution was magnetically stirred at 80°C for 2 h to obtain homogeneous solution. Simple electrospinning and coaxial electrospinning were performed for three different concentrations of PVA/water (6, 8, and 10%). Three different sizes of syringes were placed on a programmable syringe pumps electrospinning setups (Fig. 3.1).

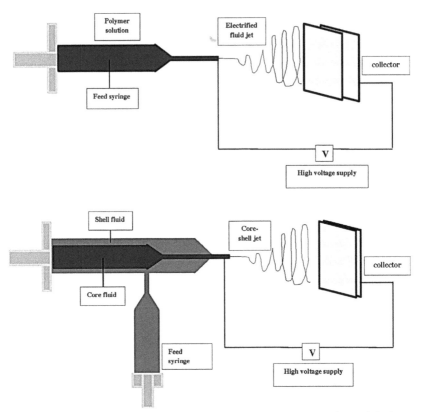

Figure 3.1 Schematic representation of simple electrospinning and coaxial electrospinning setup, respectively.

The initial spinning condition for the experimental section is adjusted similarly to the simulation part. Spinning distance, temperature, initial voltage, nozzle diameter (initial jet diameter), and the flow rate are selected as: 5–10 cm, 25°C, 0 V, 0.3–1.0 mm, and 0.1 ml·h⁻¹, respectively. Figure 3.1 shows the structure of simple electrospinning and coaxial electrospinning setups which are used to obtain experimental data in order to validate applying model.

Scanning electron microscope (SEM) (Philips XL-30) was employed to get and control nanofiber diameters. Further, diameter characterization applied for core-shell electrospun nanofibers using JEOL JEM-2010F FasTEM field emission electron microscope was operated at 100 keV. The samples were dried in a vacuum oven for 48 h at room temperature prior to transmission electron microscopy (TEM) imaging. Typical, SEM and TEM images of single and bicomponent PVA nanofibers are analyzed in next sections. Sharp boundaries of the core-shell structure along the length of the fiber can be clearly seen which indicate well produced and homogeny core-shell nanofibers.

3.4   RESULT AND DISCUSSION

Discussed governing nonlinear Equations 3.6–3.10 and 3.16 have been solved by FEniCS package software mentioned in every time steps and grids of defined finite element mesh. The setup frame in this modeling is defined in a way that the length and the nozzle dimension are simulated to be as same as the real setup. This will obtain a virtual frame similar to the real operating system. Final results such as jets diameter, electric field magnitude, solvent evaporation rate, and viscosity are defined for every time steps and grids of finite element mesh which are analyzed using ParaView software. Finally, the theoretical results validated by experimental data for both simple and coaxial electrospinning setups.

3.4.1   SIMULATION OF NANOFIBERS DIAMETER

It has been shown that fiber diameter has a direct effect on mechanical and physical properties of electrospun nanofibers which has a significant effect on their industrial application.[22,23] In this section, we focus on single and bicomponent nanofiber diameter simulation applying discussed model. Several factors may contribute to the size effect of nanofibers such as electrostatic forces, syringe feeding rate, spinning distance, and solvent evaporation.[24] The effect of these factors is clarified in following sections.

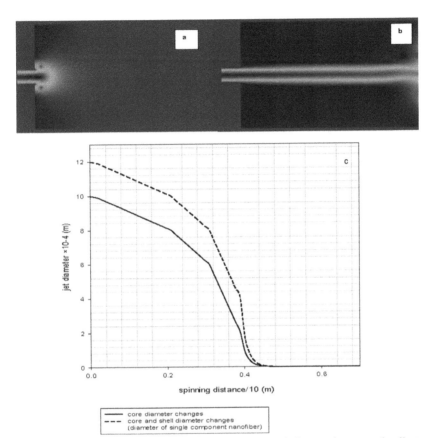

Figure 3.2   Theoretical results of jet diameter changes during motion toward collector using FEniCS simulation are presented (a) Jet initiation, (b) arriving at the collector, and (c) single and bicomponent diameter changes during electrospinning.

Figure 3.2 represents the simulation of three stages of single and bicomponent nanofibers jet diameter changes under electric fields. The continuous and dotted lines show core and core-shell or single component diameter changes during electrospinning, respectively. The initial diameter of jet and is considered equal to the electrospinning nozzle diameter. As it is recognized in the graph, the diameter reduction is sharp at first and slow at the end of the process. At the initial stage in this process, the jet is coming out of the electrospinning nozzle as a drop named Taylor cone. Finally, a jet diameter reduction is obvious in the last part of simulation. The primary jet is exerted toward collector because of attraction force between electrode connected to the collector and surface electric charges on coaxial jet. Although the force of polymers viscosity and surface tension resists against the movement of fluidic jet, the electric force between two electrodes can overcome these forces if the voltage magnitude selected correctly. As it can be seen from the vortexes around the jet, velocity is increasing during the process and the jet is further accelerated by the time. It is clear from Figure 3.2 that jet diameter of both phases is decreased, which is consistent with Feng's model.[25]

According to this analytical model, forces that determine the jet diameter during electrospinning or coaxial electrospinning can be considered as of viscosity, surface tension, electric force, and flow rate which cause jet velocity. Electric field and diameter changes are deeply interrelated, since the jet becomes thinner downstream, the increasing jet speed enlarges the surface charge density and, thus, electric field, so the electric force exerted on the jet and the diameter become smaller.[25]

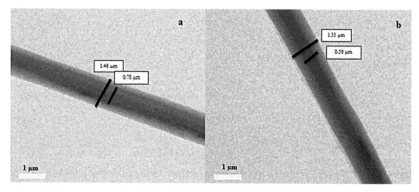

Figure 3.3   Transmission electron microscopy (TEM) results for core-shell polyvinyl alcohol (PVA) nanofibers with different polymer solution concentration while other spinning parameters are the same (nozzle diameter and spinning distances are 0.5 mm and 10 cm, respectively): (a) polymer concentration of 10%, (b) polymer concentration of 6%.

Figures 3.3 and 3.4 show TEM and SEM photographs of core-shell nanofibers of PVA with the same spinning parameters such as flow rate, voltage, nozzle diameter, and spinning distance but different polymer solution concentration (6 and 10%). As it is obvious, reduction of the concentration is lead to the finer nanofibers.

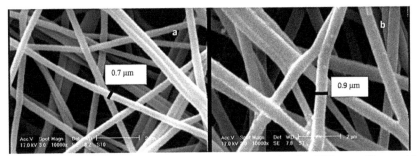

Figure 3.4 Scanning electron microscope (SEM) results for PVA nanofibers with different polymer solution concentration while other spinning parameters are the same (nozzle diameter and spinning distances are 0.3 mm and 10 cm, respectively): (a) polymer concentration of 6%, (b) polymer concentration of 10%.

Table 3.1 Comparison Between Theoretical and Experimental Nanofiber Diameter in Different Initial Jet Radius (Nozzle Diameter) and Polymer Solution Concentration for both Single and Bicomponent Nanofibers (Spinning Distance is Considered to be 10 cm).

Initial jet radius (mm)	Polymer solution concentration (wt.%)	Theoretical final jet diameter (final core-shell or single component) (μm)	Experimental final core-shell jet diameter (μm)	Experimental final single component jet diameter (μm)	Relative error (%)
0.1	6	0.54	N/A	N/A	N/A
0.3	6	0.98	1.07–2.08	0.77–1.12	8.0
0.5	6	1.28	1.35–2.03	0.99–1.23	5.0
0.1	10	0.63	0.51–2.18	0.49–0.700	19.0
0.3	10	1.19	1.20–2.71	0.93–1.27	0.8
0.5	10	1.38	1.44–3.03	1.08–1.54	4.0

Table 3.1 signifies a comparison between theoretical and experimental nanofiber diameters in different initial jet radius (nozzle diameter) for polymer solution concentration with a spinning distance of 10 cm for both single and bicomponent nanofibers. It can be found that the model results have good compatibility with the experimental data with the same initial parameters including flow rate, applied voltage, solution concentration, and nozzle diameter. The relative error shows that for the higher nozzle diameter the model can have an acceptable prediction for the final diameter of nanofibers.

3.4.2 EFFECT OF FLOWING RATE IN ELECTROSPINNING SIMULATION

The flow rate of the polymer solution within the syringe is another important process parameter. Generally, the lower flow rate is more recommended as the polymer solution will get enough time for polarization. If the flow rate is high, the Taylor cone expanded as a bog droplet and falls. If flow rate increases more, dropping will appear in which we have multi droplets instead of a continues jet, as it is depicted in Figure 3.5.[22,26]

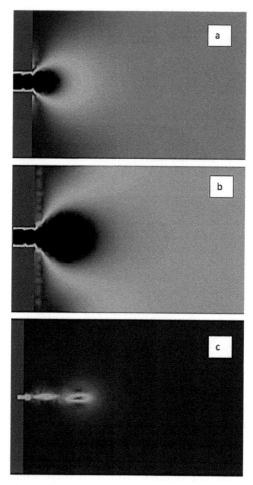

Figure 3.5 Simulation of flow rate increasing; (a) 0.2, (b) 0.4, and (c) 1 ml·h^{-1} (in higher spinning distance).

3.4.3 EFFECT OF VISCOSITY IN ELECTROSPINNING SIMULATION

Viscosity of solution is the critical key in determining the fiber morphology. It has been proven that continuous and smooth fibers cannot be obtained in very low viscosity, whereas very high viscosity results in the hard ejection of jets from solution, namely there is a requirement of suitable viscosity for electrospinning.[22]

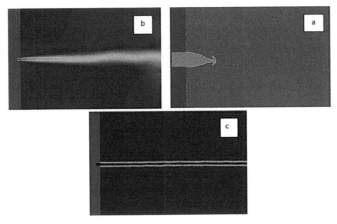

Figure 3.6   Simulation of electrospinning fluid jet with: (a) low, (b) high, and (c) suitable viscosity of the polymeric fluid.

Generally, the viscosity of solution can be tuned by adjusting the polymer concentration of the solution; thus, different products can be obtained. The viscosity range of a different polymer or oligomer solution at which electrospinning is done is different. It is important to note that viscosity, polymer concentration, and polymeric molecular weight are related to each other. For the solution of low viscosity, surface tension is the dominant factor and just beads or beaded fiber formed. If the solution is of suitable viscosity, continuous fibers can be obtained. In bicomponent nanofiber formation, the viscosity is more important due to the possibility of merging two fluids during electrospinning. Several investigation results indicated that the best condition to fabricate a homogeny core-shell nanofiber is provided when the viscosity of both fluids are near to each other.[22,23] Figure 3.6 presents the simulation of electrospinning with low, high, and suitable viscosity of polymeric fluid. As it is clear in the figure, very low viscosity cannot create a continuous electrospun jet as well as very high viscosity which stops motion of jet. In the latter case, the electrostatic forces cannot dominate surface tension and viscosity forces and as a result, the jet stops or spreads like droplets.

3.4.4   EFFECT OF APPLIED VOLTAGE IN ELECTROSPINNING SIMULATION

Within the electrospinning process, applied voltage is the crucial factor. Only the applied voltage higher than the threshold voltage, charged jets ejected from Taylor Cone can occur. Several investigations on electric voltage effects on nanofiber structures indicated that higher voltage leads to higher nanofiber diameter and the electrostatic repulsive force on the charged jet, favoring the narrowing of fiber diameter. But, the level of significance varies with the polymer solution concentration and on the distance between the tip and the collector.[22,23]

Figure 3.7 shows the electric field changing during electrospinning process. Since the electric force is small at the first time steps, surface tension and viscosity forces resist jet movement. Hence, the velocity increase is slow. After thinning of the jet in the middle of the process, electric force causes surface charge density to increase in a way that can dominate opposition forces. Therefore, the jet velocity, and as a result, electric field have a sharp rise. Electric field increases in the jet motion direction and will lead to a huge electric force which accelerates the coaxial jet towards the collector. It has a maximum value after passing the nozzle, since the shrinking radius of the jet in combination with charge conservation demands a higher rate of convection, leading to an increase in surface charge density, that is, ρe, growing. The charges then cause an increase in the mentioned axial electric field. As the jet becomes thinner downstream, the increasing jet speed will increase along with increasing electric field as Feng's model.[17]

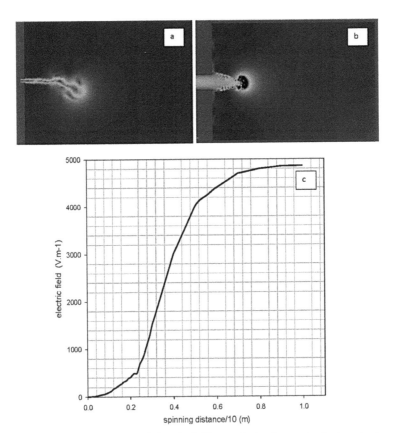

Figure 3.7 Simulation results of electric field magnitude changes in electrospinning; (a) lower magnitude of electric field, (b) higher magnitude of electric field, and (c) changing of electric field during spinning distance.

3.4.5 EFFECT OF SOLVENT EVAPORATION IN ELECTROSPINNING SIMULATION

Solvent evaporation plays a critical role in nanofiber formation and jet diameter changes in electrospinning. Applying Fick's law in this simulation, the following assumptions are used in this model to simulate solvent evaporation from thin polymer jets: (a) no temperature gradient exists inside the jet; (b) the mass diffusion and transfer is axisymmetric; (c) no chemical reactions take place inside the jet.[14] The numerical simulation demonstrates high transient inhomogeneity of solvent concentration evaporation on the surface of the spinning jet. The degree of inhomogeneity decreases when the spinning jet becomes finer. The simulated jet drying time decreases rapidly with the decreasing initial jet diameter as it has good compatibility with experimental data. Solving Fick's governing equation by different initial nozzle diameter (initial jet radius) leads to various solvent evaporation times which show some extent of homogeneity.[27]

Table 3.2 shows the variation of evaporation time of the solvent with the initial jet radius (nozzle diameter). The data presents that by decreasing the diameter of the nozzle, solvent evaporation takes more time and as a result, there will be more inhomogeneity and larger diameter. Comparing with experimental data for solvent evaporation time, simulation results show acceptable compatibility, especially in higher initial jet diameters.

Table 3.2 Variation in the Jet Drying Time (Solvent Evaporation Time) with the Initial Jet Radius.

Theoretical evaporation time (s)	Experimental evaporation time (s)	Initial Jet diameter (mm)	Final jet diameter (µm)	Relative error (%)
0.0014	N/A	0.1	0.63	N/A
0.0090	0.010	0.3	1.19	10
0.0360	0.040	0.5	1.38	10
0.2200	0.200	0.7	1.45	6
0.3100	0.300	1.0	1.80	5

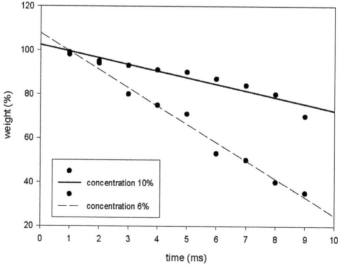

Figure 3.8 Evaporation rate of solvent in different concentration of polymer solution.

Solvent evaporation depends on solution concentration, too. As Figure 3.8 depicts, by increasing polymer solution concentration from 6 to 10%, evaporation time increases, having the same results in increasing final diameter. Figure 3.9 presents a simulation of solvent evaporation and increase in inhomogeneity of spinning jet in different concentration of the polymer solution.

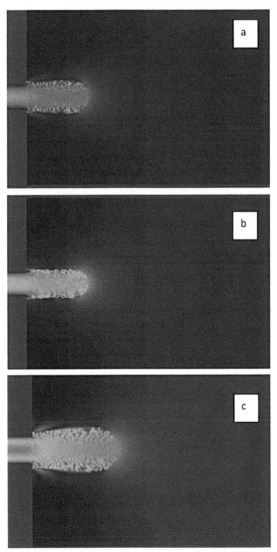

Figure 3.9 Simulation of solvent evaporation and increasing in inhomogeneity of spinning jet in different concentration of polymer solution: (a) 6, (b) 8, and (c) 10%.

3.5 CONCLUSION

Due to the rising interest in nanoscale materials and properties, research on the electrospinning process has intensified in recent years. Since the electrospinning process is dependent on a lot of different parameters, changing them will lead to significant variations in the process.

This study has focused on modeling and simulation of electrospinning process by focusing on solving the governing equations of mass, momentum conservation, Maxwell and Fick's equations together with the advection-diffusion equation of the phase-field model for electrified fluid jet using software packages including; FEniCS, GMSH, and ParaView softwares by applying finite element method. Simulation results showed that decreasing of polymeric liquid jet diameter occurs in a sinusoidal graph with acceptable compatibility with experimental data on the same condition. The electric field changed during coaxial electrospinning slowly at first and sharply in the middle. Solvent evaporation inhomogeneity for the spinning polymeric jet was also presented by this model. Solvent evaporation results indicated that the higher nozzle diameter could lead to a larger final nanofiber diameter and decrease of fluid jet homogeneity which has validated with experimental results as well. More concentrated polymer solution also had the same result in nanofiber diameter and homogeneity. Furthermore, applying the Fick's law for different initial jet diameter the final diameter of fluid jet (nanofiber) was predicted, which had good compatibility with experimental data on the same condition. The effects of other parameters such as viscosity and flowing rate are simulated, too.

3.6 ACKNOWLEDGMENT

This work is supported by MatLab department in International School for Advanced Studies (SISSA) in Trieste, Italy. The author would like to thank them for their scientific and financial support, especially, professor Luca Heltai and Dr. Alberto Sartori.

KEYWORDS

- **modeling**
- **simulation**
- **nanofiber diameter**
- **electrospinning**

REFERENCES

1. Bhardwaj, N.; Kundu, S. C. Electrospinning: A Fascinating Fiber Fabrication Technique. *Biotechnol. Adv.* **2010,** *28,* 325–347.
2. Fong, H.; Reneker, D. H. *Electrospinning and the Formation of Nanofibers, in Structure Formation in Polymeric Fibers;* Hanser Gardner Publications: Germany, 2001; p 577.
3. Taylor, G. Electrically Driven Jets. *Proceedings of the Royal Society of London A: Mathematical, Physical and Engineering Sciences*, The Royal Society, 1969.
4. Mohammadzadeh moghadam, S.; Dong, Y.; Davies, I. J. Modelling Electrospun Nanofibers: An Overview from Theoretical, Empirical, and Numerical Approaches. *Int. J. Polym. Mater. Polym. Biomater.* **2016,** (just-accepted).
5. Doshi, J.; Reneker, D. H. Electrospinning Process and Applications of Electrospun Fibers. In Industry Applications Society Annual Meeting, 1993. Conference Record of the 1993 IEEE.
6. Liu, M.; et al. Recent Advances in Directed Assembly of Nanowires or Nanotubes. *Nano-Micro Lett.* **2012,** *4*(3), 142–153.
7. Liao, I.; Chew, S.; Leong, K. Aligned Core–Shell Nanofibers Delivering Bioactive Proteins. *Nanomedicine* **2006,** *1*(4), 465–471.
8. Sill, T. J.; von Recum, H. A. Electrospinning: Applications in Drug Delivery and Tissue Engineering. *Biomaterials* **2008,** *29*(13), 1989–2006.
9. Yarin, A. L.; et al. Material Encapsulation and Transport in Core–Shell Micro/Nanofibers, Polymer and Carbon Nanotubes and Micro/Nanochannels. *J. Mater. Chem.* **2007,** *17*(25), 2585–2599.
10. Sun, Z.; et al. Compound Core–Shell Polymer Nanofibers by Co-Electrospinning. *Adv. mater.* **2003,** *15*(22), 1929–1932.
11. Zhang, Y.; et al. Preparation of Core-Shell Structured PCL-r-Gelatin Bi-Component Nanofibers by Coaxial Electrospinning. *Chem. Mater.* **2004,** *16*(18), 3406–3409.
12. Rafiei, S. Electrospinning Process: A Comprehensive Review and Update. *Appl. Methodol. Polym. Res. Technol.* **2014,** 1.
13. Rafiei, S.; et al. Advances in Electrospun Nanofibers Modeling: An Overview. In *Materials Science and Engineering. Volume I: Physical Process, Methods, and Models;* Haghi, A. K., Eds.; Apple Academic Press: Waretown, New Jersey, 2014; pp 30–50.

14. Wu, X.-F.; Salkovskiy, Y.; Dzenis, Y. A. Modeling of Solvent Evaporation from Polymer Jets in Electrospinning. *Appl. Phys. Lett.* **2011**, *98*(22), 223108.
15. Veleirinho, B.; Rei, M. F.; Lopes-DA-Silva, J. Solvent and Concentration Effects on the Properties of Electrospun Poly (Ethylene Terephthalate) Nanofiber Mats. *J. Polym. Sci. Part B: Polym. Phys.* **2008**, *46*(5), 460–471.
16. Reznik, S.; et al. Evolution of a Compound Droplet Attached to a Core-Shell Nozzle Under the Action of a Strong Electric Field. *Phys. Fluids* (1994-present), **2006**, *18*(6), 62101–62108.
17. Zussman, E.; et al. Electrospun Polyaniline/Poly (Methyl Methacrylate)-Derived Turbostratic Carbon Micro-/Nanotubes. *Adv. Mater.* **2006**, *18*(3), 348–353.
18. Li, F.; Yin, X.-Y.; Yin, X.-Z. Axisymmetric and Non-Axisymmetric Instability of an Electrified Viscous Coaxial Jet. *J. Fluid Mech.* **2009**, *632*(2), 199–225.
19. Gómez, H.; et al. Isogeometric Analysis of the Cahn–Hilliard Phase-Field Model. *Comput. Methods Appl. Mech. Eng.* **2008**, *197*(49), 4333–4352.
20. Jacqmin, D. Calculation of Two-Phase Navier–Stokes Flows Using Phase-Field Modeling. *J. Comput. Phys.* **1999**, *155*(1), 96–127.
21. Dayal, P.; Kyu, T. Porous Fiber Formation in Polymer-Solvent System Undergoing Solvent Evaporation. *J. Appl. Phys.* **2006**, *100*(4), 043512.
22. Thompson, C.; et al. Effects of Parameters on Nanofiber Diameter Determined from Electrospinning Model. *Polymer* **2007**, *48*(23), 6913–6922.
23. Beachley, V.; Wen, X. Effect of Electrospinning Parameters on the Nanofiber Diameter and Length. *Mater. Sci. Eng.* **2009**, *29*(3), 663–668.
24. Nasir, M.; et al. Control of Diameter, Morphology, and Structure of PVDF Nanofiber Fabricated by Electrospray Deposition. *J. Polym. Sci. Part B: Polym. Phys.* **2006**, *44*(5), 779–786.
25. Feng, J. Stretching of a Straight Electrically Charged Viscoelastic Jet. *J. Non-Newtonian Fluid Mech.* **2003**, *116*(1), 55–70.
26. Deitzel, J.M.; et al. The Effect of Processing Variables on the Morphology of Electrospun Nanofibers and Textiles. *Polymer* **2001**, *42*(1), 261–272.
27. Wannatong, L.; Sirivat, A.; Supaphol, P. Effects of Solvents on Electrospun Polymeric Fibers: Preliminary Study on Polystyrene. *Polym. Int.* **2004**, *53*(11), 1851–1859.

CHAPTER 4

AN EXPERIMENTAL INVESTIGATION ON ELECTROSPINNING PROCESS TO PRODUCE PAN NANOFIBERS

SHIMA MAGHSOODLOU*

University of Guilan, Rasht, Iran

Corresponding author. E-mail: sh.maghsoodlou@gmail.com

CONTENTS

ABSTRACT

The objective of this work is to apply genetic programming (GP) as a machine learning technique in the context of nanofiber diameter detection. Defined functions automatically enable GP to describe the useful and reusable subroutines dynamically during a run. By using this method, a mathematical equation can be obtained without any restriction. Several relationships based on GP to determine the nanofiber diameter were proposed. The result showed that all solutions had high regression coefficient which indicted the accuracy of solution predictions. Solution E had the best accuracy percentage and the lowest error percentage.

4.1 INTRODUCTION

Electrospinning is a useful process for producing nanofibers. In this technology, the interrelationships among the relevant variables are unknown or poorly understood. If the problem domain is well understood, there maybe analytical tools that will provide quality solutions without the uncertainty inherent in a stochastic search process. Some researchers have tried to estimate the influence of each parameter on the fiber diameter. For this purpose, a polynomial equation has been discovered by using different methods such as response surface methodology, neural network, and genetic algorithm.[1-4] Recently, the polynomial relationships by designing response surface models have been expressed. They used a typical response surface function for their input variables and then optimized it for the parameters' mutual interaction on the diameter of produced fibers.[5-8]

The objective of this study was to investigate the effective electrospinning parameters, solution concentration, applied voltage, feeding rate, and collector distance on electrospun polyacrylonitrile (PAN) nanofibers diameter by a new method which expresses different solutions with complete mathematical equations. One of the efficient techniques for obtaining optimum conditions in a multivariable system is genetic programming (GP), an intelligent data mining method.

GP and other evolutionary algorithms that explore the interrelationships among the relevant variables are unknown or poorly understood. It can help discover variables and operations that are important; provide novel solutions to individual problems; unveil unexpected relationships among variables; and, sometimes, GP can discover new concepts that can be applied in a wide variety of circumstances.[9,10]

4.2 TAGUCHI METHOD

4.2.1 MATERIALS

PAN powder (Mw = 10^5 g/mol), consisting of 93.7 wt.% acrylonitrile (AN) and 6.3 wt.% methyl acrylate (MA) was supplied with Polyacryl Co. (Isfahan, Iran) and N,N-dimethylformamide (DMF) was obtained from Merck, respectively, as polymer and solvent.

4.2.2 DESIGN OF EXPERIMENTS

In this study, the design of experiments was conducted using Taguchi method. Four factors of design were considered as the most significant parameters in electrospinning process. For solution concentration, six levels were considered in the range of 8–18 (%w/v). For each of three other factors, three levels were considered which include 10, 14, and 18 kV as applied voltage, 0.5, 0.75, and 1 ml/h as solution feeding rate, and 10, 13 and 15 cm as spinning distance. Therefore, the L_{18} arrays of Taguchi method were used to design electrospinning experiments for preparing PAN nanofibers (Table 4.1). The average fiber diameter of each case was determined by analyzing scanning electron microscope (SEM) images with Image J software.

TABLE 4.1 Applied Amount of Factors According to L18 Arrays of Taguchi Method for Electrospinning Process.

Trial number	Solution concentration (%w/v)	Applied voltage (kV)	Feeding rate (ml/h)	Spinning distance (cm)
1	8	10	0.5	10
2	8	14	0.75	13
3	8	18	1	15
4	10	10	0.5	13
5	10	14	0.75	15
6	10	18	1	10
7	12	10	0.75	10
8	12	14	1	13
9	12	18	0.5	15
10	14	10	1	15
11	14	14	0.5	10
12	14	18	0.75	13
13	16	10	0.75	15
14	16	14	1	10
15	16	18	0.5	13
16	18	10	1	13
17	18	14	0.5	15
18	18	18	0.75	10

4.3 METHODOLOGY

GP is a form of evolutionary computation in which the individuals in the evolving population are computer programs rather than bit strings. These computer programs can be solutions for any problems. In GP, programs are usually expressed as syntax trees rather than as lines of a code. The variables and constants in the program are leaves of the tree. In GP, they are called terminals, while the arithmetic operations are internal nodes called functions. The sets of allowed functions and terminals together form the primitive set of a GP system. In this case, the representation used in GP is a set of trees. All the basic steps are indicated in Figure 4.1.

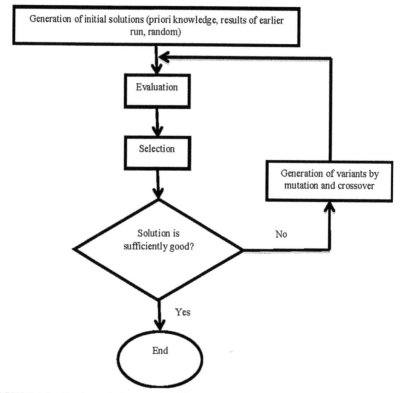

FIGURE 4.1   Basic evolutionary algorithm.

Similar to other evolutionary algorithms, the individuals in the initial population are typically randomly generated in GP. There are a number of different approaches to generate this random initial population. In this case, we used the combination of two simple methods of full and grow to generate initial population known as ramped half-and-half. At first, initial population of solutions should be generated.[11,12] Population sizing has been one of the most important topics to consider in evolutionary computation. Researchers usually argue that a "small" population size could guide the algorithm to poor solutions and that a "large" population size could make the algorithm expend more computation time in finding a solution. Therefore, here we are facing a trade-off that needs to be approximated, feeding the algorithm with "enough" chromosomes, to obtain "good" solutions. "Enough," for us, is directly related to instances in the search space and diversity. Genetic operators in GP are applied to individuals that are probabilistically selected based on fitness. That is, better individuals are more likely to have more child programs than inferior individuals.[13,14]

As mentioned earlier, it is necessary to be able to evaluate how "good" a potential solution is relative to other potential solutions. The "fitness function" is responsible for performing this evaluation and returning a positive integer number, or "fitness value" that reflects how optimal the solution is: the higher the number, the better the solution.[15,16]

Selection is the stage of a GP in which individual genomes are chosen from a population for breeding later. GP departs significantly from other evolutionary algorithms in the implementation of the operators of crossover and mutation. The most commonly used form of crossover is subtree crossover. Given two parents, subtree crossover randomly (and independently) selects a crossover point (a node) in each parent tree.[17,18]

The most commonly used form of mutation in GP (which we will call subtree mutation) randomly selects a mutation point in a tree and substitutes the subtree rooted there with a randomly generated subtree. Subtree mutation is sometimes implemented as a crossover between a program and a newly generated random program; this operation is also known as "headless chicken" crossover.[19]

After the fitness of each individual has been found in the population, 10% of individuals, who have the best performance, get copied without alteration into the next generation. One percent of the remaining individuals get randomly selected for mutation and the rest of them generate offsprings by using subtree crossover operation.

In this method, the goodness of each solution on the training data has been used as a fitness measure.

These solutions have been expressed by tree views (Fig. 4.2). Terminal set for diameter prediction is: $\{x_1, x_2, x_3, x_4\}$:

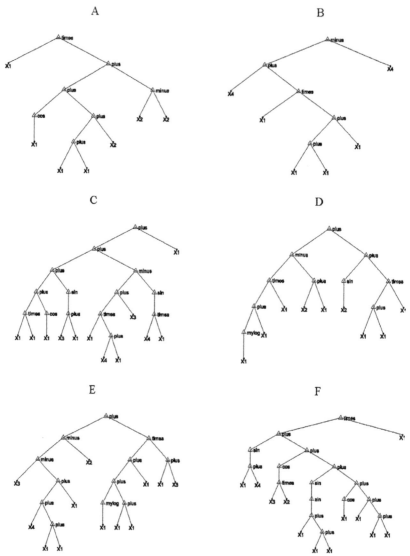

FIGURE 4.2   Tree generation for different iterations (A–F).

Where x_1 is the concentration, x_2 is the voltage, x_3 is the feeding rate, and x_4 is spinning distance.

TABLE 4.2 GP Parameters in the Study.

Max generation	10–45	Crossover rate	0.89
Population size	200	Mutation rate	0.01
Grow method	Ramped half-and-half	Elitism	0.1
Selection method	Roulette wheel/tournament	Function set	$\{+,-,*,\log, \sin, \cos\}$

To explore the mathematical interrelationship of these parameters, mathematical operators have been used as functions set (Table 4.2).

In following, we called solutions which have been discovered by running GP algorithm for 10 iterations as solution A, for 15–25–30 iterations as solution B, for 20 iterations as solution C, for 35 iterations as solution D, for 40 iterations as solution E, and for 45 iterations as solution F (Fig. 4.2).

As it can be seen in Figure 4.2, after running the GP algorithm several times, we surprisingly reached the same trees by running the algorithm for 15, 25, and 30 iterations (Table 4.3).

TABLE 4.3 GP Equations for Different Tree Generation.

Tree generation	Equation
A	$x_1(\cos(x_1) + 2x_1 + x_2$
B	$3x_1^2$
C	$x_1^2 + \cos(x_1) + \sin(x_1 + x_3) + x_1 + (x_1(x_1 + x_4)) + x_3 - \sin(x_1 x_4)$
D	$x_1(\log(x_1) + x_1) - x_2 - x_1 + \sin(x_2) + 2x_1^2$
E	$x_3 - x_4 - 3x_1 - x_2 + ((\log(x_1) + 3x_1)(x_1 + x_3))$
F	$x_1*(\sin(x_1 + x_4) + \cos(x_2 + x_3) + \sin(\sin(3x_1)) + \cos(x_1) + 3x_1)$

4.4 RESULTS AND DISCUSSION

At first, by using equations which were obtained from the GP processor, prediction for each case was obtained (Table 4.4). By the results of each prediction formula, the best solution must be selected. Therefore, statistical analysis methods including errors and accuracy will be calculated as follows.

Error measurement statistics play a critical role in tracking forecast accuracy, monitoring for exceptions, and benchmarking of forecasting process. Interpretation of these statistics can be tricky, particularly when working with low-volume data or when trying to assess accuracy across multiple items.[20]

TABLE 4.4   Actual and Solution Results Report.

Number of cases	Experiment results	Computations					
		A	B	C	D	E	F
1	106.88	206.836	192	218.1469	180.6807	168.1763	174.7295
2	106.66	238.836	192	241.5508	178.2153	167.652	186.6335
3	120.07	270.836	192	256.686	172.4737	168.1278	185.6816
4	267.18	291.6093	300	339.7113	289.456	273	270.0423
5	255.43	331.6093	300	359.6558	286.9906	275	276.1845
6	294.48	371.6093	300	309.6673	281.249	284	302.2763
7	440.55	418.1262	432	421.1958	422.4062	417.5096	429.0587
8	468.35	466.1262	432	459.1458	419.9408	420.0294	421.3778
9	431.33	514.1262	432	482.0787	414.1992	394.9898	454.833
10	590.52	533.9143	588	617.3193	579.5018	581.1919	569.5766
11	572.45	589.9143	588	546.5914	577.0364	560.1189	561.1584
12	638.75	645.9143	588	589.9155	571.2948	564.1554	606.1254
13	825.16	656.6774	768	765.9835	760.7219	751.919	731.2049
14	769.66	720.6774	768	686.8615	758.2565	765.47	741.6053
15	718.26	784.6774	768	734.2215	752.5149	733.368	745.9736
16	956.58	839.8857	972	900.8114	966.0509	973.8502	967.1498
17	1152.29	911.8857	972	936.9939	963.5855	939.7225	985.9526
18	986.89	983.8857	972	848.1121	957.8439	954.7864	997.13

Mean absolute error (MAE) measures the average magnitude of the errors in a set of forecasts, without considering their direction. It measures the accuracy for continuous variables. MAE is the average over the verification sample of the absolute values of the differences between forecast and the corresponding observation. It is a linear score which means that all the individual differences are weighted equally in the average.[21]

$$\mathrm{MAE} = \frac{1}{n}\sum_{i=1}^{n}|F_i - A_i| \tag{4.1}$$

Mean absolute percent error (MAPE) measures the size of the error in percentage terms. It is calculated as the average of the unsigned percentage error, and is given by Equation 4.2:[21,22]

$$\mathrm{MAPE} = \frac{1}{n}\sum_{i=1}^{n}\left|\frac{F_i - A_i}{A_i}\right| \tag{4.2}$$

where A_1 is the actual value and F_1 is the forecast value.

MAE and MAPE methods have been used to choose the best solutions among the different solutions which have been generated using GP method. Obtained results are shown in Table 4.5. A regression approximation has also been applied to the solutions data with experimental data. The fundamentals for the least squares procedure can be found in Table 4.5.

TABLE 4.5 Solution Results Comparison with MAP, MAPE, Accuracy Percent, and Regression Coefficient.

Solutions	MAP	MAPE	Accuracy percent	R^2
A	77.40553	0.286713	71.32872	0.9437
B	42.15833	0.169415	83.05852	0.9766
C	44.80421	0.283897	71.61034	0.9774
D	42.71605	0.150551	84.94492	0.9767
E	42.48063	0.134651	86.53492	0.9731
F	40.47469	0.149517	85.04826	0.9783

As seen in Table 4.5, the solutions D, E, and F have the best performance in forecasting nanofiber diameter based on the highest accuracy and lowest error. Among them, solution E has the maximum accuracy percent and minimum MAPE. This solution equation includes all initial factors that make it more applicable (Table 4.5). The regression analysis for these solutions is shown in Figure 4.3. High regression coefficient between the actual and predicted data (Table 4.5) indicates excellent solution evaluation results.

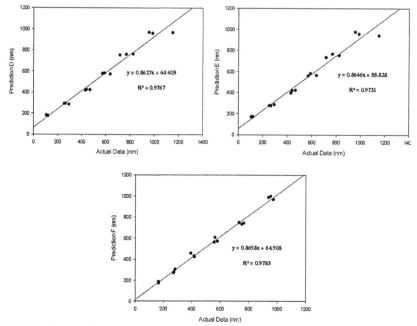

FIGURE 4.3   Predicted versus actual mean fiber diameter plots for D, E, and F solutions.

4.5   CONCLUSION

An experimental investigation of the electrospinning process to produce PAN nanofibers was carried out. The goal was to investigate the parameters interrelationship effects on the diameter and to establish a prediction scheme of this influence by using a complete equation, not a typical polynomial formula. GP was used for modeling and optimizing the average nanofiber diameters, and different solutions were obtained. High regression coefficients and accuracy percent with low error quantities indicated an acceptable GP performance. Some of the equations did not contain all of the factors but solution E that had the maximum accuracy percent included all of them.

KEYWORDS

- **electrospinning process**
- **nanofiber diameter estimation**
- **genetic programming**

REFERENCES

1. Fridrikh, S. V.; Yu, J. H.; Brenner, M. P.; Rutledge, G. C. *Phys. Rev. Lett.* **2003**, *90*, 144502–1.
2. Maleki, M.; Latifi, M.; Amani-Tehran, M. *World. Acad. Sci. Eng. Technol.* **2010**, *64*, 389.
3. Ziabari, M.; Mottaghitalab, V.; Haghi, A. K. *Korean J. Chem. Eng.* **2010**, *27*, 340.
4. Saehana, S.; Iskandar, F.; Abdullah, M. *World Acad. Sci. Eng. Technol. Int. J. Chem. Mol. Nucl. Mater. Metall. Eng.* **2013**, *7*, 86.
5. Vargas, J. H.; Campos, J. B. G.; Romero, J. L.; Ortega, J. M. P. *ACS Sustain. Chem. Eng.* **2013**, *2*, 454.
6. Yazdanpanah, M.; Khanmohammadi, M.; Aghdam, R.; Shabani, K.; Rajabi, M. *Curr. Chem. Lett.* **2014**, *3*, 175.
7. Yördem, O. S.; Papila, M.; Menceloğlu, Y. Z. *Mater. Design.* **2008**, *29*, 34.
8. Nasouri, K.; Bahrambeygi, H.; Rabbi, A.; Shoushtari, A. M.; Kaflou, A. *J. Appl. Polym. Sci.* **2012**, *126*, 127.
9. Koza, J. R.; Poli, R. *Search Methodologies;* Burke, E. K., Kendall, G., Eds.; Springer: US, 2005; p 127.
10. Koza, J. R.; Keane, M. A.; Streeter, M. J.; Mydlowec, W.; Yu, J.; Lanza, G. *Genetic Programming IV;* Springer Science and Business Media: USA, 2006; p 590.
11. Koza, J. R. *Stat. Comput.* **1994**, *4*, 87.
12. Koza, J. R. *Genetic Programming;* Cambridge MA: USA, 1992; p 680.
13. Deb, K.; Pratap, A.; Agarwal, S.; Meyarivan, T. *IEEE. T. Evolut. Comput.* **2002**, *6*, 182.
14. Sivanandam, S. N.; Deepa, S. N. *Introduction to Genetic Algorithms*; Springer: USA, 2008; p 131.
15. Altenberg, L. *Adv. Genet. Program.* **1994**, *3*, 47.
16. Burke, E. K.; Gustafson, S.; Kendall, G. *IEEE Trans. Evolut. Comput.* **2004**, *8*, 47.
17. Yang, J.; Honavar, V. *Feature Extraction, Construction and Selection*, Liu, H., Motoda, H., Eds.; Springer: US, 1998; p 136.
18. Muni, D. P.; Pal, N. R.; Das, J. *IEEE Trans. Syst. Man Cybern. B.* **2006**, *36*, 106.
19. Luke, S.; Spector, L. *Genet. Program.* **1997**, *97*, 240.
20. Fornell, C.; Larcker, D. F. *J. Mark. Res.* **1981**, *18*, 382.
21. Li, G.; Shi, J. *Appl. Energy.* **2010**, *87*, 2313.
22. Swanson, D. A.; Tayman, J.; Bryan, T. M. *J. Popul. Res.* **2011**, *28*, 225.

CHAPTER 5

NANO-ZINC BORATES AS POLYVINYL CHLORIDE THERMAL STABILIZERS

SEVDIYE ATAKUL[1] AND DEVRIM BALKÖSE[2,*]

[1]*Akzo Nobel Boya AŞ İzmir, Turkey*

[2]*İzmir Institute of Technology Department of Chemical Engineering, Gulbahce Urla İzmir, Turkey*

Corresponding author. E-mail: devrimbalkose@gmail.com

CONTENTS

ABSTRACT

Different zinc borate species were tested for their use as thermal stabilizers for polyvinyl chloride plastigels. PVC thermomat tests, thermal gravimetric analysis (TGA) and color tests were made for measuring thermal stability. PVC thermomat tests indicated that the induction time was shorter and the stability time was longer for plastigel without any zinc borate than the plastigels having different kinds of zinc borates. Zinc borates prevented dehydrochlorination of PVC at early times, but after accumulation of $ZnCl_2$ to a critical level, the samples underwent accelerated dehydrochlorination due to strong Lewis acidity of $ZnCl_2$. Zinc borates had a smoke suppressant effect leading to increase in the solid residue formed at 600°C as indicated by TGA. The discoloration of the samples with $2ZnO$ $3B_2O_3$ $7H_2O$, $7ZnO$ B_2O_3, and $4ZnO$ B_2O_3 H_2O was prevented at early heating periods as indicated by their yellowness index values. Thus zinc borates can be recommended as thermal stabilizers for short heating periods at low temperatures.

5.1   INTRODUCTION

Polyvinyl chloride (PVC) is one of the most frequently used polymers and a significant amount is processed in the form of plastisols. It is extensively used as a thermoplastic material because it has superior mechanical and physical properties, high chemical resistance, easy modification, low production cost, and widely utilized in durable applications, for example, pipes, window profiles, house siding, wire cable insulation, and flooring.[9] PVC plastisol is commonly used as a textile ink for screen printing and as a coating, particularly in outdoor applications (roof and furniture).

PVC processing and handling has one serious problem which is its rather low thermal stability. The dehydrochlorination of PVC starts at about 100°C and the formation of polyene sequences by dehydrochlorination reaction causes discoloration. Unplasticized PVC products do not contribute to fire propagation and pass most flammability tests without flame retardants due to the high chlorine content of PVC.[15] The exposure of the PVC to the UV component of sunlight results in discoloration of the PVC molecules. For example, white PVC pipe will turn yellow to brown color.

A PVC compound may contain polymer, stabilizers, plasticizers, extenders, lubricants, fillers, pigments, polymeric processing aids, and impact modifiers. Other miscellaneous materials used in PVC occasionally include fire retardants, optical bleaches, and blowing agents.

Organic flame retardants and smoke suppressants used in plasticized PVC include members of the families of phosphate esters and brominated phthalate plasticizers. Inorganic flame retardants and smoke suppressants include alumina trihydrate,[11,12] antimony trioxide, barium metaborate, huntite (hydrous magnesium calcium carbonate), transition metal oxides,[12] magnesium hydroxide, ammonium octamolybdate, various mixed metal complexes, natural zeolite, zinc stearate,[1] zinc borate (ZB),[5] zinc hydroxystannate,[2,6] and zinc phosphate.[4]

Dioctyl phthalate (DOP) is considered by most to be the PVC industry standard general purpose plasticizer due to its long use history and excellent balance of properties in nondemanding applications. DOP is sanctioned by Food and Drug Administration (FDA) and corresponding organizations in many other countries for use in numerous food-packaging applications. DOP has a history of safe usage in medical devices such as blood bags and tubing used in kidney dialysis machines.[2] However, EU has classified DOP as a hazardous substance for reproduction (Category 2: toxic to reproduction and development) and included it in the so-called SVHC Candidate List as a "substance of very high concern" (SVHC) in October 2008. Then, it was announced in the list in 2015.

There is a number of different ZB species defined in the literature, and each of them has a specific application based on their unique properties. The features of ZB are usually determined by several factors, such as B_2O_3/ZnO molar ratio, the amount of water, and bonding type of water molecules to the main structure.[8] $2ZnO\ 3B_2O_3\ 7H_2O$ is formed when borax is added to aqueous solutions of soluble zinc salts at temperatures below 70°C and $2ZnO\ 3B_2O_3\ 3.5H_2O$ is generally produced with the reaction between zinc oxide and boric acid.[3] Inverse emulsion technique was used to produce ZB for lubrication.[14] Mergen et al.[10] have investigated the production of $4ZnO\ B_2O_3\ H_2O$ and its effect on PVC. Gonen et al.[8] have investigated supercritical ethanol (SCE) drying of ZBs for obtaining nanoparticles. Commercial and synthesized $2ZnO\ 3B_2O_3\ 3H_2O$ and $ZnO\ B_2O_3\ 2H_2O$ samples were dried by both conventional and SCE drying methods and the products were compared by Gonen et al.[8] It was found that $ZnO\ B_2O_3\ 2H_2O$ decomposed completely into zinc oxide and boric acid. However, ZB having oxide formula of $2ZnO \cdot 3B_2O_3\ 3H_2O$ decomposed partially to form anhydrous ZB, zinc oxide, water, and boric acid during the SCE drying carried out at 250°C and 6.5 MPa.[8]

In this study, different commercial and synthesized ZB additives were used as a thermal stabilizer in PVC plastigels obtained by gelation of PVC-DOP plastisols. Dehydrochlorination of PVC on heating and the mass loss from plastisols were measured by 763 PVC Thermomat and thermal gravimetric analysis (TGA). Colors of the plastigels were measured by Avantes Avamouse color measurement instrument.

5.1.1 THEORY

According to the Frye–Horst mechanism,[4] anions exchange with labile chlorine atoms next to double bonds or carbonyl substituents. The ZB in the system reacts with the HCl released from the system during the heating by exchanging their borate ions.

$$MX_2 + \quad \sim\!\!\!\!\sim CH - \overset{\overset{\textstyle Cl}{|}}{\underset{\underset{\textstyle H}{|}}{C}} \sim\!\!\!\!\sim \quad \longrightarrow \quad ClMX + \quad \sim\!\!\!\!\sim \overset{\overset{\textstyle X}{|}}{C}\!-\!H - \overset{\overset{\textstyle H}{|}}{\underset{\underset{\textstyle H}{|}}{C}} \sim\!\!\!\!\sim \qquad (5.1)$$

where X is the borate anion. The bond between C and borate anions is more stable than the C–Cl bond. At long heating times, this bond will be broken too and HCl gas will be released. The reactions occurring in the PVC system are as follows:

$$PVC\ (s) \rightarrow dePVC\ (s) + HCl\ (g) \qquad (5.2)$$

$$2ZnO \cdot 3B_2O_3 + 12HCl \rightarrow$$
$$Zn(HO)Cl + ZnCl_2 + 3BCl_3 + 3HBO_2 + 4H_2O \qquad (5.3)$$

Presence of yellowness is associated with scorching, soiling, and general product degradation by light, chemical exposure, and processing of PVC. Moreover, it is an important property in many industries for several reasons. First of all, processing of various materials may cause yellowing. Then, the purity of some products may be determined based on the amount of yellowness present. Moreover, some products degrade and turn yellow with exposure to sunlight, temperature, or other environmental factors during their use. Yellowness indices are used chiefly to quantify these types of degradation with a single value. They can be used when measuring clear, near-colorless liquids or solids in transmission, and near-white, opaque solids in reflectance. Thus, yellowness has become an important variable to measure in industries such as textiles, paints, and plastics. According to yellowness indices application note from Hunter Lab Company (2008), yellowness index (Y) of plastigels was prepared by using Equation 5.4 for measurements at D65 light source and 2° standard observer angle.[17]

$$YI = \frac{100 \times (1.2985X - 1.1335Z)}{Y} \qquad (5.4)$$

where X, Y, and Z values are the tristimulus values measured by the spectrophotometer.

5.2 MATERIALS AND METHODS

5.2.1 MATERIALS

In this study, PVC (Petvinil P 38/74, from PETKIM), DOP with 0.98 g/cm^3 density from Merck, Viscobyk 5025 which is a liquid with 0.88 g/cm^3 density and consisted of low-volatile carboxylic acid derivatives with acidic wetting and dispersing components from BYK chemicals,[16] and ZB are used to investigate thermal stability of PVC.

PVC was emulsion type and was the product of Petkim (Petvinil P 38/74). Savrik et al.[13] have characterized PVC powder in a previous study. It was found that it composed of spheres of a large particle size range from 10 nm to 20 µm. Fourier-transform infrared spectroscopy (FTIR) indicated that it contained carbonyl and carboxylate groups belonging to additives such as surface active agents, plasticizers, and antioxidants used in the production of PVC. These additives were 1.6% in mass of PVC as determined by ethanol extraction. Energy dispersive X ray analysis showed that the surfaces of PVC particles were coated with carbon-rich materials. These additives from polymerization process made PVC powder more thermally stable as understood from Metrom PVC Thermomat tests as well. Viscobyk 5025 was used as a wetting and dispersing agent to reduce the viscosity of the liquid phase of the plastisols.

ZBs were used as thermal stabilizers for PVC. Two commercial ZBs from US Borax (Fire Brake 2235 and Fire Brake 411) and ZB prepared from zinc nitrate and borax,[8] from zinc nitrate, borax, and Span 60[14] and their derivatives obtained by SCE extraction[8] were used as thermal stabilizers. The plastigel sample numbers and their empirical formulas are shown in Table 5.1.

TABLE 5.1 Sample Number, Zinc Borate (ZB) Codes, and Description.

Sample no.	Zinc borate code	Raw material	Chemical formula	Produced by
1	–	–	–	–
2	ZB 1	Commercial fire brake 2335	$2ZnO\ 3B_2O_3\ 3.5H_2O$	US Borax
3	ZB 2	Zinc nitrate, borax deca-hydrate, span 60	$ZnO\ 3B_3O_3(OH)_5\ H_2O$	[14]

TABLE 5.1 *(Continued)*

Sample no.	Zinc borate code	Raw material	Chemical formula	Produced by
4	ZB 3	Zinc nitrate, borax decahydrate	$2ZnO\ 3B_2O_3\ 7H_2O$	[8]
5	ZB 4 SCE	Supercritical ethanol (SCE) extracted zinc borate (ZB) produced from ZnO and boric acid	$70ZnO\ B_2O_3$	[8]
6	ZB 5 SCE	SCE extracted ZB produced from commercial fire brake 2335	$7ZnO\ B_2O_3$	[8]
7	ZB 6	Commercial fire brake 415	$4ZnOB_2O_3\ H_2O$	US Borax
8	ZB 7 SCE	SCE extracted ZB-3	$58ZnO\ B_2O_3$	[8]

5.2.2 METHODS

5.2.2.1 PREPARATION OF PVC PLASTISOL

The mixture of PVC plastisols was prepared with seven different ZBs, DOP, viscobyk, and PVC. In the first part, a powder form of PVC was put in the air-circulating oven at 60°C for 2 h to dry. Then, 100 g PVC, 80 ml DOP, and 5 ml Viscobyk 5025 were mixed in a beaker and they were stirred with a mechanical mixer at room temperature for 5 min. Next, 10 g of mixture was put in eight different small beakers and 0.135 g of different ZB samples were added to each of them. Thus, plastisols having 53.98% PVC, 42.32% DOP, 2.37% Viskobyk 5025, and 1.33% ZB were obtained.

All of the prepared plastisols were spread on a cardboard to obtain 90 μm film thicknesses by using film applicator (Sheen 113 N) with 110 mm/s speed. Air-circulating oven was used to gel plastisol films at 140°C for 15 min.

5.2.2.2 THERMAL STABILITY TESTS

5.2.2.2.1 PVC Thermomat

Thermal stabilization of PVC can be determined by Metrohm 763 PVC Thermomat equipment. This equipment has eight different cells to put samples and measurement can be done at the same temperature for these eight different samples simultaneously. Plastigel films were cut into small pieces nearly having 0.5 cm sides. Small sample pieces of 0.5 g were put into tubes and placed into the heating blocks of the thermomat. When PVC plastigels are heated in PVC thermomat in the presence of nitrogen gas, the conductivity of water, through which nitrogen gas is passed, changes with respect to time. The period when the conductivity starts to increase is called induction time, and the period when the conductivity value reaches 50 s/cm is called stability time as shown in Figure 5.1. This value is the maximum acceptable level of degradation. The rate of dehy-drochlorination can be determined by the conductivity of the solution where the evolved HCl gas is transferred.[1] Conductivity time measure-ments of plastigel films heated at 140, 160, and 180°C was made by ther-momat apparatus.

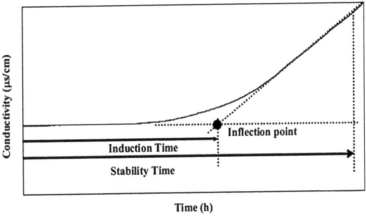

FIGURE 5.1 A representative conductivity versus time curve for polyvinyl chloride (PVC) thermomat results.

5.2.2.2.2 Thermal Gravimetric Analysis (TGA)

TGA was achieved to observe loss of mass caused by the elimination of HCl and the evaporation of plasticizer from plastigels. Setaram Labsys TGA instrument was used for this observation. Plastigel samples were heated from room temperature up to 1000°C with 10°C/min heating rate in dry air flow.

5.2.2.3 COLOR TEST

The photographs of plastigel samples before and after PVC thermomat tests were taken with an Olympus digital camera. The yellowness index values of the prepared samples were measured with Avantes Avamouse color measurement instrument with illuminant D65 and 2° standard observer value.

5.3   RESULTS AND DISCUSSION

5.3.1 THERMAL STABILITY TEST

5.3.1.1 PVC THERMOMAT TEST

PVC thermomat instrument investigated the thermal stability of PVC. The kinetic parameters of dehydrochlorination of PVC were determined by heating them at 140, 160, and 180°C. Conductivity versus time curves of PVC samples are shown in Figures 5.2–5.4 and induction and stability times are reported in Table 5.2. The difference in induction and stability times of pure PVC was higher than that of other PVC's which contains different kinds of ZBs. The induction time of the samples with ZBs was longer than that of the control plastigel at 140 and 160°C indicating the stabilizing effect of ZB before accumulation of $ZnCl_2$ to a critical value. However, the opposite behavior was observed at 180°C, indicating faster accumulation of $ZnCl_2$ to a critical value at this high temperature. In general, induction time of the samples with ZB was determined between 4–8 h at 140°C, whereas it was 1 h at 160°C and 0.25 h at 180°C. If induction and stability times were compared to the results of Erdogdu et al.[4] study, it can be seen that the results are different. In Erdoğdu et al.[4] study, induction and stability times of plastigel without ZB were 2.99 and 15.21 h at 140°C and 1.49 and 3.35 h at 160°C, respectively. In the present study, they are 3.43 and 9.4 h at 140°C and 0.8 and 2.41 h at 160°C. In Erdogdu et al.[4] study, the plastigel with 1.36% ZnO $3B_2O_3$ $3.5H_2O$ had induction and stability times of 7.04 and 7.18 h at 140°C. In the present study, for the sample with 1.33% ZnO $3B_2O_3$ $3.5H_2O$, induction and stability times at 140°C were 4.43 and 4.91 h, respectively. Both studies indicated that ZB addition increased the induction time period with respect to the control sample and the difference in induction and stability times get very close to each other which indicates thermal stabilization effect up to induction time and the zinc burning effect is observed after induction time.

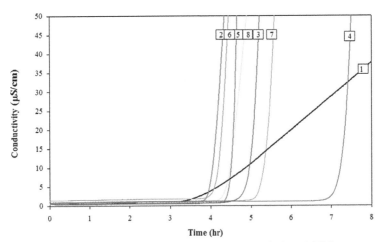

FIGURE 5.2   Conductivity versus time curves of PVC plastigels at 140°C.

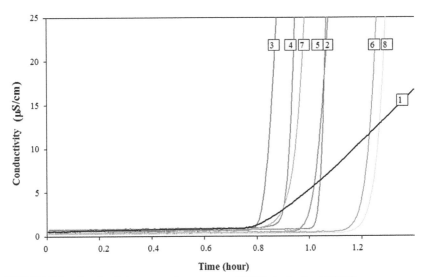

FIGURE 5.3   Conductivity versus time curves of PVC plastigels at 160°C.

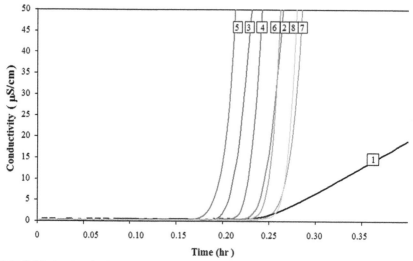

FIGURE 5.4   Conductivity versus time curves of PVC plastigels at 180°C.

Considering induction times at different temperatures, the reaction rate constant maybe calculated to investigate the kinetic behavior of the PVC samples. This phenomenon is formulated as Arrhenius equation.[7]

$$k = Ae^{-E/RT} \qquad (5.5)$$

In this equation, k is the reaction rate constant, A is the pre-exponential factor, E represents activation energy, R is the gas constant, and T is the temperature. The units of variables depend on the reaction order.

Table 5.2   Induction and Stability Time Values of the PVC Samples.

Sample		Induction time (h)			Stability time (h)		
		140°C	160°C	180°C	140°C	160°C	180°C
1	–	3.43	0.8	0.25	9.4	2.41	0.63
2	$2ZnO\ 3B_2O_3\ 3.5H_2O$	4.43	1.04	0.24	4.91	1.12	0.27
3	$ZnO\ 3B_3O_3(OH)_5\ H_2O$	4.00	0.85	0.22	4.35	0.92	0.23
4	$2ZnO\ 3B_2O_3\ 7H_2O$	5.07	0.92	0.23	5.21	0.97	0.25
5	$70ZnO\ B_2O_3$	7.72	1.1	0.22	7.5	1.08	0.22
6	$7ZnO\ B_2O_3$	4.68	1.32	0.27	4.66	1.28	0.26
7	$4ZnO\ B_2O_3\ H_2O$	4.22	1.09	0.29	4.46	1.02	0.29
8	$58ZnO\ B_2O_3$	5.77	1.34	0.28	5.59	1.32	0.28

Pre-exponential factor, A, is generally determined from experiment, described as collision frequency and R is the gas constant which can be found in literature. Only if the A is known for this study, the reaction rate constant can be calculated.

First of all, the general kinetic model can be shown below for this study at Equation 5.6 where c is the concentration of groups available for dehydrochlorination in solid PVC at time t.

$$-\frac{dc}{dt} = kc \qquad (5.6)$$

If Equation 5.6 is rearranged and integrated as shown by Equations 5.7 and 5.8,

$$-\int \frac{dc}{c} = k \int dt \qquad (5.7)$$

$$-\ln(c/c_0) = kt \qquad (5.8)$$

where c_0 is the initial concentration of groups available for dehydrochlorination. If C is defined as

$$-\ln(c/c_0) = C \qquad (5.9)$$

$$\text{then} \qquad C = kt \qquad (5.10)$$

From Equations 5.5 and 5.10, we get Equation 5.11.

$$\ln\frac{1}{t} = -\frac{E}{RT} + \ln A + \ln C \qquad (5.11)$$

The final form of the kinetic model is shown in Equation 5.11 in which $\ln A$ and $\ln C$ cannot be known initially. $\ln A$ can be observed experimentally and $\ln C$ is very small compared to the first and second term and can be neglected.

$$\ln\frac{1}{t} = -\frac{E}{RT} + \ln A \qquad (5.12)$$

Thus, the activation energy can be found from the slopes of the lines $\ln t$ versus $1/T$ using Equation 5.12.

The same extent of dehydrochlorination occurs in all samples at induction times.

The activation energy of PVC dehydrochlorination and R^2 values of the lines for ln t versus $1/T$ are shown in Table 5.3, and the activation energy for induction time data in Figure 5.5. The activation energy was found from the slopes of the lines using Equation 5.12.

Table 5.3 Activation Energies of PVC Dehydrochlorination Reaction for the Samples.

Sample no.	From induction time	
	Activation energy (kJ/mo)l	R^2
1	105.11	0.9986
2	112.83	0.9991
3	114.42	0.9996
4	118.28	0.9989
5	137.44	0.9992
6	112.05	0.9916
7	106.40	0.9996
8	116.35	0.9978

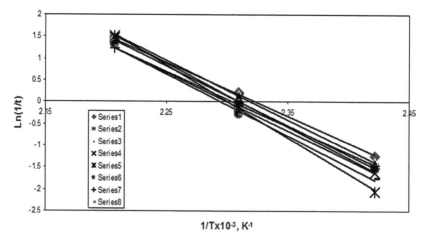

FIGURE 5.5 Activation energy determination using induction times at different temperatures.

The activation energy of the plastigel without any ZB was found as 105 kJ/mol from induction time. The activation energy for the film without any ZB was found at the same order as 135.4 kJ/mol by Erdoğdu et. al.[4] by using the kinetic data obtained from conductivity time curves. The activation energy values of all the samples having ZBs were higher than the activation energy of PVC without ZB, indicating the slowing down the dehydrochlorination reaction by ZB additive according to Equation 5.3. However, the samples with ZBs were dehydrochlorinated at a very fast rate after the induction period, indicating the concentration of strong Lewis acid $ZnCl_2$ reached to the critical value to accelerate the dehydrochlorination.

5.3.1.2 THERMAL GRAVIMETRIC ANALYSIS

The thermal decomposition of PVC is a two-stage process: thermal decomposition in the first stage is mainly the evolution of HCl, and in the second stage is mainly cyclization of conjugated polyene sequences to form aromatic compounds. It is known that zinc-containing compounds can greatly reduce the smoke formation. Zinc chloride, which is formed along with the thermal decomposition of PVC, acts as an effective catalyst for the ionic dehydrochlorination of PVC due to its strong Lewis acidity, resulting in an increase in char formation and a decrease in smoke production.[11]

Mass loss caused by the elimination of HCl and the evaporation of plasticizer from plastigels and degradation of PVC to a solid residue by evolving organic volatile components was determined with TGA by dynamic heating up to 1000°C. Figure 5.6 shows the TG curves of the samples. In Table 5.4, the onset temperature of the mass loss and residual mass percentage values at 225, 400, and 600°C are shown.

The mass loss of the plastigel without any ZB started at 225°C. However, for the samples with ZB, it started around 200°C. At 225°C, the remaining mass of the samples with ZB had 77–97%. The mass remained at 400°C after the first step of TG curve was higher for samples with ZB (35–38%) than without ZB (29%). This result shows that adding ZB to PVC decreases the amount of volatiles formed. Further, heating the samples up to 600°C also showed the solid residue increase from 5.5 to 5.7–17% with ZB addition. Thus, Erdogdu et al.[4] and Ning and Guo[11] observed that ZB acted as a smoke suppressant.

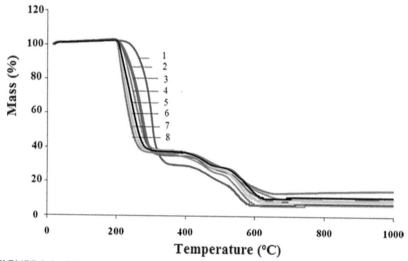

FIGURE 5.6 TG curves of the PVC plastigels: (1) control sample, (2) with 2ZnO 3B$_2$O$_3$ 3.5H$_2$O, (3) with ZnO 3B$_3$O$_3$(OH)$_5$ H$_2$O, (4) with 2ZnO 3B$_2$O$_3$ 7H$_2$O, (5) with70ZnO B$_2$O$_3$, (6) with 7ZnO B$_2$O$_3$, (7) with 4ZnO B$_2$O$_3$ H$_2$O, and (8) with 58ZnOB$_2$O$_3$.

TABLE 5.4 Onset Temperature of Mass Loss and the Residual Mass of PVC Plastigels at 225, 400, and 600°C.

Sample		T of onset mass loss (°C)	Residual mass (%)		
			225°C	400°C	600°C
1	–	222.0	99.9	29.0	5.5
2	2ZnO 3B$_2$O$_3$ 3.5H$_2$O	199.9	96.2	38.0	15.5
3	ZnO 3B$_3$O$_3$(OH)$_5$ H$_2$O	200.0	97.0	37.8	17.0
4	2ZnO 3B$_2$O$_3$ 7H$_2$O	200.0	96.5	36.0	12.5
5	70ZnO B$_2$O$_3$	199.9	77.5	37.8	9.99
6	7ZnO B$_2$O$_3$	199.9	82.0	37.8	10.0
7	4ZnO B$_2$O$_3$ H$_2$O	199.9	77.0	35.0	5.7
8	58ZnO B$_2$O$_3$	200.0	88.9	38.0	12.8

5.3.2 COLOR TEST

The photographs of the plastigel films prepared by heating the plastisols at 140°C for 15 min and the samples after PVC thermomat tests are seen in

Figures 5.7 and 5.8, respectively. As seen in the figures, the prepared films did not have any discoloration, but after the thermal stability test, they were all blackened. The change in color of a test sample from clear or white toward yellow was described with yellowness index by Equation 5.4. The yellowness index values of the samples as prepared are shown in Table 5.5. The yellowness index of the samples with $2ZnO\ 3B_2O_3\ 7H_2O$, $7ZnO\ B_2O_3$, and $4ZnO\ B_2O_3\ H_2O$ were lower than that of the sample without ZB. This indicated the number of conjugated dienes formed during the gelation period at $140°C$ was lower for the samples with this type of ZBs.

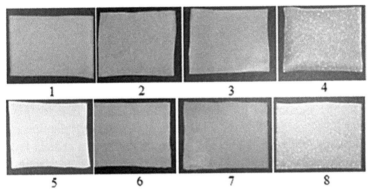

FIGURE 5.7   Photographs of the plastigels at $140°C$ for 15 min.

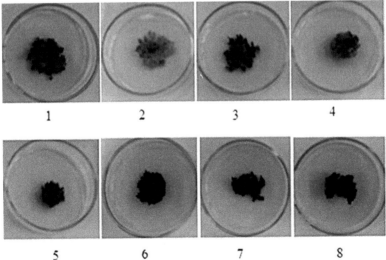

FIGURE 5.8   Photographs of the plastigels after PVC thermomat test at $140°C$.

TABLE 5.5　The Yellowness Index of the Films as Prepared.

Sample	Additive type	Plastigels as prepared at 140°C
1	–	5.09
2	$2ZnO\ 3B_2O_3\ 3.5H_2O$	6.18
3	$ZnO\ 3B_3O_3(OH)_5\ H_2O$	7.15
4	$2ZnO\ 3B_2O_3\ 7H_2O$	2.59
5	$70ZnO\ B_2O_3$	10.42
6	$7ZnO\ B_2O_3$	4.44
7	$4ZnO\ B_2O_3\ H_2O$	2.61
8	$58ZnO\ B_2O_3$	5.15

5.4　CONCLUSION

The present study indicated that the induction time was shorter and the stability time was longer for plastigel without any ZB than the plastigels having different kinds of ZBs. ZBs prevent dehydrochlorination of PVC at early times, but after the accumulation of $ZnCl_2$ to a critical level, the samples undergo accelerated dehydrochlorination due to strong Lewis acidity of $ZnCl_2$. The solid residues formed were also higher for the samples with ZBs showing the smoke suppressant effect. Thus, ZBs can be recommended as thermal stabilizers for short heating periods at low temperatures. The presence of $2ZnO\ 3B_2O_3\ 7H_2O$, $7ZnO\ B_2O_3$, and $4ZnO$ $B_2O_3\ H_2O$ prevents discoloration of PVC during this period as indicated by yellowness index values.

KEYWORDS

- PVC
- dehydrochlorination
- zinc borates
- yellowness index
- thermal degradation

REFERENCES

1. Atakul, S.; Balköse, D.; Ülkü, S. Synergistic Effect of Metal Soaps and Natural Zeolite on Poly(vinyl Chloride) Thermal Stability. *J. Vinyl Addit. Technol.* **2005,** *11*(2), 47–56.

2. Coaker, A. W. Poly(vinyl Chloride). In *Applied Polymer Science, 21st Century;* Craver, C. D., Carraher, C. E., Eds.; Elsevier: Oxford, 2000.

3. Eltepe, H. E.; Balkose, D.; Ulku, S. Effect of Temperature and Time on Zinc Borate Species Formed from Zinc Oxide and Boric Acid in Aqueous Medium. *Ind. Eng. Chem. Res.* **2007,** *46*(8), 2367–2371.

4. Erdoğdu, C. A.; Atakul, S.; Balköse, D.; Ülkü, S. Development of Synergistic Heat Stabilizers for PVC from Zinc Borate-Zinc Phosphate. *Chem. Eng. Commun.* **2009,** *196,* 148–160.

5. Fang, Y. Q.; Wang, Q.; Guo, C.; Song, Y.; Paul, A.; Cooper, P. A. Effect of Zinc Borate and Wood Flour on Thermal Degradation and Fire Retardancy of Polyvinyl Chloride (PVC) Composites. *J. Anal. Appl. Pyrolysis* **2013,** *100,* 230–236.

6. Ferm, D. J.; Shen, K. K. The Effect of Zinc Borate in Combination with Ammonium Octamolybdate or Zinc Stannate on Smoke Suppression in Flexible PVC. *J. Vinyl Addit. Technol.* **1997,** *3*(1), 33–41.

7. Fogler, H. S. *Elements of Chemical Reaction Engineering;* Prentice-Hall: New Jersey, 1986.

8. Gonen, M.; Balkose, D.; Ulku, S. Supercritical Ethanol Drying of Zinc Borates of $2ZnO\ 3B(2)O(3)\ 3H(2)O$ and $ZnO\ B_2O_3\ 2H(2)O$. *J. Supercrit. Fluids* **2011,** *59,* 43–52.

9. Gong, F.; Feng, M.; Zhao, C.; Zhang, S.; Yang, M. Thermal Properties of Poly(vinyl Chloride)/Montmorillonite Nanocomposites. *Polym. Degrad. Stab.* **2004,** *84,* 289–294.

10. Mergen, A.; Ipek, Y.; Bölek, H.; Öksüz, M. Production of Nano Zinc Borate (4ZnO $B_2O_3H_2O$) and Its Effect on PVC. *J. Eur. Ceram. Soc.* **2012,** *32*(9), 2001–2005.

11. Ning, Y.; Guo, S. Y. Frame-Retardant and Smoke-Suppressant Properties of Zinc Borate and Aluminum Trihydrate-Filled Rigid PVC. *J. Appl. Polym. Sci.* **2000,** *77*(14), 3119–3127.

12. Pi, H.; Guo, S. Y.; et al. Mechanochemical Improvement of the Flame-Retardant and Mechanical Properties of Zinc Borate and Zinc Borate—Aluminum Trihydrate-Filled Poly(vinyl Chloride). *J. Appl. Polym. Sci.* **2003,** *89*(3), 753–762.

13. Savrik, S. A.; Balkose, D.; Ulutan, S.; Ulku, S. Characterization of Poly(vinyl Chloride) Powder Produced by Emulsion Polymerization. *J. Therm. Anal. Calorim.* **2010,** *101*(2), 801–806.

14. Savrik, S. A.; Balkose, D.; Ulku, S. Synthesis of Zinc Borate by Inverse Emulsion Technique for Lubrication. *J. Therm. Anal. Calorim.* **2011,** *104*(2), 605–612.

15. Yong-zhong, B.; Zhi-ming, H.; Shen-xing, L.; Zhi-xue, W. Thermal Stability, Smoke Emission and Mechanical Properties of Poly(vinyl Chloride)/Hydrotalcite Nanocomposites. *Polym. Degrad. Stab.* **2008,** *93,* 448–455.

16. http://www.byk.com/en/additives/additives-by-name/viscobyk-5025.php. (Access date 2.2.2017)

17. https://www.hunterlab.se/wp-content/uploads/2012/11/Yellowness-Indices.pdf. (Access date: 2.2.2017)

CHAPTER 6

NANOFIBER PRODUCTION CAPABILITY OF ELECTRO-CENTRIFUGE TECHNIQUE

S. RAFIEI*

University of Guilan, Rasht, Iran

Corresponding author. E-mail: saeedeh.rafieii@gmail.com

CONTENTS

ABSTRACT

In this chapter, nanofiber production capability of the electro-centrifuge technique is compared with that of the conventional electrospinning method.

6.1 INTRODUCTION

When the diameters of polymer fibers are shrunk from micrometers to submicro- or nanometers, several new characteristics emerge, including a very high ratio of surface area to volume, flexibility in surface functionalities, and superior mechanical performance (e.g., stiffness and tensile strength). These unique properties make polymeric fibers ideal candidates for many important applications.[1-3]

Although several methods have been proposed for nanofiber manufacturing so far, an efficient and cost-effective procedure of production is still a challenge and is debated by many experts. Electrospinning has been the most successful method to produce nanofibers so far. It is a well-known and prominent procedure which is based on the electrostatic force that provides the possibility for spinning nanofibers from many types of polymers using melt or solution spinning.[4-6]

The use of mechanical methods to apply high rates of tension to a polymer solution or melt is only possible for polymers with high extension ability, to prevent uneven transmission of tension and stress concentration. In order to apply high rates of tension to a polymer solution or melt, methods which can apply even distribution of stress during the tensioning process are required. When the centrifugal force acts upon a substance, the particles of that matter experience a force proportional to their distance from the center of rotation. Thus, this force can be used to apply high rates of tension on a polymer solution. During the tensioning process, if the polymer solution has sufficient viscosity, it will be stretched as a string, transforming to a polymeric fiber after drying.[7,8]

Electro-centrifuge spinning has been recently introduced as a beneficial method for nanofiber production. It is necessary to specify the effective parameters on nanofiber production and diameters. In this research, the effective parameters of the process and the influence of each parameter on fiber diameter are addressed. Also, the fiber production capability of the electro-centrifuge technique is compared with that of the conventional electrospinning method.[9-11]

6.1.1 NANOFIBERS

Nanofibers are defined as fibres with diameters less than 100 nm which can be produced by different methods such as melt processing, interfacial polymerization, electrospinning, electro-centrifuge spinning, and antisolvent-induced polymer precipitation.[5] The fibers which are produced by these methods have a very high specific surface area, unusual strength, high surface energy, surface reactivity, and high thermal and electric conductivity that could be used in products by special application. These unique properties make the polymeric nanofibers ideal candidates for many important applications, such as nanofiltration,[12] nanocatalysis, tissue scaffolds,[13] protective clothing, filtration, nanoelectronics,[14] high-performance nanofibers,[15,16] drug-delivery systems, wound dressings, and composites. Although several methods have been proposed for nanofiber manufacturing so far, an efficient and cost-effective procedure of production is still a challenge and is debated by many experts.[17–20]

6.1.2   ELECTROSPINNING METHOD

Electrospinning is an efficient and simplicity mean of producing nanofibers by solidification of a polymer solution stretched by an electric field.[21] It is a well-known and prominent procedure based on electrostatic forces that provides the possibility for spinning nanofibers from many types of polymers using melt or solution spinning.[22,23]

In typical electrospinning process, an electrical potential is applied between droplets of polymer solution or melt held through a syringe needle and a grounded target. Afterward the electrostatic field stretches the polymer solution into fibers as the solvent is evaporating. During the process, the polymer jet undergoes instabilities, which together with the solution properties, determine the morphology of the forming nano-sized structures obtained onto the collector. The electrical forces elongate the jet thousands of times causing the jet to become very thin. Ultimately, the solvent evaporates or the melt solidifies, and a web containing very long nanofibers is collected on the target. The fiber morphology is controlled by the experimental design and is dependent on the solution properties and system conditions such as conductivity, solvent polarity, solution concentration, polymer molecular weight, viscosity, and applied voltage.[4,24]

To overcome various limitations of the typical electrospinning method, researchers have applied some modifications to the original setup.[24,25] Some of these reformed techniques which have been done by now include: electrospinning using a collector consisting of two pieces of electrical conductive plates separated by a gap, collecting spun nanofibers on a rotating thin wheel with a sharp edge, fabricating aligned yarns of nylon 6 nanofibers by rapidly oscillating a grounded frame within the jet,[26] and using a metal frame as a collector to generate parallel arrays of nanofibers.[27] Matthews et al.[26] have applied a rotating drum to collect aligned nanofibers at a very high speed up to thousands of revolution per minute (RPM). In another approach, Deitzel et al.[28] have used a multiple field technique which can straighten the polymer jet to some extent. Luming, et al.[29] have used this technique to collect aligned nanofibers by increasing the surface velocity of the drum.

Increasing the production rate of nanofibers has always been one of the most important challenges of researchers. Most efforts at this point have been focused on increasing the number of nozzles.[30] Yarin et al. have introduced the concept of "upward needleless electrospinning of multiple nanofibers".[31] A similar idea was discussed by Liu and He[32,33] by the name of "bubble electrospinning" method.

6.1.3 CENTRIFUGAL FORCE

In Newtonian mechanics, the term centrifugal force is used to refer to an inertial force (also called a "fictitious" force) directed away from the axis of rotation that appears to act on all objects when is viewed in a rotating reference frame. The concept of centrifugal force can be applied in rotating devices such as centrifuges, centrifugal pumps, centrifugal governors, centrifugal clutches, and so forth, as well as in centrifugal railways, planetary orbits, banked curves, and so forth, when they are analyzed in a rotating coordinate system.[34]

Centrifugal force is an outward force apparent in a rotating reference frame, it does not exist when measurements are made in an inertial frame of reference. All measurements of position and velocity must be made relative to some frame of reference. For example, if we are studying the motion of an object in an airliner traveling at great speed, we could calculate the motion of the object with respect to the interior of the airliner, or to the surface of the earth. An inertial frame of reference is one that does not accelerate (including rotation). The use of an inertial frame of reference, which will be the case for all elementary calculations, is often not explicitly stated but maybe generally assumed.[35,36]

In terms of an inertial frame of reference, centrifugal force does not exist. All calculations can be performed using only Newton's laws of motion and the real forces. In its current usage, the term "centrifugal force" has no meaning in an inertial frame. According to Newton's first law, in an inertial frame, an object that has no force acting on it travels in a straight line. When measurements are made with respect to a rotating reference frame, however, the same object would have a curved path, because the frame of reference is rotating. If it is desired to apply Newton's law in the rotating frame, it is necessary to introduce new, fictitious forces to account for this curved motion. In the rotating reference frame, all objects, regardless of their state of motion, appear to be under the influence of a radially (from the axis of rotation) outward force that is proportional to their mass, the distance from the axis of rotation of the frame, and to the square of the angular velocity of the frame. This is the centrifugal force.[37,38]

6.1.4 CENTRIFUGAL SPINNING METHODS

Hooper patented the centrifugal spinning in 1924. Classical centrifugal spinning involves supplying centripetal force using a rotary distribution disc with side nozzle holes. The rotation shears the spinning dope, classical thermoplastic materials, for example, molten mineral or glass to form fibers such as mineral wool. Recent advances in centrifugal nanospinning have demonstrated the potential of this technique for nanofibers production. The initial work done of merging electrospinning with centrifugal concepts indicated a competitive edge compared with the technique alone in terms of its ability to produce homogeneous nanofibers less than 100 nm in size and a simultaneous one-step post spin-draw possibility to enhance molecular alignment and fiber crystallinity for improved tensile strength of the resultant fiber.[8,9]

Nanofibers which are produced by traditional electrospinning usually have a wide diameter distribution, whereas the fiber diameter has an important effect on the performance of nanofiber mats for many important applications. Several variables can influence the electrospun nanofiber diameter, diameter uniformity, and nanofiber quality such as volumetric charge density, distance from nozzle to the collector, initial jet/orifice radius, relaxation time, viscosity, concentration, solution density, electric potential, perturbation frequency, and solvent vapor pressure. Furthermore, fiber diameter increases when surface tension and flow rate are increased, but it decreases when electric current increases. Solvent–polymer interactions critically influence the diameter and morphology of the electrospun fibers. A method for obtaining finer nanofiber is to reduce the concentration of the polymer solution. Also, it has been found that with the increase in electrical conductivity of the polymer solution, there is a significant decrease in the electrospun nanofiber diameter. Also, the control of the process parameters such as applied voltage and temperature can influence electrospun nanofiber diameter. The diameter of the nanofibers increases when electrospinning voltage is increased.[2,39]

Although all these approaches can influence nanofiber diameter, there are some limitations due to the addition of unwanted components, such as insufficient and difficult controllability and strict thinning effect. Therefore, it is necessary to provide a method that can fabricate nanofibers with high uniformity and fineness without any additional procedures.

6.1.5 CENTRIFUGAL ELECTROSPINNING

Centrifugal electrospinning has been attempted and the setup typically comprises of rotary spinnerets similar to centrifugal spinning. Dosunmu et al. have reported centrifugal electrospinning using a rotary spinneret in 2006, by using a porous ceramic tube spinneret. Andrady et al. have reported using a rotatable spray head with four individual extrusion elements, which in turn can be made of bundles of multiple nozzles.

Reiter Oberflachen Technik GmbH developed the Hyper Bell centrifugal electrospinning technology, which was subsequently acquired by Dienes Apparatebau GmbH. A centrifugal electrospinning unit with three spin heads currently supplied by Dienes Apparatebau is able to increase the throughput of conventional nozzle electrospinning by thousand fold with minimum achievable nanofiber diameter of 80 nm. Coupling centripetal force with electrostatic force, highly aligned PLA electrospun fibers with improved modulus of 3.3 GPa (PLA in chloroform and tetrahydrofuran) were produced.

6.1.6 CENTRIFUGAL NANOSPINNING

A spinning method using only centripetal force to produce nanofibers was recently developed. Force spinning by FibeRio Technology Corporation achieved a minimum as-spun fibers diameter of 45 nm. The first polyethylene oxide (PEO) nanofibers were obtained by force spinning demonstrated homogeneity with an average diameter of 300–105 nm.

Furthermore, aligned PLA nanofibrous scaffolds were prepared using a similar method by Badrossamay et al. under the name "Rotary-Jet Spinning" and were used to seed cardiomyocytes in mice. The result showed a good tolerance of the nanofibers and the cell seedings successfully developed into pulsating multicellular tissues. The temperature, the rotational speed of the spinneret, and the collection distance are parameters influencing the geometry and morphology of centrifugally spun nanofibers.

Centrifugal nanospinning is a more versatile processthan electrospinning with respect to the broad range of the materials that can be processed, including melts, solutions, and emulsions. The process offers higher productivity and simplicity in an equipment setup without the complication of high-voltage in upscaled processing and the material constraint on electrical conductivity or relative permittivity compared to electrospinning. Nevertheless, centrifugal nanospinning is an extrusion process limited by challenges related to the material properties and the designs of the spinneret, which can lead to large differences in fiber quality and productivity. The fiber diameter in a single PLA sample can vary from 50 nm to 3.5 mm. The degree of complexity in the spinneret design is also proportional to the cost incurred. Although centrifugal nanospinning is a facile technique to generate 3D scaffolds with a moderately high degree of uniaxial alignment. Formation of more complex 3D nanostructures with functionalized nano features or alignment of fibers in more than one direction maybe difficult and has been demonstrated with this technique to date.

There has been much research and progress in the development of various designs and modification to the electrospinning process over the last century. Nevertheless, there are still many areas where further refinement of the process will be welcomed. To begin with, although there are several setups designed to achieve fiber alignment, there is still a serious shortcoming in getting highly aligned nanofibers over a large area of substantial thickness. Generally, a drum collector is not able to get highly aligned fibers even though it is able to get a larger area of fibrous mesh. Aligned nanofibers have been shown to induce cell elongation and proliferation in the direction of the fiber alignment. The ability to fabricate highly aligned fibers in large quantity over a large area will allow more investigation in cellular response to the fiber alignment in terms of gene expression and cell interaction. Typically, electrospun assemblies are in a two-dimensional form and in the case of yarn one-dimensional form. The only three-dimensional electrospun structure with significant length, width, and height is a fibrous tube. However, researchers have yet to find a way to consistently fabricate a solid three-dimensional structure through electrospinning. With the ability to fabricate three-dimensional structure, other applications such as bone replacement scaffold can be considered. Recently, Smit et al and Khil et al have demonstrated the fabrication of continuous yarn made out of pure nanofibers. However, the spinning speed of 30 m·min^{-1} is still much slower than the industrial fiber spinning process, which runs from 200 to 1500 m·min^{-1} for dry spinning. Yarn made out of electrospun fibers has many applications especially, when fabricated into textiles. However, for electrospun yarn to be adopted by the textile industry, its yarn-spinning speed has to be improved significantly.

Although Sun et al have made a breakthrough in the electrospinning process by creating controlled patterning using electrospun nanofibers, the small volume of solution that can be spun at one time significantly reduced the practicality of the process. An advantage of electrospinning is its ability to spin long continuous fibers at high speed. However, it is still not possible to form highly ordered structures rapidly. Arrayed nanofiber assemblies created thus far are based on arranging aligned nanofibers with very little control on the distance between each fiber.

In the following experimental part of this study, besides exploring the effects of centrifugal force on the nanofibers' diameter, the fiber production capability of the electro-centrifuge technique is compared with that of the conventional electrospinning method.

6.2 EXPERIMENTAL STUDY BY FOCUS ON FLOW RATE FOR BOTH DEVICES

6.2.1 MATERIALS

Commercial polyacrylonitrile (PAN) polymer powder with $\overline{Mw} = 100{,}000$ g/mol and $\overline{Mn} = 70{,}000$ g/mol was supplied by Poly Acryl, Iran. The used solvent was dimethylformamide (DMF) from Merck Company. The polymer solutions of PAN in DMF with the concentration of 13–16 wt.% were prepared using a digital scale (Libror AEU-210, Shimadzu) with accuracy to the gage of 0.0001 g. Dissolving and stirring of the mixture was performed with constant speed at room temperature after that, the solution was kept at 70°C for 2 h to complete the dissolution.

Scanning electron microscope (SEM) images were used to measure the nanofibers' diameter of different samples using a field emission scanning electron microscope (Philips SEM, XL-30). At least 100 fibers were chosen to compare each sample layer and diameter with the image scale. The average of the results was applied as the diameter of fibers produced within the process.

In order to measure the flow rate, $0.3 + x$ mg of the solution with four decimal places accuracy was poured into the needle container. Then the needle was located in the centrifuge apparatus and centrifugal action was continued until the solution weight reached $0.3 - x$ mg. The difference between the two weights indicate the amount of pumped solution in grams. By measuring the density of solvent and polymer, the volume of the pumped solution can be calculated.

6.2.2 ELECTRO CENTRIFUGE METHOD AND SETUP

Figure 6.1 represents a schematic picture of the centrifuge apparatus. In order to exert centrifugal force, a rotating disk with the capability of controlling speed in the range of 0–10,000 rounds per minute has been used. A tube with an inner diameter of around 4 mm is eccentrically placed within the cylinder body as a polymer container which is connected to a needle with geometric characteristics of 0.165 mm inner diameter, 0.3 mm outer diameter and 17 mm of length ofnozzle. The assembly of the polymer solution container and the nozzle inside the disk reduces the effects of airstream on the nozzle tip during the rotation. The intense impact between the air and the needle causes the polymer solution to dry at the needle tip and block the flow of the polymer solution through the nozzle.

The centrifuge method for PAN solution with the concentration of 12–16 wt.% at a rotational speed in the range of 0–9540 rpm can produce defected nanofiber with drop sprinkling and bead. To improve the centrifuge method in order to obtain intact nanofibers, electrical force was used.

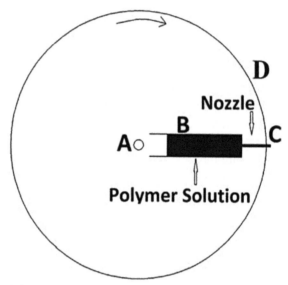

FIGURE 6.1   Schematic representation of centrifuge spinning setup, (**A**) axle of rotation, (**B**) polymer solution container, (**C**) nozzle tip, and (**D**) cylinder.

A high-voltage power supply was used to apply an electrostatic force which is able to generate DC voltages up to 22 kV. A metallic cylinder of 26.6 cm diameter and 10 cm height which is connected to a negative electrode of high-voltage supply was used as the fiber collector and the nozzle was connected to the positive electrode as shown in Figure 6.2. The distances between the disk center and the needle tip, and the interior surface of the collector cylinder and the needle tip are 5.3 and 8 cm, respectively. The volume of the container filled by the solution was only 0.30 ml in all experiments.[37,38]

Applying centrifugal force causes the polymer solution to flow out of the nozzle. A jet is formed at the nozzle exit if the viscosity of the solution is high enough; the electrical and centrifugal forces elongate the jet thousands of times and it becomes very thin. Ultimately, by evaporating the solvent, very long nanofibers are collected on the interior surface of the collector.

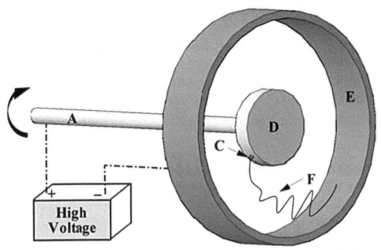

FIGURE 6.2   Schematic representation of electro-centrifuge setup, (**A**) axle of rotation, (**C**) nozzle tip, (**D**) cylinder, (**E**) collector, and (**F**) polymeric jet.

6.2.3   THEORY OF FLOW RATE

It should be noted that the flow rate is controlled only by the centrifugal force in the present experimental setup. When the free surface of the polymer solution in the container and the nozzle tip is at the same atmospheric pressure, the pressure within the container increases as the centrifugal force is exactly equal to the pressure loss across the nozzle due to the viscosity effect that is, $\Delta P_{1,2} = \Delta P_{2,3}$. On the other hand, the pressure rise in the container can be estimated as:

$$\Delta P_{1,2} = \int_{x_0}^{L_0} \rho l \omega^2 dl \text{ or } \Delta P_{1,2} = \frac{1}{2} \rho \left(L_0^2 - x_0^2 \right) \omega^2 \qquad (6.1)$$

where ρ is the liquid density, L_0 is the distance between the center of rotation and the nozzle entrance, x_0 is the distance between the center of rotation and the liquid free surface, and ω is the angular speed.

Equation 6.1 shows that the pressure loss across the nozzle depends on the length of the liquid column $(L_0 - x_0)$ within the container. Since the flow rate is also proportional to the same pressure difference, we can control the flow rate by changing x_0 and ω.

6.2.4   RESULTS AND DISCUSSION

Coulombic electrical forces and external field forces play an important role in electrospinning method. The electro-centrifuge method has been established with the intention of adding centrifugal force to the above forces. Therefore, the effect of centrifugal force on the production rate and fiber diameter is a matter of importance and has been investigated in this study.

6.2.5   EFFECT OF ROTATIONAL SPEED AT CONSTANT FLOW RATE ON THE NANOFIBER DIAMETER

In order to investigate the effect of the rotational speed at the constant flow rate on the fibers' diameter, fibers were produced from a 15 wt.% solution at five different quantities of the solution. For each run, the rotational speed was adjusted by trial and error in a way that the flow rate at all rotational speeds remains unchanged (1.2140 ml/h). As reported in Table 6.1, the obtained results reveal that the fibers' diameter increases by increasing the rotational speed at the constant flow rate. This can be interpreted in this way by increasing the rotational speed, the aerodynamic effects of the airstream formed in between of rotating cylinders become more prominent around the fiber. As the velocity of the airflow increases, the polymer jet gets dry before it can be elongated further by the centrifugal force. Therefore, it seems that at every flow rate there is a rotational speed in which the minimum fiber diameter can be spun.

TABLE 6.1   Nanofiber Diameters using Different Rotational Speeds at a Constant Flow Rate 1.2140 ml/h.

Nanofiber production method	Rotational speed (rpm)	Nanofiber diameter (nm)
Electro-centrifuge	6360	344
Electro-centrifuge	7950	396
Electro-centrifuge	9540	470
Electrospinning	0	352

6.2.6 COMPARISON BETWEEN THE NANOFIBER PRODUCTION RATES IN ELECTROSPINNING AND ELECTRO-CENTRIFUGE SPINNING SYSTEMS

Before comparing the two systems of electro-centrifuge and electrospinning, it is essential to know that all the effective variables (excluding the centrifugal force) of the two systems have the same conditions.

The variation of surface tension cannot be significant when the polymer concentration is in a range of 13–16% and therefore, we assume that this parameter is constant for all experiments. Comparisons between the two systems are made at two different applied voltages of 10 and 15 kV, and for 13–16 wt.% solutions of PAN at the same conditions. The most effective voltage which is possible to be applied to an 8 cm gap is equal to 15 KV. At higher voltage, the surrounding air will be ionized, which results in the generation of electric current between the positive and negative poles. The setup parameters are exactly the same for both systems, except for the flow rate. In the electrospinning system, a syringe pump provides the necessary flow rates, while in the electro-centrifuge system the flow rate is adjusted by the centrifugal force. For the specified parameters, the mean diameter of the nanofibers obtained from the two systems was in a range of 200–600 nm. Figure 6.3. shows typical SEM images of nanofibers produced from a 15 wt.% polymer solution with a rotational speed of 6360 rpm and an average diameter of 410 nm.

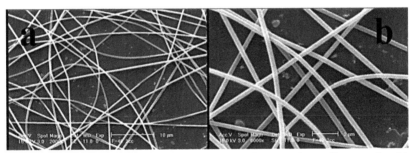

FIGURE 6.3 Scanning electron microscope (SEM) images of nanofibers produced by electro-centrifuge method with 6360 rpm and 0.3 cc polymer solution in the container, (a) low magnification (2000×); (b) high magnification (6000×).

Keeping other parameters constant, by increasing the flow rate to a specific limit, bead formation and drop sprinkling occur in electrospinning.

The flow rate of the solution at the beginning of the bead formation is defined as the maximum rate of solution pumping for electrospinning. However in the electro-centrifuge case, increasing the rotational speed, will increase the flow rate of the solution when other parameters are constant. Therefore, the production rate of fibers by the electro-centrifuge system is calculated by measuring the solution flow rate at the maximum possible rotational speed that leads to the production of fibers which are free of defects and beads. The increase in the production rate of fibers by the electro-centrifuge method is defined as the ratio of flow rates of the two systems when the produced fibers have the same diameter.

As it can be seen, the electro-centrifuge method can increase the rate of fiber production to a large extent. For the concentration of 16 wt.%, the rpm of 9540 is not the maximum possible rotational speed and it is feasible to produce fibers with this concentration at higher velocities. This case was not investigated here due to the apparatus limitations.

TABLE 6.2   Effect of Polymer Concentration on the Production Rate at Constant Voltage of 15 KV.

Nanofiber production method	Rotational speed (rpm)	Polymer solution concentration (wt.%)	Production rate percent
Electro-centrifuge	6360	13	1200
Electro-centrifuge	7950	14	723
Electro-centrifuge	9540	15	450
Electro-centrifuge		16	195
electrospinning	0	13	197

Tables 6.2 shows a decline in the production rate index as the concentration of the polymer solution are increased in constant voltage. The observations show that at a constant rotational velocity and applied voltage, upon increasing the concentration, the flow rate decreases and the production rate of the electro-centrifuge system rapidly approaches that of the electrospinning method. Interestingly, increasing the applied voltage it reduces the production rate when other parameters are constant. The effect of the centrifugal force on the production rate increase is more prominent for lower voltages due to the lower electrospinning production rate in that range.

It can be concluded from the experiment that at lower polymer concentrations, the centrifugal forces are more effective. For example, at the concentration of 13 wt.% and voltage of 10 kV, the production rate is 12 times larger than the conventional electrospinning approach (Table 6.3).

TABLE 6.3   Effect of Applied Voltage on the Production Rate in Constant Concentration of 13%.

Nanofiber production method	Rotational speed (rpm)	Applied voltage (KV)	Production rate percent
Electro-centrifuge	6360	10	187
Electro-centrifuge	7950	12	444
Electro-centrifuge	9540	14	678
Electro-centrifuge	–	15	1200
Electrospinning	0	10	177

6.3   CONCLUSION

Centrifugal and electrical forces have been employed simultaneously for nanofiber production and the effect of adding the centrifugal force to the electrical forces on the nanofiber diameter and production rate (which are the most important factors for nanofiber production) has been investigated in this study.

At first, the effect of increasing the rotational speed at a constant flow rate on the fiber diameter was investigated which causes an increase in the diameter of the produced fibers upon increasing the rotational speed. As a result of an increase in the impact intensity of the air surrounding the nozzle with the exiting jet, the jet dries quickly.

In addition, the nanofiber production capability of the novel method was investigated and compared with the conventional electrospinning. The results demonstrated a significant increase in the production rate of the electro-centrifuge method compared to electrospinning which is depended on the concentration of the polymer solution and the applied voltage. Employing the electro-centrifuge technique can overcome the low production rate of electrospinning method. Therefore, this approach can be a satisfactory alternative to electrospinning.

KEYWORDS

- nanofibers
- electrospinning
- production techniques

REFERENCES

1. Ramakrishna, S.; et al. *An Introduction to Electrospinning and Nanofibers*; World Scientific: USA, 2005; Vol. 8.
2. Frenot, A.; Chronakis, I. S. Polymer Nanofibers Assembled by Electrospinning. *Curr. Opin. Colloid Interface Sci.* **2003**, *8*(1), 64–75.
3. Ding, B.; Yu, J. *Electrospun Nanofibers for Energy and Environmental Applications*. Springer: USA, 2014.
4. Doshi, J.; Reneker, D. H. *Electrospinning Process and Applications of Electrospun Fibers*. in *Industry Applications Society Annual Meeting, 1993., Conference Record of the 1993 IEEE*. 1993. IEEE.
5. Reneker, D. H.; Chun, I. Nanometre Diameter Fibres of Polymer, Produced by Electrospinning. *Nanotechnology* **1996**, *7*(3), 216.
6. Greiner, A.; Wendorff, J. H. Electrospinning: A Fascinating Method for the Preparation of Ultrathin Fibers. *Angew. Chem. Int. Ed.* **2007**, *46*(30), 5670–5703.
7. Rist, R. C. *Influencing the Policy Process with Qualitative Research*. 1994.
8. Voelker, H.; et al. *Production of Fibers by Centrifugal Spinning*. 1996, Google Patents.
9. Edmondson, D.; et al. Centrifugal Electrospinning of Highly Aligned Polymer Nanofibers Over a Large Area. *J. Mater. Chem.* **2012**, *22*(35), 18646–18652.
10. Liu, S. L.; et al. Assembly of Oriented Ultrafine Polymer Fibers by Centrifugal Electrospinning. *J. Nanomater.* **2013**, *2013*(2514103), 8.
11. Mary, L. A.; et al. Centrifugal Spun Ultrafine Fibrous Web as a Potential Drug Delivery Vehicle. *eXPRESS Polym. Lett.* **2013**, *7*(3), 238–248.
12. Subramanian, S.; Seeram, R. New Directions in Nanofiltration Applications—Are Nanofibers the Right Materials as Membranes in Desalination? *Desalination* **2013**, *308*, 198–208.
13. Pham, Q. P.; Sharma, U.; Mikos, A. G. Electrospinning of Polymeric Nanofibers for Tissue Engineering Applications: A Review. *Tissue Eng.* **2006**, *12*(5), 1197–1211.
14. Lee, J.; Feng, P. X.-L.; Kaul, A. B. Characterization of Plasma Synthesized Vertical Carbon Nanofibers for Nanoelectronics Applications. in *MRS Proceedings*. 2012. Cambridge Univ Press.
15. Zhang, W.; Pintauro, P. N. High-Performance Nanofiber Fuel Cell Electrodes. *Chem. Sus. Chem.* **2011**, *4*(12), 1753–1757.

16. Tan, K.; Obendorf, S. K. Fabrication and Evaluation of Electrospun Nanofibrous Antimicrobial Nylon 6 Membranes. *J. Membr. Sci.* **2007,** *305*(1), 287–298.

17. Kenawy, E.-R.; et al. Processing of Polymer Nanofibers Through Electrospinning as Drug Delivery Systems. *Mater. Chem. Phys.* **2009,** *113*(1), 296–302.

18. Chen, J.-P.; Chang, G.-Y.; Chen, J.-K. Electrospun Collagen/Chitosan Nanofibrous Membrane as Wound Dressing. *Colloids Surf. A.* **2008,** *313*, 183–188.

19. Huang, Z.-M.; et al. A Review on Polymer Nanofibers by Electrospinning and Their Applications in Nanocomposites. *Compos. Sci. Tech.* **2003,** *63*(15), 2223–2253.

20. Jayaraman, K.; et al. Recent Advances in Polymer Nanofibers. *J. Nanosci. Nanotechnol.* **2004,** *4*(1–2), 52–65.

21. Frenot, A.; Chronakis, I. S. Polymer Nanofibers Assembled by Electrospinning. *Curr. Opin. Colloid Interface Sci.* **2003,** *8*, 64–75.

22. Persano, L.; et al. Industrial Upscaling of Electrospinning and Applications of Polymer Nanofibers: A Review. *Macromol. Mater. Eng.* **2013,** *298*(5), 504–520.

23. Rafiei, S.; et al. Mathematical Modeling in Electrospinning Process of Nanofibers: A Detailed Review. *Cellul. Chem. Technol.* **2013,** *47,* 323–338.

24. Luo, C.; et al. Electrospinning Versus Fibre Production Methods: From Specifics to Technological Convergence. *Chem. Soc. Rev.* **2012,** *41*(13), 4708–4735.

25. Sahay, R.; Thavasi, V.; Ramakrishna, S. Design Modifications in Electrospinning Setup for Advanced Applications. *J. Nanomater.* **2011,** *2011*, 17.

26. Fong, H.; et al. Generation of Electrospun Fibers of Nylon 6 and Nylon 6-Montmorillonite Nanocomposite. *Polymer* **2002,** *43*(3), 775–780.

27. Dersch, R. et al.; Electrospun Nanofibers: Internal Structure and Intrinsic Orientation. *J. Polym. Sci. Part A.* **2003,** *41*(4), 545–553.

28. Deitzel, J.; et al. Controlled Deposition of Electrospun Poly (ethylene oxide) Fibers. *Polymer* **2001,** *42*(19), 8163–8170.

29. Pan, H.; et al. Continuous Aligned Polymer Fibers Produced by a Modified Electrospinning Method. *Polymer* **2006,** *47*(14), 4901–4904.

30. Fang, D.; Hsiao, B.; Chu, B. Multiple-Jet Electrospinning of Non-Woven Nanofiber Articles. *Polym. Prepr.* **2003,** *44*(2), 59–60.

31. Yarin, A.; Zussman, E. Upward Needleless Electrospinning of Multiple Nanofibers. *Polymer* **2004,** *45*(9), 2977–2980.

32. Liu, Y.; He, J.-H. Bubble Electrospinning for Mass Production of Nanofibers. *Int. J. Nonlinear Sci. Numer. Simul.* **2007,** *8*(3), 393–396.

33. Dosunmu, O.; et al. Electrospinning of Polymer Nanofibres from Multiple Jets on a Porous Tubular Surface. *Nanotechnology* **2006,** *17*(4), 1123.

34. Arya, A. P. *Introduction to Classical Mechanics;* Allyn and Bacon: Boston, MA, 1990.

35. Badrossamay, M. R.; et al. Nanofiber Assembly by Rotary Jet-Spinning. *Nano Lett.* **2010,** *10*(6), 2257–2261.

36. Truesdell, C. *Rational Mechanics;* Academic Press: New York, 1983.

37. Synge, J. L. *Principles of Mechanics;* Read Books Ltd.: USA, 2013.

38. Gantmakher, F. R. *Lectures in Analytical Mechanics;* Mir publishers: Russia, 1970.

39. Sandou, T.; Oya, A. Preparation of Carbon Nanotubes by Centrifugal Spinning of Coreshell Polymer Particles. *Carbon* **2005,** *43*(9), 2015–2017.

PART III
Chemical Theory and Computation

QUANTUM CHEMICAL EVALUATION OF CARBONATE ION EFFECTS ON THE ANTICANCER ACTIVITY OF PT(II) AND PD(II) COMPLEXES

ANA M. AMADO*

Química-Física Molecular, Departamento de Química, FCTUC, Universidade de Coimbra, P3004-535 Coimbra, Portugal, Tel./Fax: +351-239826541

Corresponding author. E-mail: amado.am@gmail.com

CONTENTS

ABSTRACT

Despite many years of intense research, the mechanisms underlying cisplatin (*cis*-Pt(NH$_3$)$_2$Cl$_2$, cDDP) anticancer activity, toxic side effects, and natural/acquired tumor resistance are still quite elusive. Understanding the cellular, molecular, and submolecular aspects of the drug action while of utmost relevance remain a challenge for researchers. There are some key points unanimously accepted, such as nuclear DNA being the main biological target and activation step being a hydrolysis reaction of a chloride ligand. A balance between kinetic liability and thermodynamic stability of the metal-ligand bonds has been pointed as a determining issue of cDDP biological activity.

Understanding the mechanisms of cDDP toxicity and resistance are of utmost relevance in chemotherapy. Moreover, the knowledge of potential drug interactions with physiological fluids constituents during the drugs's course from the site of administration to the biological target is extremely important.

Among the inorganic components of physiological fluids, bicarbonate ion stands out as a potential interfering agent in the cDDP anticancer activity. The present report intends to evaluate the influence of bicarbonate on the normal activation pathway of cDDP and on the stability of the active aqua forms of the drug at a molecular level. For that, quantum chemical calculations are performed for simulating the hydration and carbonation reactions of cDDP. Attempting to understand the anticancer inactivity of the parent complexes cDDPd, tDDP, and tDDPd, the calculations were extended to those systems.

Overall, the results showed that the presence of bicarbonate can contribute to a decrease of complex bioavailability due to the occurrence of diverting side reactions—carbonation of both original and aqua forms of the drug.

7.1 INTRODUCTION

In cDDP clinical formulations, the presence of 9 mg/ml of NaCl (154 mM) partially suppresses the chloride hydrolysis process.[1,2] Nevertheless, about 15% of the platinum drug transforms into its monoaqua form $(Pt(NH_3)_2(H_2O)Cl)^+$.[3,4] After entering the bloodstream, where the chloride concentration drops to about 105 mM,[5] further hydrolysis of cDDP to the monoaqua species takes place. The formation of the aqua species is activated even further after entrance into the cell, where the chloride concentration dropsvery drastic (4–80 mM depending on the cell type).

Along the path to its biochemical target (nuclear DNA), cDDP and the aqua species formed, come into contact with the different physiological components of extra- and intracellular fluids, namely inorganic ions. This allows the occurrence of side reactions that may divert the platinum species from or hindering its binding to the cellular target. After chloride ions, carbonate species are among the most abundant inorganic species of extracellular fluids. The reaction of CO_2 produced by the cells with H_2O leads to the formation of carbonic acid, H_2CO_3, which rapidly ionizes bicarbonate (HCO_3^-) at physiological pH conditions (pKa=6.1).

The formation of cDDP carbonated species has been demonstrated experimentally,[6] but the resultant biochemical impact remains poorly understood. Binter et al.[5] suggested that the presence of carbonate induces the formation of monofunctional cisplatin-DNA adducts which in turn dominate over the bifunctional adducts. Accordingly, a reduction in cDDP activity would be promoted, as monoadducts are unable to induce local DNA kinking and unwinding, effects that have been correlated to cDDP anticancer activity. Todd et al.[7] however, argued that carbonate species suppress the overall binding efficiency of cDDP to DNA rather than affecting the binding mode preferences. In accord to the authors, the carbonated species have either a neutral or a negative charge, which prevents the interaction with DNA. More recently, however, it was suggested that the carbonate interference with cDDP anticancer activity is mostly due to a combination between the two effects (prevalence of monoadducts and cut in binding efficiency).[8]

Aiming to provide some clues that might help to understand the carbonate effects on cDDP anticancer activity at a molecular level, quantum chemical calculations were performed to estimate the kinetic and thermodynamic parameters associated to the potential reactions of cDDP with HCO_3^-. The parent systems *trans* counterpart of cDDP (tDDP) and the respective palladium (II) derivatives (cDDPd and tDDPd), which do not bear anticancer activity,[9–11] were also considered. Hopefully, the results will contribute to an understanding of the carbonate effect with regard to cDDP anticancer activity and, tDDP, cDDPd, and tDDPd inactivity.

7.2   COMPUTATIONAL DETAILS

Calculations were performed using Gaussian 09 program (G09w)[12] installed on a PC machine. The standard all-electron 6-31G* basis set and the LANL2DZ effective core potential (ECP) were used to describe the nonmetal atoms and the metal centers (Pt and Pd), respectively, in combination with the mPW1PW91 DFT. The author is aware of the smallness and limitations of the theory level used. However, the mPW1PW/6-31G*/

LANL2DZ combination has been proven to yield a good balance between accuracy and computational demands in prior works.[13–17]

All calculations were performed within the self-consistent reaction field (SCRF) methodology to account for the solvent effect (aqueous solution; $\varepsilon = 78.39$). The newly implemented density-based solvation model (SMD), recommended for computing solvation ΔG values,[18] was used in all calculations along with the default parameters, including radii model for building up the solvent cavity.[18,19] Within this SCRF model, the solvation free energy is separated into two main components[20]: (i) the electrostatic contributions, determined within the polarizable continuum model (PCM) formalism, and (ii) the nonelectrostatic terms, inferred within the cavity-dispersion-solvent-structure protocol. The SMD model has proven to be a good choice for determining molecular properties, such as vibrational frequencies and pKa values.[19]

In all cases, full geometry optimizations were performed without any symmetry constraints. The obtained geometries were subjected to the frequency calculations (at 298.15 K) in order to estimate the energy corrections (e.g., zero-point vibrational energy (zpve), thermal, enthalpy, and Gibbs free energy) and to confirm the convergence either to a minimum (no negative eigenvalues) or to a first-order saddle point (one negative eigenvalues).

7.2.1 SEARCH FOR THE REACTION INTERVENING SPECIES

It is widely accepted that the activation step of cDDP corresponds to a hydrolysis reaction of a chloride ligand via an associative substitution pathway (S_N2 mechanism). It starts with the entering water molecule approaching the Pt(II) centers from one side of the coordination plane and proceeds via a trigonal bipyramidal transition state to a final product, in which a chloride ligand is displaced on the opposite side of the coordination plane. This mechanism is assumed to stand for the parent systems tDDP, cDDPd, and tDDPd.

In order to simulate the reaction pathways by quantum chemical calculations, the first step is to set up the initial guesses for the trigonal bipyramidal transition state (TS) geometries. The TS geometries present elongated bonds between the metal (M) and both leaving (L_{leav}) and entering (L_{ent}) ligands. The assembled initial guesses were submitted to full optimization, using the Gaussian OPT=TS keyword.[12] Confirmation of correspondence to true TS within the reaction path was obtained through a frequency calculation—a negative eigenvalue, corresponding to a state where the L_{leav} and L_{ent} alternately turn away and close to the metal center, is computed.

The optimized TS geometries were used, within the intrinsic reaction coordinate (IRC) formalism of G09w,[12] to assemble the initial guesses of their action intermediate geometries: of reactant (I_1), presenting the L_{leav} still linked to the Pt(II) center and the L_{ent} already interacting with the metal, and of product (I_2), presenting the L_{ent} linked to the Pt(II) center and the L_{leav} still interacting with the metal. The assembled I_1 and I_2 geometries were then submitted to the optimization and vibrational frequency calculations, in order to confirm the correspondence to real minima. Finally, in order to confirm that the assembled geometries I_1, TS, and I_2 are related within the same reaction coordinate patha final check was performed through a QST3 calculation.[12]

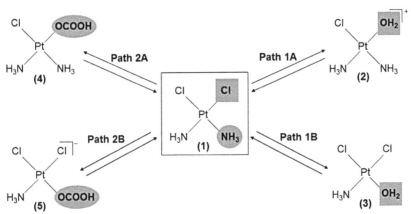

FIGURE 7.1 Schematic representation and nomenclature of the hydration and direct carbonation reactions possible for the dichlorodiamine complexes, taking cDDP as a model.

The described sequence of calculations was followed for each hydration and carbonation reactions considering the four complexes. The different reactions gathered are depicted in Figure 7.1 (hydration and direct carbonation, taking cDDP as a model), 2 (carbonation of cis-aqua species, taking cDDP as a model) and 3 (carbonation of trans-aqua species, taking tDDP as a model).

7.2.2 DETERMINATION OF THERMODYNAMIC AND KINETIC PARAMETERS

The forward and reverse rate constants of each reaction (k_i and k_{-i}, respectively) were calculated within the transition state theory[21] using Equation (7.1)

$$k_i = \frac{k_B T}{h c^0} \times e^{\left(\frac{-\Delta G_i^a}{RT}\right)}$$ (7.1)

Where k_B is the Boltzmann constant, T is the absolute room temperature (298.15 K), h is the Planck's constant, R is the ideal gas constant, and c^0 is the standard concentration (1 mol dm^{-3}). ΔG_i^a stands for the activation

Gibbs free energy predicted, being determined as

$$\Delta G_{i(-i)}^a = \Delta G_{TS} - \Delta G_{I_1(I_2)}$$ (7.2)

The ΔG value of each species (TS, I_1, or I_2) was determined as

$$\Delta G = E_{0K} + zpve + \Delta \Delta G_{0 \to TK}$$ (7.3)

where the three terms correspond to the calculated total energy at 0 K, the zpve, and the Gibbs free thermal energy corrections, respectively.

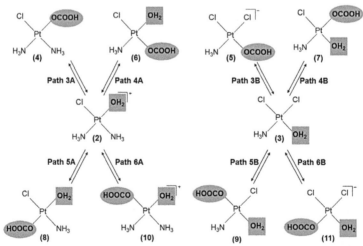

FIGURE 7.2 Schematic representation and nomenclature of the carbonation reactions possible for the aqua species of the *cis*-complexes formed by prior Cl$^-$ or NH$_3^-$ hydration (left and right side, respectively). cDDP is taken as a model.

Reactions spontaneity and thermal (endothermic or exothermic) character are inferred based on the predicted reaction Gibbs free energy (ΔG_i^r) and enthalpy change (ΔH_i^r), respectively. The two quantities are determined as

$$\Delta G_i^r = \Delta G_{I2} - \Delta G_{I1} \qquad (7.4)$$

and

$$\Delta H_i^r = \Delta H_{I2} - \Delta H_{I1} \qquad (7.5),$$

respectively. Each of the ΔG and ΔH values are determined according to the Equation (7.3) (in the case of ΔH, the last term corresponding to $\Delta\Delta H_{0\to TK}$, the enthalpy thermal corrections, instead of $\Delta\Delta G_{0\to TK}$).

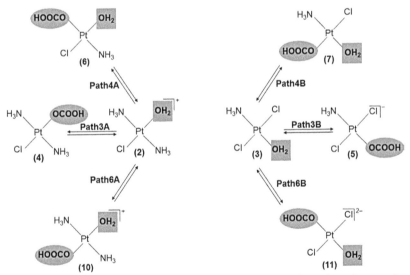

FIGURE 7.3 Schematic representation and nomenclature of the carbonation reactions possible for the aqua species of the *cis*-complexes formed by prior Cl⁻ or NH₃⁻ hydration (left and right side, respectively). tDDP is taken as a model.

7.3 RESULTS AND DISCUSSION

Notwithstanding, the high number of theoretical studies reported for the hydration reactions of cDDP and tDDP,[22–35] the quantum chemical reassessment of those reactions justifies for the reasons that follow.

Contrasting to cDDP and tDDP, information regarding the hydration of the palladium systems is very scarce.[28-36] On the other hand, there are no theoretical results on the reactions of bicarbonate with either of the complexes. Finally, the parameters used to define the calculations in the existing works are very different, which makes the comparisons dubious. In order to make the comparative analysis conclusive, it is important that the highest sameness between calculation parameters is ensured.

7.3.1 THERMODYNAMIC ANALYSIS

Figures 7.4 and 7.5 show a graphical representation of calculated ΔH_i^r and ΔG_i^r values in function of the hydration and carbonation reactions of dichlorodiamine and monoaqua complexes, respectively. The calculated ΔH_i^r and ΔG_i^r values are listed in Table 7.1 of the supplementary material.

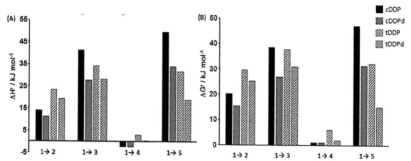

FIGURE 7.4 ΔH_i^r and ΔG_i^r values obtained for the hydration and direct carbonation reactions of the four dichlorodiamine complexes (see text and Figs. 7.1–7.3).

TABLE 7.1 Thermodynamic Parameters (kJ mol⁻¹) Predicted for the 1st Hydration and Possible Carbonation Reactions as a Function of Metal Complex.

Reaction[a]		Parameter	cDDP	cDDPd	tDDP	tDDPd
Path 1	(1)↔(2)	ΔH_{1A}^r	13.7	10.9	23.1	19.2
		ΔG_{1A}^r	20.2	15.4	29.6	25.3
	(1)↔(3)	ΔH_{1B}^r	40.9	27.4	33.9	28.0
		ΔG_{1B}^r	38.4	26.9	37.7	31.0

TABLE 7.1   *(Continued)*

Reaction[a]		Parameter	cDDP	cDDPd	tDDP	tDDPd
Path 2	(1)↔(4)	ΔH_{2A}^r	−2.6	−2.6	2.7	0.2
		ΔG_{2A}^r	1.1	1.1	6.2	2.0
	(1)↔(5)	ΔH_{2B}^r	49.3	33.8	31.5	18.9
		ΔG_{2B}^r	46.9	31.3	32.2	14.9
Path 3	(2)↔(4)	ΔH_{3A}^r	−43.1	−38.0	−53.4	−42.9
		ΔG_{3A}^r	−41.5	−40.0	−54.0	−42.2
	(3)↔(5)	ΔH_{3B}^r	3.4	0.8	−11.7	−14.2
		ΔG_{3B}^r	−0.5	−1.8	−9.6	−14.2
Path 4	(2)↔(6)	ΔH_{4A}^r	43.9	50.8	79.8	56.1
		ΔG_{4A}^r	37.7	45.0	73.6	53.5
	(3)↔(7)	ΔH_{4B}^r	78.8	51.5	−9.2	−7.4
		ΔG_{4B}^r	71.8	50.1	−2.5	−5.9
Path 5	(2)↔(8)	ΔH_{5A}^r	51.5	36.6	–	–
		ΔG_{5A}^r	44.9	31.0	–	–
	(3)↔(9)	ΔH_{5B}^r	−2.2	0.9	–	–
		ΔG_{5B}^r	4.1	5.9	–	–
Path 6	(2)↔(10)	ΔH_{6A}^r	−2.2	−4.1	−1.8	−0.6
		ΔG_{6A}^r	0.6	−0.4	4.4	2.7
	(3)↔(11)	ΔH_{6B}^r	12.5	11.3	30.5	48.0
		ΔG_{6B}^r	15.2	13.2	29.8	41.0

[a]See Figures 7.1–7.3.

7.3.1.1 *HYDRATION AND CARBONATION OF THE DICHLORODIAMINE COMPLEXES*

Regarding the thermal character, most of the reactions are predicted as strongly endothermic ($\Delta H_i^r > 0$; Fig. 7.4). Exceptions are observed for the carbonation reactions at the chloride position, for which the predicted ΔH_i^r values are either slightly negative (*cis* complexes) or slightly positive (*trans* complexes). On the other hand, the generality of the reactions is predicted to be nonspontaneous ($\Delta G_i^r > 0$; Fig. 7.4). Yet, the spontaneity evaluation of the carbonation reactions involving a Cl displacement (1→4) demands caution given the smallness of the related ΔG_i^r values. This is particularly true for the *cis* complexes as in these cases the ΔG_i^r predicted values are around 1 kJ mol⁻¹ (Table 7.1). Still, this result is particularly interesting in that it evidences the bicarbonate ion as a potential interfering agent, able to divert part of the drug from its normal physiological activating path still while traveling in the bloodstream. In other words, direct carbonation reactions of the dichlorodiamine complexes may contribute to lowering of the drug's bioavailability, affecting the drug's required dose to achieve therapeutic activity desired.

7.3.1.2   CARBONATION OF THE MONOAQUA COMPLEXES

For the monoaqua species obtained by chloride hydration (Fig. 7.5(A)), the calculations predict the carbonation reaction via H_2O displacement (2→4) as significant exothermic and spontaneous process, regardless of metal and complex configuration. The equivalent carbonation reaction onthe aqua species obtained by NH_3–hydration (3→5; Fig. 7.5(B)) is also ranked as spontaneous (Fig. 7.5(B)), though the smallness of the ΔH_i^r and ΔG_i^r values advise a careful characterization of its thermodynamic character, particularly regarding cDDP and cDDPd. In turn, the carbonation reactions involving NH_3 displacement (2→6, 2→8 and 3→11) are all predictably endothermic and nonspontaneous processes.

The same uniformity does not verify, however, when the carbonation reactions involve the displacement of a Cl ligand. For example, for the aqua species obtained by NH_3 displacement, it is found that substitution of the Cl ligand *cis*-oriented to H_2O (3→7; Fig. 7.5(B)) is either extremely endothermic and nonspontaneous or exothermic and spontaneous, depending on the original configuration of the complex (either *cis* or *trans*, respectively). On the other hand, in the cases where the reaction involves the displacement of the Cl *trans*-oriented with respect to H_2O (3→9; Fig. 7.5(B)), it only possible for the *cis* complexes, the calculations seem to indicate some dependency on the metal. Evaluation of the thermodynamic nature of equivalent reactions in the aqua species obtained by Cl hydration (2→10; Fig. 7.5(A)) is hampered by the smallness of the associated ΔH_i^r and ΔG_i^r values. Nevertheless, the results seem to evidence some dependency of the thermal character on the original configuration of the complex.

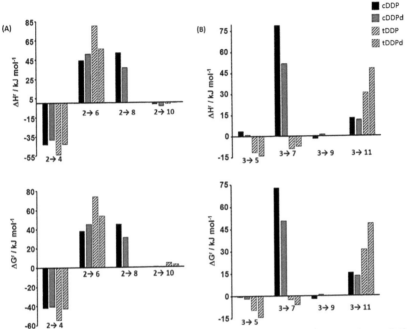

FIGURE 7.5 ΔH_i^r and ΔG_i^r values obtained for the carbonation reactions of the different monoaqua species obtained by either Cl⁻ (A) or NH_3^- (B) hydration of the four dichlorodiamine complexes (see text and Figs. 7.1–7.3).

In summary, the thermodynamic results suggest that carbonation processes may divert the complexes from their regular accepted therapeutic pathway in more than a single point. The first detour route may occur soon in the bloodstream and in the interstitial fluid, where the high concentration of bicarbonate can favor the displacement of a chloride ligand. The second very significant diverting route seems to occur after the drug activation by hydration.

7.3.2   KINETIC ANALYSIS

7.3.2.1   CARBONATION OF THE MONOAQUA COMPLEXES

Figure 7.6 compares the activation energy barriers determined for the direct (ΔG_i^a) and reverse (ΔG_{-i}^a) reactions gathered in the present work (Figs. 7.1–7.3). The calculated values are listed in Table 7.2 of the supplementary material.

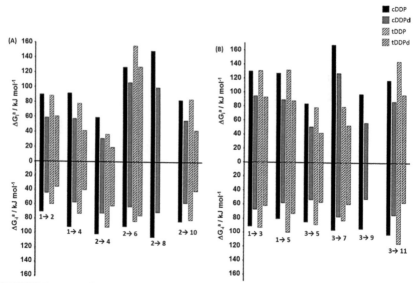

FIGURE 7.6 ΔG_i^a and ΔG_{-i}^a values predicted for the different hydration and carbonation reactions evaluated: (A) hydration and carbonation via Cl-displacement and carbonation of the aqua species thus formed; (B) hydration and carbonation via NH$_3$-displacement and carbonation of the aqua species thus formed (see text and Figs. 7.1–7.3).

Starting with the dichlorodiamine complexes, the results show that, regardless of complex configuration, both hydration and direct carbonation via NH$_3$ displacement (1→3 and 1→5) present larger activation energy barriers than via Cl replacement (1→2 and 1→4, respectively). The same applies to the generality of the carbonation reactions on the aqua species (Fig. 7.6(B) vs. Fig. 7.6(A)). However, there are two exceptions to this general rule, one regarding tDDP and tDDPd and the other cDDP and cDDPd. In the case of the *trans* complexes, the activation energy barrier associated to the carbonation reaction via displacement of the ligand *cis*-oriented in relation to H$_2$O (2→6 vs. 3→7, Fig. 7.6) is about 2.0–2.4 times larger for the aqua species obtained by Cl hydration. A similar result is obtained for the *cis* complexes, but now regarding the displacement of the ligand *trans*-oriented to H$_2$O (2→8 vs. 3→9, Fig. 7.6). In this case, the ΔG_i^a values are about 1.5–1.8 times larger for the aqua species obtained by Cl hydration than for those obtained by NH$_3$ hydration.

In general, the activation energy barriers related to the direct reactions are either greater than or of the same magnitude as those associated with the respective reverse reactions (Table 7.2 and Fig. 7.6). An exception is found for the carbonation reactions involving the displacement of the H_2O in the aqua species obtained by Cl hydrolysis (reactions 2→4). In this case, the ΔG_{-i}^a values are predicted to be about 1.7 (cDDP), 2.5 (cDDPd and tDDP), and 3.3 (tDDPd) folds larger than the respective ΔG_i^a values. This result suggests that in this case the equilibrium is shifted toward the direct reaction (carbonation by hydrolysis), the extent of the effect following the relative order cDDP<cDDPd≈tDDP<tDDPd.

TABLE 7.2 Selected Kinetic Parameters (kJ mol^{-1}) Predicted for the 1st Hydration and Possible Carbonation Reactions as a Function of Metal Complex.

Reaction[a]		Parameter	cDDP	cDDPd	tDDP	tDDPd
Path 1	(1)↔(2)	ΔH_{1A}^a	90.1	59.1	89.0	60.7
		ΔG_{1A}^a	69.9	43.7	59.4	35.4
	(1)↔(3)	ΔH_{1B}^a	130.0	94.3	130.8	93.1
		ΔG_{1B}^a	91.6	67.4	93.1	62.1
Path 2	(1)↔(4)	ΔH_{2A}^a	92.0	57.8	78.5	41.6
		ΔG_{2A}^a	90.8	56.7	72.3	39.6
	(1)↔(5)	ΔH_{2B}^a	127.5	89.2	131.8	88.0
		ΔG_{2B}^a	80.6	57.9	99.7	73.1
Path 3	(2)↔(4)	ΔH_{3A}^a	59.3	31.4	37.0	19.4
		ΔG_{3A}^a	100.8	71.4	90.9	61.6
	(3)↔(5)	ΔH_{3B}^a	84.2	51.2	78.9	42.7
		ΔG_{3B}^a	84.7	53.0	88.5	56.9
Path 4	(2)↔(6)	ΔH_{4A}^a	127.7	107.1	156.2	128.2
		ΔG_{4A}^a	90.1	62.0	82.6	74.8
	(3)↔(7)	ΔH_{4B}^a	168.0	127.7	79.8	53.3
		ΔG_{4B}^a	96.2	77.0	82.2	59.2
Path 5	(2)↔(8)	ΔH_{5A}^a	149.4	100.8	–	–
		ΔG_{5A}^a	104.5	69.8	–	–
	(3)↔(9)	ΔH_{5B}^a	98.3	57.4	–	–
		ΔG_{5B}^a	94.2	51.5	–	–

TABLE 7.2 *(Continued)*

Reaction[a]		Parameter	cDDP	cDDPd	tDDP	tDDPd
Path 6	(2)↔(10)	ΔH_{6A}^a	83.3	56.1	84.7	42.7
		ΔG_{6A}^a	82.7	56.5	80.3	40.0
	(3)↔(11)	ΔH_{6B}^a	117.7	87.5	144.9	97.3
		ΔG_{6B}^a	102.5	74.2	115.1	56.7

[a]See Figures 7.1–7.3.

7.3.2.2 REACTION RATE CONSTANTS

Based on the calculated ΔG_i^a and ΔG_{-i}^a values it is possible to estimate the reaction rate (k_i and k_{-i}) and equilibrium constants (K_i) associated with the different reactions. Table 7.3 lists the predicted values calculated as described in the computational details section.

TABLE 7.3 Calculated Reaction Rate Constants (k_i and k_{-i}) and Equilibrium (K_i) Constants[a].

Reaction[a]		Parameter	cDDP	cDDPd	tDDP	tDDPd
Path 1	(1)↔(2)	k_{1A}	1.0×10^{-3}	2.8×10^2	1.6×10^{-3}	1.4×10^2
		k_{-1A}	3.5	1.4×10^5	2.4×10^2	3.9×10^6
		K_{1A}	2.9×10^{-4}	2.0×10^{-3}	6.5×10^{-6}	3.7×10^{-5}
	(1)↔(3)	k_{1B}	1.0×10^{-10}	1.9×10^{-4}	7.6×10^{-11}	3.0×10^{-4}
		k_{-1B}	5.6×10^{-4}	9.7	3.0×10^{-4}	8.2×10^1
		K_{1B}	1.9×10^{-7}	1.9×10^{-5}	2.5×10^{-7}	3.7×10^{-6}
Path 2	(1)↔(4)	k_{2A}	4.7×10^{-4}	4.6×10^2	1.1×10^{-1}	3.2×10^5
		k_{-2A}	7.7×10^{-4}	7.2×10^2	1.3	7.2×10^5
		K_{2A}	6.2×10^{-1}	6.4×10^{-1}	8.2×10^{-2}	4.5×10^{-1}
	(1)↔(5)	k_{2B}	2.9×10^{-10}	1.5×10^{-3}	5.0×10^{-11}	2.4×10^{-3}
		k_{-2B}	4.7×10^{-2}	4.5×10^2	2.1×10^{-5}	9.7×10^{-1}
		K_{2B}	6.1×10^{-9}	3.3×10^{-6}	2.4×10^{-6}	2.5×10^{-3}
Path 3	(2)↔(4)	k_{3A}	2.5×10^2	2.0×10^7	2.0×10^6	2.5×10^9
		k_{-3A}	1.4×10^{-5}	1.9	7.4×10^{-4}	1.0×10^2
		K_{3A}	1.9×10^7	1.0×10^7	2.8×10^9	2.5×10^7
	(3)↔(5)	k_{3B}	1.1×10^{-2}	6.7×10^3	9.3×10^{-2}	2.1×10^5

TABLE 7.3   *(Continued)*

Reaction[a]		Parameter	cDDP	cDDPd	tDDP	tDDPd
		k_{-3B}	9.0×10^{-3}	3.2×10^{3}	1.9×10^{-3}	6.7×10^{2}
		K_{3B}	1.2	2.1	4.8×10	3.1×10^{2}
Path 4	$(2) \leftrightarrow (6)$	k_{4A}	2.6×10^{-10}	1.1×10^{-6}	2.7×10^{-15}	2.2×10^{-10}
		k_{-4A}	1.0×10^{-3}	8.5×10^{1}	2.1×10^{-2}	4.9×10^{-1}
		K_{4A}	2.6×10^{-7}	1.3×10^{-8}	1.3×10^{-13}	4.4×10^{-10}
	$(3) \leftrightarrow (7)$	k_{4B}	2.3×10^{-17}	2.6×10^{-10}	6.5×10^{-2}	2.9×10^{3}
		k_{-4B}	8.7×10^{-5}	2.0×10^{-1}	2.5×10^{-2}	2.6×10^{2}
		K_{4B}	2.6×10^{-3}	1.3×10^{-9}	2.6	1.1×10
Path 5	$(2) \leftrightarrow (8)$	k_{5A}	4.2×10^{-14}	1.4×10^{-5}	–	–
		k_{-5A}	3.1×10^{-6}	3.7	–	–
		K_{5A}	1.4×10^{-8}	3.7×10^{-6}	–	–
	$(3) \leftrightarrow (9)$	k_{5B}	3.7×10^{-5}	5.5×10^{2}	–	–
		k_{-5B}	2.0×10^{-4}	5.9×10^{3}	–	–
		K_{5B}	1.9×10^{-1}	9.3×10^{-2}	–	–
Path 6	$(2) \leftrightarrow (10)$	k_{6A}	1.6×10^{-2}	9.2×10^{2}	9.0×10^{-3}	2.1×10^{5}
		k_{-6A}	2.0×10^{-2}	7.9×10^{2}	5.3×10^{-2}	6.1×10^{5}
		K_{6A}	7.9×10^{-1}	1.2	1.7×10^{-1}	3.4×10^{-1}
	$(3) \leftrightarrow (11)$	k_{6B}	1.5×10^{-8}	2.9×10^{-3}	2.6×10^{-13}	5.6×10^{-5}
		k_{-6B}	6.9×10^{-6}	6.2×10^{-1}	4.3×10^{-8}	7.2×10^{2}
		K_{6B}	2.2×10^{-3}	4.7×10^{-3}	6.0×10^{-6}	7.7×10^{-8}

[a] k_i in s^{-1} and k_{-i} in $M^{-1} \cdot s^{-1}$; see text and Figures 7.1–7.3.

Regarding the hydration of cDDP and tDDP via Cl displacement, the predicted rate constants are in line with the experimentally reported values.[3,4,37–41]. As stated previously information regarding the remaining complexes and reactions are either scarce or inexistent. Nevertheless, it is possible to verify that the predicted data for the hydration reactions of the palladium complexes (by Cl displacement) are in accordance with the results of the literature. In fact, the presented results establish that the Cl hydration reactions of the Pd (II) complexes are about 10^{5}–10^{5} times more reactive than the respective platinum complexes.[42,43]

The presented results reinforce the earlier conclusion that bicarbonate ion may actually contribute to a decrease in the drug's bioavailability, predicting that carbonation via hydrolysis of the aqua species is not only an extremely exothermic spontaneous process but also an expectedly extended process. Moreover, the relative ordering of the complexes based on the ratio of the ΔG_i^a and ΔG_{-i}^a associated to those reactions can explain,

at least in part, the inactivity of the parent complexes cDDPd, tDDP, and tDDPd when compared to the cDDP.

   In resume, the results presented show that the possibility of interaction of the drug with bicarbonate ions along with the pathway to its therapeutic target should be considered and evaluated further, at a molecular level, since such interaction can significantly affect the bioavailability of the drug. In what particularly concerns the rate constants of the carbonation reactions, the values obtained may be good starting estimates. In view of the presented results, the experimental evaluation of the carbonation kinetics of the complexes under physiological conditions seems to be of extreme relevance, since these reactions can significantly condition the bioavailability of the drug.

7.4   ACKNOWLEDGMENTS

The author acknowledges financial support from the Portuguese Foundation for Science and Technology—Unidade de Química-Física Molecular (UID/Multi/00070/2013). Acknowledgements are also due to Laboratório-Associado CICECO, University of Aveiro, Portugal, for the free access to computational facilities (G09w).

KEYWORDS

- Pt(II) and Pd(II) complexes
- anticancer activity
- hydrolysis kinetics
- carbonation kinetics
- quantum chemical calculations

REFERENCES

1. DBL™ Cisplatin Injection. http://www.medsafe.govt.nz/profs/Datasheet/d/DBLCisplatininj. pdf. (accessed March 22, 2017).

2. Cisplatin Injection. http://www.pfizer.com/files/products/uspi_cisplatin.pdf. (accessed March 20, 2017).

3. Miller, S. E.; Gerard, K. J.; House, D. A. The Hydrolysis Products of Cis-Diamminedichloroplatinum(II) .6. a Kinetic Comparison of the Cis-Isomers and Trans-Isomers and Other Cis-di(amine)di(chloro)platinum(II) Compounds. *Inorg. Chim. Acta* **1991**, *190*, 135–144.

4. Miller, S. E.; House, D. A. The Hydrolysis Products of Cis-Dichlorodiammineplatinum(II) 3. Hydrolysis Kinetics at Physiological pH. *Inorg. Chim. Acta* **1990**, *173*, 53–60.

5. Binter, A.; Goodisman, J.; Dabrowiak, J. C. Formation of Monofunctional Cisplatin-DNA Adducts in Carbonate Buffer. *J. Inorg. Biochem.* **2006**, *100*, 219–1224.

6. Centerwall, C. R.; Goodisman, J.; Kerwood, D. J.; Dabrowiak, J. C. Cisplatin Carbonato Complexes. Implications for Uptake, Antitumor Properties, and Toxicity. *J. Am. Chem. Soc.* **2005**, *127*, 12768–12769.

7. Todd, R. C.; Lovejoy, K. S.; Lippard, S. J. Understanding the Effect of Carbonate Ion on Cisplatin Binding to DNA. *J. Am. Chem. Soc.* **2007**, *129*, 6370–6371.

8. Park, J.-S.; Kim, S. H.; Lee, N.-K.; Lee, K. J.; Hong, S.-C. In Situ Analysis of Cisplatin Binding to DNA: The Effects of Physiological Ionic Conditions. *Phys. Chem. Chem. Phys.* **2012**, *14*, 3128–3133.

9. Coluccia, M.; Natile, G. Trans-Platinum Complexes in Cancer Therapy. *Anti-Cancer Agents Med. Chem.* **2007**, *7*, 111–123.

10. Kasparkova, J.; Marini, V.; Bursova, V.; Brabec, V. Biophysical Studies on the Stability of DNA Intrastrand Cross-Links of Transplatin. *Biophys. J.* **2008**, *95*, 4361–4371.

11. Marchán, V.; Pedroso, E.; Grandas, A. Insights into the Reaction of Transplatin with DNA and Proteins: Methionine-Mediated Formation of Histidine–Guanine trans-Pt(NH$_3$)$_2$ Cross-Links. *Chem. A Eur. J.* **2004**, *10*, 5369–5375.

12. Frisch, M. J.; Trucks, G. W.; Schlegel, H. B.; Scuseria, G. E.; Robb, M. A.; Cheeseman, J. R.; Scalmani, G.; Barone, V.; Mennucci, B.; Petersson, G. A.; Nakatsuji, H.; Caricato, M.; Li, X.; Hratchian, H. P.; Izmaylov, A. F.; Bloino, J.; Zheng, G.; Sonnenberg, J. L.; Hada, M.; Ehara, M.; Toyota, K.; Fukuda, R.; Hasegawa, J.; Ishida, M.; Nakajima, T.; Honda, Y.; Kitao, O.; Nakai, H.; Vreven, T.; Montgomery, J. A., Jr.; Peralta, J. E.; Ogliaro, F.; Bearpark, M.; Heyd, J. J.; Brothers, E.; Kudin, K. N.; Staroverov, V. N.; Kobayashi, R.; Normand, J.; Raghavachari, K.; Rendell, A.; Burant, J. C.; Iyengar, S. S.; Tomasi, J.; Cossi, M.; Rega, N.; Millam, J. M.; Klene, M.; Knox, J. E.; Cross, J. B.; Bakken, V.; Adamo, C.; Jaramillo, J.; Gomperts, R.; Stratmann, R. E.; Yazyev, O.; Austin, A. J.; Cammi, R.; Pomelli, C.; Ochterski, J. W.; Martin, R. L.; Morokuma, K.; Zakrzewski, V. G.; Voth, G. A.; Salvador, P.; Dannenberg, J. J.; Dapprich, S.; Daniels, A. D.; Farkas, O.; Foresman, J. B.; Ortiz, J. V.; Cioslowski, J.; Fox, D. J. Gaussian 09, Revision A.02. In Gaussian 09, Revision A.02, Gaussian, Inc.: Wallingford CT, 2009.

13. Amado, A. M.; Fiuza, S. M.; Marques, M. P. M.; Batista de Carvalho, L. A. E. Conformational and Vibrational Study of Platinum(II) Anticancer Drugs:

Cis-Diamminedichloroplatinum(II) as a Case Study. *J. Chem. Phys.* **2007,** *127,* article n° 185104.

14. Fiuza, S. M.; Amado, A. M.; Dos Santos, H. F.; Marques, M. P. M.; Batista de Carvalho, L. A. E. Conformational and Vibrational Study of Cis-Diamminedichloropalladium(II). *Phys. Chem. Chem. Phys.* **2010,** *12,* 14309–14321.

15. Padrão, S.; Fiuza, S. M.; Amado, A. M.; Amorim da Costa, A. M.; Batista de Carvalho, L. A. E. Validation of the mPW1PW Quantum Chemical Calculations for the Vibrational Study of Organic Molecules—Re-Assignment of the Isopropylamine Vibrational Spectra. *J. Phys. Org. Chem.* **2012,** *24,* 110–121.

16. Amado, A. M.; Fiuza, S. M.; Batista de Carvalho, L. A. E.; Ribeiro-Claro, P. J. A. On the Relevance of Considering the Intermolecular Interactions on the Prediction of the Vibrational Spectra of Isopropylamine. *J. Chem.* **2013.** Article ID 682514.

17. Fiuza, S. M.; Silva, T. M.; Marques, M. P. M.; de Carvalho, L. A. E. B.; Amado, A. M. On the Correction of Calculated Vibrational Frequencies for the Effects of the Counterions α,ω-Diamine Dihydrochlorides. *J. Mol. Model.* **2015,** *21,* 1–13, C7–266.

18. http://www.gaussian.com/g_tech/g_ur/k_scrf.htm (accessed March1, 2015).

19. Amado, A. M.; Fiuza, S. M.; Batista de Carvalho, L. A. E.; Ribeiro-Claro, P. J. A. On the Effects of Changing Gaussian Program Version and SCRF Defining Parameters: Isopropylamine as a Case Study. *Bull. Chem. Soc. Jan.* **2012,** *85,* 962–975.

20. Marenich, A. V.; Cramer, C. J.; Truhlar, D. G. Universal Solvation Model Based on Solute Electron Density and on a Continuum Model of the Solvent Defined by the Bulk Dielectric Constant and Atomic Surface Tensions. *J. Phys. Chem. B* **2009,** *113,* 6378–6396.

21. Costa, L. A. S.; Hambley, T. W.; Rocha, W. R.; De Almeida, W. B.; Dos Santos, H. F. Kinetics and Structural Aspects of the Cisplatin Interactions with Guanine: A Quantum Mechanical Description. *Int. J. Quantum Chem.* **2006,** *106,* 2129–2144.

22. Burda, J. V.; Zeizinger, M.; Leszczynski, J. Hydration Process as an Activation of Trans- and Cisplatin Complexes in Anticancer Treatment. DFT and abInitio Computational Study of Thermodynamic and Kinetic Parameters. *J. Comput. Chem.* **2005,** *26,* 907–914.

23. Raber, J.; Zhu, C. B.; Eriksson, L. A. Activation of Anti-Cancer Drug Cisplatin—Is the Activated Complex Fully Aquated? *Mol. Phys.* **2004,** *102,* 2537–2544.

24. Burda, J. V.; Zeizinger, M.; Sponer, J.; Leszczynski, J. Hydration of Cis- and Trans-Platin: A Pseudopotential Treatment in the Frame of a G3-Type Theory for Platinum Complexes. *J. Chem. Phys.* **2000,** *113,* 2224–2232.

25. Chval, Z.; Sip, M. Pentacoordinated Transition States of Cisplatin Hydrolysis—abInitio Study. *J. Mol. Struct.: THEOCHEM* **2000,** *532,* 59–68.

26. 26.Tsipis, A. C.; Sigalas, M. P. Mechanistic Aspects of the Complete Set of Hydrolysis and Anation Reactions of Cis- and Trans-DDP Related to Their Antitumor Activity Modeled by an Improved ASED-MO Approach. *J. Mol. Struct.: THEOCHEM* **2002,** *584,* 235–248.

27. Zhang, Y.; Guo, Z. J.; You, X. Z. Hydrolysis Theory for Cisplatin and its Analogues Based on Density Functional Studies. *J. Am. Chem. Soc.* **2001,** *123,* 9378–9387.

28. Burda, J. V.; Zeizinger, M.; Leszczynski, J. Activation Barriers and Rate Constants for Hydration of Platinum and Palladium Square-Planar Complexes: An abinitio Study. *J. Chem. Phys.* **2004,** *120,* 1253–1262.

29. Robertazzi, A.; Platts, J. A. Hydrogen Bonding, Solvation, and Hydrolysis of Cisplatin: A Theoretical Study. *J. Comput. Chem.* **2004,** *25,* 1060–1067.

30. Lau, J. K. C.; Deubel, D. V. Hydrolysis of the Anticancer Drug Cisplatin: Pitfalls in the Interpretation of Quantum Chemical Calculations. *J. Chem. Theory Compu.* **2006,** *2,* 103–106.

31. Lau, J. K.-C.; Ensing, B. Hydrolysis of Cisplatin-a First-Principles Metadynamics Study. *Phys. Chem. Chem. Phys.* **2010,** *12,* 10348–10355.

32. Zimmermann, T.; Leszczynski, J.; Burda, J. V. Activation of the Cisplatin and Transplatin Complexes in Solution with Constant pH and Concentration of Chloride Anions; Quantum Chemical Study. *J. Mol. Model.* **2011,** *17,* 2385–2393.

33. Melchior, A.; Sanchez Marcos, E.; Pappalardo, R. R.; Martinez, J. M. Comparative Study of the Hydrolysis of a Third- and a First-Generation Platinum Anticancer Complexes. *Theo. Chem. Acc.* **2011,** *128,* 627–638.

34. Kuduk-Jaworska, J.; Chojnacki, H.; Janski, J. J. Non-Empirical Quantum Chemical Studies on Electron Transfer Reactions in Trans- and Cis-Diamminedichloroplatinum(II) Complexes. *J. Mol. Model.* **2011,** *17,* 2411–2421.

35. Melchior, A.; Tolazzi, M.; Manuel Martinez, J.; Pappalardo, R. R.; Sanchez Marcos, E. Hydration of Two Cisplatin Aqua-Derivatives Studied by Quantum Mechanics and Molecular Dynamics Simulations. *J. Chem. Theory Comput.* **2015,** *11,* 1735–1744.

36. Zeizinger, N.; Burda, J. V.; Sponer, J.; Kapsa, V.; Leszczynski, J. A Systematic abInitio Study of the Hydration of Selected Palladium Square-Planar Complexes. A Comparison with Platinum Analogues. *J. Phys. Chem. A* **2001,** *105,* 8086–8092.

37. Appleton, T. G.; Hall, J. R.; Ralph, S. F.; Thompson, C. S. M. NMR-Study of Acid-Base Equilibria and Other Reactions of Ammineplatinum Complexes with Aqua and Hydroxo Ligands. *Inorg. Chem.* **1989,** *28,* 1989–1993.

38. Banerjea, D.; Basolo, F.; Pearson, R. G. Mechanism of Substitution Reactions of Complex Ions. XII.1 Reactions of Some Platinum(II) Complexes with Various Reactants. *J. Am. Chem. Soc.* **1957,** *79,* 4055–4062.

39. Bodenner, D. L.; Dedon, P. C.; Keng, P. C.; Borch, R. F. Effect of Diethyldithiocarbamateon Cis-Diamminedichloroplatinum(II)-Induced Cytotoxicity, DNA Cross-Linking, and Gamma-Glutamyl-Transferase Transpeptidase Inhibition. *Cancer Res.* **986,** *46,* 2745–2750.

40. Bose, R. N.; Viola, R. E.; Cornelius, R. D. Phosphorus-31 NMR and Kinetic Studies of the Formation of Ortho-, Pyro-, and Triphosphato Complexes of Cis-Dichlorodiammineplatinum(II). *J. Am. Chem. Soc.* **1984,** *106,* 3336–3343.

41. Reishus, J. W.; Martin, D. S. Cis-Dichlorodiammineplatinum (II). Acid Hydrolysis and Isotopic Exchange of the Chloride Ligands. *J. Am. Chem. Soc.* **1961,** *83,* 2457–2462.

42. Butour, J. L.; Wimmer, S.; Wimmer, F.; Castan, P. Palladium(II) Compounds with Potential Antitumour Properties and Their Platinum Analogues: A Comparative Study of the Reaction of Some Orotic Acid Derivatives with DNA in Vitro. *Chem.-Biol. Interact.* **1997,** *104,* 165–178.

43. Hofer, T. S.; Randolf, B. R.; Adnan Ali Shah, S.; Rode, B. M.; Persson, I. Structure and Dynamics of the Hydrated Palladium(II) Ion in Aqueous Solution A QMCF MD Simulation and EXAFS Spectroscopic Study. *Chem. Phys. Lett.* **2007,** *445,* 193–197.

CHAPTER 8

TRIPLE SYSTEMS, BASED ON NICKEL COMPOUNDS, AS EFFECTIVE HYDROCARBONS OXIDATION CATALYSTS AND MODELS OF NI-ACIREDUCTONE DIOXYGENASE

LUDMILA I. MATIENKO[1,*], LARISA A. MOSOLOVA[1], VLADIMIR I. BINYUKOV[1,2], ELENA M. MIL[1,3], AND GENNADY E. ZAIKOV[1,4]

[1]N M Emanuel Institute of Biochemical Physics, Russian Academy of Sciences, 4 Kosygin str., Moscow 119334, Russia

[2]E-mail: bin707@mail.ru

[3]E-mail: elenamil2004@mail.ru

[4]E-mail: chembio@sky.chph.ras.ru

*Corresponding author. E-mail: matienko@sky.chph.ras.ru

CONTENTS

ABBREVIATIONS

AFM method   Atomic-Force Microscopy method
(Acac)⁻   Acetylacetonate ion
DN   donor number
GLC   Gas-liquid chromatography method
His   L-Histidine
HMPA   hexamethylphosphorotriamide
MP   N-methylpirrolidone-2
MSt   stearates of alkaline metals (M= Li)
Ni(Fe)-ARD   Ni(Fe) Acireductone Dioxygenase
PhOH   phenol
Tyr   L-Tyrosine

ABSTRACT

In this chapter, we research the kinetics and mechanism of selective ethyl-benzene oxidation with dioxygen catalyzed by triple catalytic systems $\{Ni^{II}(acac)_2 + L^2 + PhOH, PhOH = phenol\}$ ($L^2 = HMPA$, MP (MP=N-meth-ylpyrrolidone-2)). With atomic force microscopy (AFM) method, the possibility of forming the stable supramolecular nanostructures based on triple systems, $\{Ni^{II}(acac)_2 + HMPA(MP) + PhOH\}$, due to intermolecular H-bonds was researched. We also used AFM method to study the possibility of formation of stable supramolecular nanostructures based on triple system $\{Ni^{II}(acac)_2 + L^2 + Tyr\}$ (Tyr=L-Tyrosine, $L^2 = MP$ or L-Histidine), that is, structural and functional model of Ni-Acireductone Dioxygenase (Ni-ARD),—with the assistance of intermolecular H-bonds. The bibliography includes 22 references.

8.1   INTRODUCTION

The method of modifying the Ni^{II} and $Fe^{II,III}$ complexes with mono- or multidentate modifying ligands, used in the selective oxidation of alkyl-arens (ethylbenzene and cumene) with molecular oxygen to afford the corresponding hydroperoxides which aimed at increasing their selectivity was first proposed by L.I. Matienko.[1,2] The mechanism of action of such modifying ligands was elucidated, and new efficient catalysts of selective oxidation of ethylbenzene to α-phenyl ethyl hydroperoxide (PEH) were developed.[1,2] The problem of selective oxidation of alkylarens to hydroperoxides is economically sound, in connection with the use of hydroperoxides as intermediates in the large-scale production of important monomers. For instance, propylene oxide and styrene are synthesized from α-phenyl ethyl hydroperoxide, and cumyl hydroperoxide is the precursor in the synthesis of phenol and acetone.[3,4]

The phenomenon of a substantial increase in the selectivity (S) and conversion (C) of ethylbenzene oxidation to PEH upon the addition of phenol (PhOH) together with alkali metal stearate MSt (M=Li, Na), or the other electron donating compounds L^2, namely, monodentate ligands MP (N-metylpirrolidon-2), HMPA to metal complexes Ni(acac)$_2$ was discovered in our works[1,2] These results by L. I. Matienko and L. A. Mosolova are protected by the Russian Federation patent (2004).

The high efficiency of three-component systems {Ni (acac)$_2$+L^2 +PhOH} that introduced the compound of redox-inactive metal, (L^2=metalloligands MSt (M=Na, Li)), in the reaction of selective oxidation of ethylbenzene to α-phenyl ethyl hydroperoxide, and maybe associated with the formation of extremely stable binuclear hetero-ligand complexes {Ni (acac)$_2$L^2·PhOH}$_n$ due to intermolecular H-bonds.[2,5]

We have proposed a new approach for researching the possibility of formation of supramolecular structures based on catalytically active nickel (and iron) complexes,[6] due to H-bonds, and at first, we have received the data in favor of this hypothesis with the AFM method.[6,7] In the present work, we have examined the kinetics and mechanism of catalysis with triple system {Ni(acac)$_2$+L^2+PhOH} (L^2=HMPA, MP), in ethylbenzene oxidation to PEH in order to investigate the effect of ligands with various electron donating properties (DN) on the activity of triple systems comprising PhOH as ligand-modifier. We researched the possibility of formation of supramolecular structures based on triple system {Ni(acac)$_2$+L^2+PhOH} (L^2=HMPA, MP), in comparison with earlier researched catalysts and catalytic systems—{Ni(acac)$_2$+L^2+PhOH} (L^2=MSt (M=Na, Li)), Ni$_2$(OAc)$_3$(acac)·MP·2H$_2$O.[5-7] We also discuss the possible role of macrostructures and Tyr-fragment as regulatory factors in the mechanism of action of Ni-Acireductone Dioxygenase (Ni-ARD), based on experimental data, which we received with AFM on model systems{Ni(acac)$_2$+MP+Tyr} (Tyr=L-Tyrosine). AFM data for the case of triple system {Ni(acac)$_2$+L^2+Tyr} (L^2=His=L-Histidine), we received for the first time.

8.2 MATERIALS AND METHODOLOGY

Ethylbenzene (RH) was oxidized with dioxygen at 120°C in glass bubbling-type reactor[5] in the presence of {Ni^{II}(acac)$_2$+L^2+PhOH} (L^2=HMPA) systems.

Analysis of Oxidation Products   α-Phenyl ethyl hydroperoxide (PEH) was analyzed by iodometry. By-products, including methylphenylcarbinol (MPC), acetophenone (AP), and PhOH as well as the RH content in the oxidation process were examined by GLC.[5] The selectivity S$_{PEH}$ and conversion C of ethylbenzene into PEH oxidation were determined using the formulas:

$$S_{PEH}=[PEH]/\Delta[RH]\cdot 100\% \text{ and } C=\Delta[RH]/[RH]_0\cdot 100\%.$$

An order in which PEH, AP, and MPC formed was determined from the time dependence of product accumulation rate ratios at t→0. The variation of these ratios with time was evaluated by graphic differentiation, (see Figs. 8.4–8.6)[5]. Experimental data processing was done using special computer programs Mathcad and Graph2Digit.

AFM SOLVER P47/SMENA/ with Silicon Cantilevers NSG11S (NT MDT) with curvature radius 10 nm, tip height: 10–15 μm, and cone angle ≤22° in taping mode on resonant frequency 150 KHz was used.[6]

As a substrate, the polished special chemically modified silicone surface was used.

Waterproof modified silicone surface was exploited for the self-assembly-driven growth due to H-bonding of triple systems {Ni(acac)$_2$+L^2+PhOH} (L^2=HMPA, MP), {Ni(acac)$_2$+MP(His) +Tyr} and binary systems {Ni(acac)$_2$+MP(His)}, {Ni(acac)$_2$+Tyr} with silicone surface. The saturated H$_2$O solutions of triple systems were put on a surface, maintained for some time, and then the solvent was removed from the surface by means of special method—spin-coating process.

In the course of scanning of investigated samples, it has been found that the structures are fixed on a surface strongly enough due to H-bonding. The self-assembly-driven growth of the supramolecular structures on modified silicone surface, based on systems {Ni(acac)$_2$+L^2+PhOH} (L^2=HMPA, MP) {Ni(acac)$_2$+His+Tyr} and the other triple systems (as compared to corresponding binary systems {Ni(acac)$_2$+MP(His)}, {Ni(acac)$_2$+Tyr}, due to H-bonds and possibly other non-covalent interactions, was researched.

8.3 RESULTS AND DISCUSSION

8.3.1 MECHANISM OF CATALYSIS WITH TRIPLE SYSTEM {Ni(acac)$_2$+L^2+PhOH} (L^2=HMPA, MP) IN ETHYLBENZENE OXIDATION WITH DIOXYGEN

As mentioned before, the phenomenon of a substantial increase in the conversion (C) of ethylbenzene oxidation to PEH at the selectivity (S_{PEH}) level of not less than 90% upon the addition of PhOH together with alkali metal stearates MSt (M=Li, Na) as metalloligands or monodentate ligands MP, HMPA to metal complexes Ni(acac)$_2$ was discovered in our works (see, for example, Fig. 8.1).[1,2,5] But kinetics and mechanism of ethylbenzene oxidation to α-phenyl ethyl hydroperoxide, catalyzed with triple systems {Ni(acac)$_2$+HMPA(MP)+PhOH}, were not researched enough.

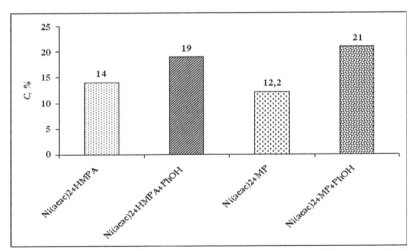

FIGURE 8.1 Conversion C (%) to α-phenyl ethyl hydroperoxide (PEH) in the ethylbenzene oxidation, catalyzed with the triple systems {Ni(acac)$_2$+HMPA(or MP) +PhOH}, in comparison with catalysis by binary systems {Ni(acac)$_2$+HMPA(or MP)} in the ethylbenzene oxidation. [Ni(acac)$_2$]=3 × 10^{-3} mol/l, [HMPA]=2 × 10^{-2} mol/l, [MP]=7 × 10^{-2} mol/l, [PhOH]=3 × 10^{-3} mol/l at 120°C.

The catalysis with binary system $\{Ni(acac)_2 + HMPA\}$ is characterized by maximum initial rates of accommodation of all products (PEH, AP, MPC) ($t \leq 1$ h), and similar to the catalytic action of a binary system $\{Ni(acac)_2 + MP\}$. During ethylbenzene oxidation, the rates of products accommodation are reduced to some values (compare Figs. 8.2a, 8.3a). At $t > 20$ h, the rates of AP, MPC accommodation $w_{AP(MPC)}$ increase, the rate of accumulation of phenol w_{PhOH}, which is a product of heterolysis of PEH, significantly increases, but w_{PEH} decreases (these data at $t > 15$ or 20 h are not presented in the Figs. 8.2a and 8.3a).

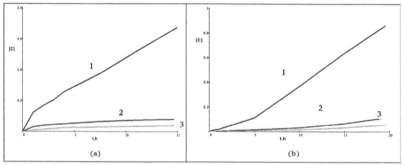

FIGURE 8.2 Kinetics of accumulation of (1) PEH, (2) acetophenone (AP), (3) methylphenylcarbinol (MPC) in ethylbenzene oxidation with dioxygen, catalyzed by binary system $\{Ni(acac)_2 + HMPA\}$ (a), and catalyzed by triple system $\{Ni(acac)_2 + HMPA + PhOH\}$ (b). $[Ni(acac)_2] = 3 \times 10^{-3}$ mol/l, and $[HMPA] = 2 \times 10^{-2}$ mol/l, $[PhOH] = 3 \times 10^{-3}$ mol/l at 120°C.

In the study of the mechanism of catalysis, we found that the additives of PhOH to binary system $\{Ni(acac)_2 + HMPA\}$ make changes in the kinetics of the products in ethylbenzene oxidation.

In the presence of $\{Ni(acac)_2 + HMPA + PhOH\}$, the rates of accommodation of oxidation products PEH, AP, MPC increase rapidly as compared to the initial stage and then they reach constant values and retain the same during the oxidation process (≤ 20 (HMPA) or ≤ 30 h (MP)) (see for example Figs. 8.2b, 8.3b, new data) by analogy with catalysis by $\{Ni(acac)_2 + MSt + PhOH\}$.[5,7]

In the case of catalysis with triple systems $\{Ni(acac)_2 + L^2 + PhOH\}$, the value of S_{PEH} is high (about 90%) both at the beginning of the reaction ($t \leq 1$ h) and at the significant depth of the process, unlike the catalysis by $\{Ni(acac)_2 + L^2\}$. In the latter case, the dependence of S_{PEH} on C has a well-defined extremum. For example, in the case of catalysis with $\{Ni(acac)_2 + MP\}$ $S_{PEH,max} = 85 - 87\%$ at $C \sim 8 - 10\%$.

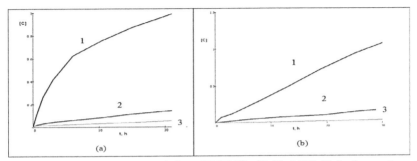

FIGURE 8.3   Kinetics of accumulation of PEH (1), AP (2), MPC (3) in the reaction of ethylbenzene oxidation with dioxygen, catalyzed by binary system {Ni(acac)$_2$+MP} (a), and catalyzed by triple system {Ni(acac)$_2$+MP+PhOH} (b). [Ni(acac)$_2$]=3 × 10^{-3} mol/l, and [MP]=7 × 10^{-2} mol/l, [PhOH]=3 × 10^{-3} mol/l at 120°C (new data).

The mechanism of catalysis with binary systems {Ni(acac)$_2$+L^2} (L^2=HMPA, MP, [Ni(acac)$_2$]=3 × 10^{-3} mol/l) was researched by us earlier.[1,2] We have found (with the inhibitor method) that the increase in the initial rate (t≤1 h) of the ethylbenzene oxidation with dioxygen, catalyzed with Ni(acac)$_2$ (3 × 10^{-3} mol/l) in the presence of additives of electron donating ligands L^2, is due to higher activity of formed complexes Ni(acac)$_2$·L^2 in the micro stages of chain initiation (O$_2$ activation) and/ or decomposition of PEH with free radical formation. So, in this case, the selectivity ($S_{PEH,max}$ ~75–80%) is not high.

The rate of free radical formation in the micro stages of chain initiation (O$_2$ activation) increases in a row of catalysts, that is, Ni(acac)2 <{Ni(acac)2·MP}<{Ni(acac)2·HMPA}, as far as there is increase of donor ability of the added ligands (MP·(DN=27.30)<HMPA·(DN=37.07) (DN (kkal/mol)). MP and HMPA additives increase the rate of chain initiation (O$_2$ activation) 2.4 and 4.16 times, accordingly.

The absence of correlation between rates of catalytic ethylbenzene oxidation and free radical formation in stages of chain initiation (O$_2$ activation), and also in catalytic PEH decomposition seems to be due to the influence of complex Ni(acac)$_2$·L^2 not only on the steps of the free radical formation, but also on the activity of RO$_2$• radicals in the micro step of chain propagation Cat+RO$_2$•→.

With process development, the increase in S_{PEH} (from 80% to~90%) and decrease in w are observed (II macro stage of the ethylbenzene oxidation).[1,2] We established that ligands L^2 (HMPA, MP) control the transformation of Ni(acac)$_2$·L^2 complexes into more active and selective hetero-ligand complexes Ni$_x$(acac)$_y$(AcO)$_z$(L^2)$_n$(H$_2$O)$_m$·(Ni$_2$(AcO)$_3$(acac)·MP·2H$_2$O, L^2

=MP, HMPA) (by analogy with action of Ni-ARD[2,6]). Here, the increase in S_{PEH} is due to the expense of catalyst participation in O_2 activation and inhibition of chain and heterolytic decomposition of PEH. Besides this, we proposed that the direction of formation of side products, acetophenone (AP), and methylphenylcarbinol (MPC), may be changed from consequent (under hydroperoxide decomposition) to parallel at the expense of modified catalyst in the chain propagation (Cat+RO_2•→). At the III macro stage of the catalytic ethylbenzene oxidation, the sharp fall of the S_{PEH} is associated with the heterolysis of PEH to phenol and acetaldehyde, catalyzed with completely transformed catalytic form, that is, Ni(OAc)$_2$.

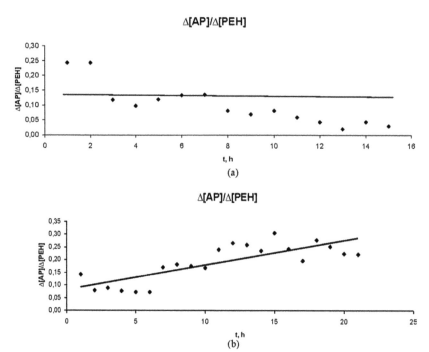

FIGURE 8.4 Dependences $\Delta[AP]_{ij}/\Delta[PEH]_{ij}$ on time t_j in the course of ethylbenzene oxidation, catalyzed with {Ni(acac)$_2$+HMPA}(a) or {Ni(acac)$_2$+MP}(b) systems. [Ni(acac)$_2$]=3×10^{-3} mol/l, [HMPA]=2×10^{-2} mol/l, [MP]=7×10^{-2} mol/l at 120°C. This dependences $\Delta[AP]_{ij}/\Delta[PEH]_{ij}$ on t_j, during catalysis of ethylbenzene oxidation with systems {Ni(acac)$_2$+HMPA (or MP)} are presented at first.

We observed parallel formation products PEH, MPC, AP in the II macro stage of the ethylbenzene oxidations, catalyzed by binary systems

{Ni(acac)$_2$+HMPA(MP)}, $w_{AP(MPC)}/w_{PEH} \neq 0$ at t→0, $w_{AP}/w_{MPC} \neq 0$ at t→0. For example, in Figures 8.4a and 8.4b, one can see parallel formation of AP and PEH: $w_{AP}/w_{PEH} \neq 0$ at t→0) for these catalytic systems.

During catalysis with triple system {Ni(acac)$_2$+HMPA(MP)+PhOH}, the parallel formation of PEH, AP, and MPC was found ($w_{AP(MPC)}/w_{PEH} \neq 0$ at t→0, $w_{AP}/w_{MPC} \neq 0$ at t→0) at the earliest stages of ethylbenzene oxidation and throughout the reaction of ethylbenzene oxidation at t≤20–30 h (see, for example Figures 8.5 and 8.6).

$$\Delta[AP]/\Delta[PEH]$$

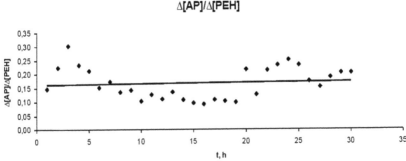

FIGURE 8.5   Dependence of $\Delta[AP]_{ij}/\Delta[PEH]_{ij}$ on time t_j in the course of ethylbenzene oxidation, catalyzed with {Ni(acac)$_2$+MP+PhOH} system. [Ni(acac)$_2$]=3 × 10^{-3} mol/l, [MP]=7 × 10^{-2} mol/l, [PhOH]=3 × 10^{-3} mol/l at 120°C.

We have observed the analogical kinetics earlier, in the case of the catalysis with triple system {Ni(acac)$_2$+L^2+PhOH} L^2 =MSt (M=Na, Li).[2,5]

So, AP and MPC form in parallel with PEH rather than as a result of PEH decomposition during ethylbenzene oxidation, catalyzed with {Ni(acac)$_2$+MP(HMPA)+PhOH} system. This is analogous to the catalysis by triple system {Ni(acac)$_2$+L^2+PhOH} (L^2=MSt)[5] but is unlike non-catalytic oxidation and catalysis by binary systems {Ni(acac)$_2$+MP(HMPA)} at the early stages of ethylbenzene oxidation.[2] Data shown in Figures 8.2–8.6 have been obtained and presented for the first time.

There are common featured characteristic of the triple systems including L^2=HMPA, MP (and metalloligand–modifiers NaSt, LiSt, too), compared with the most active binary systems.[2] The advantage of these triple systems is the long-term activity (ethylbenzene oxidation rate w≅constant) of the in situ-formed complexes Ni(acac)$_2$·L^2·PhOH.[2] This indicates that unlike binary systems, the acac$^-$ ligand in the structure of triple nickel complex does not undergo transformation in the course of ethylbenzene oxidation.

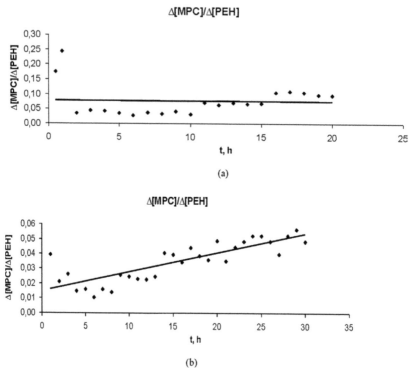

FIGURE 8.6 Dependence of $\Delta[MPC]_{ij}/\Delta[PEH]_{ij}$ on time t_j in the course of ethylbenzene oxidation, catalyzed with {Ni(acac)$_2$+HMPA+PhOH}(a) and {Ni(acac)$_2$+MP+PhOH} (b) system. [Ni(acac)$_2$]=3 × 10^{-3} mol/l, [HMPA] MP 2 × 10^{-2} mol/l, [MP] , 7 × 10^{-2} mol/l [PhOH] =3 × 10^{-3} mol/l at 120°C.

The formation of triple complexes Ni(acac)$_2$·MP·PhOH at the very early stages of oxidation was established earlier with kinetic methods.[1,2,5]

Analogy in kinetics of the ethylbenzene oxidations, catalyzed with two systems {Ni(acac)$_2$+L^2+PhOH} (L^2=HMPA, MP), testify in favor of the formation of triple complexes Ni(acac)$_2$·HMPA·PhOH. Earlier we have established that the concentration of PhOH during catalysis with systems {Ni(acac)$_2$+MP(MSt)+PhOH(3 × 10^{-3} mol/l)}, decreases during the first hours of oxidation (see, for example, Fig. 8.7b).[2,5,7] We received analogous results in the case of ethylbenzene oxidation with molecular oxygen, catalyzed by triple systems {Ni(acac)$_2$+HMPA+PhOH} (Fig. 8.7a). These changes in PhOH concentrations seem to be due to the triple complexes

Ni(acac)$_2$·MP(MSt)·PhOH formation,[2,5,7] and also the formation point of triple complexes {Ni(acac)$_2$·HMPA·PhOH}.

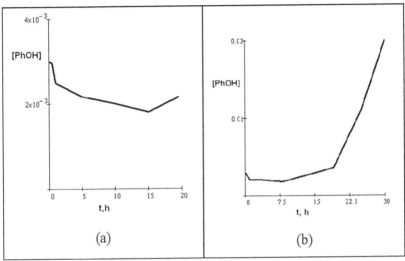

(a)   (b)

FIGURE 8.7 PhOH kinetics in reactions of ethylbenzene oxidation catalyzed by triple systems: (a){Ni(acac)$_2$ (3.0 × 10^{-3} mol/l))IHMPA(2.0 × 10^{-2} mol/l))IPhOH (3 × 10^{-3} mol/l)}, (b){Ni(acac)$_2$ (3.0 × 10^{-3} mol/l))IMP(7.0 × 10^{-2} mol/l))IPhOH(3 × 10^{-3} mol/l) at 120°C. Data in the Figure 8.7 is presented for the first time.

8.3.2 ROLE OF INTERMOLECULAR H-BONDING IN STABILIZATION OF TRIPLE CATALYTIC COMPLEXES {Ni(acac)$_2$ L^2 PhOH}. SELF-ORGANIZATION OF TRIPLE COMPLEXES IN MACROSTRUCTURES

The nature commonly uses H-bonds for the fabrication of supramolecular assemblies because they are directional and have a wide range of interaction energies that are tunable by adjusting the number of H-bonds, their relative orientation, and their position in the overall structure. Due to cooperative dipolar interactions, H-bonds in the center of protein helices can be 20 kcal/mol.[8,9]

The porphyrin linkage through H-bonds is the generally observed binding type in nature. One of the simplest artificial self-assembling supramolecular porphyrin systems is the formation of a dimer based on carboxylic acid functionality.[10,11]

The high efficiency of three-component systems $\{Ni(acac)_2 + L^2 + PhOH\}$ $(L^2 = HMPA, MP (MSt (M = Na, Li))$ in the reaction of selective oxidation of ethylbenzene to PEH on parameters S, C, w = constant is associated with the formation of extremely stable hetero-ligand complexes $\{Ni(acac)_2 \cdot L^2 \cdot PhOH\}$. We assumed that the stability of complexes $\{Ni(acac)_2 \cdot L^2 \cdot PhOH\}$ during ethylbenzene oxidation can be associated with one of the reasons with the formation of supramolecular structures due to intermolecular H-bonds (phenol–carboxylate)[12–14] and, possibly, the other non-covalent interactions:

$$\{Ni(acac)_2 + L^2 + PhOH\} \rightarrow \{Ni(acac)_2 \cdot L^2 \cdot PhOH\}$$
$$\rightarrow \{Ni(acac)_2 \cdot L^2 \cdot PhOH\}_n$$

AFM-microscopy data has shown to be in favor of formation of supramolecular macrostructures due to intermolecular (phenol–carboxylate) H-bonds and, possibly, other non-covalent interactions,[12–14] based on the triple complexes $\{Ni(acac)_2 \cdot L^2 \cdot PhOH\}$ $(L^2 = HMPA, MP)$, (Fig. 8.8a, b), that we have received for the first time in this article and also in our earlier works for other nickel structures.[7,15] Spontaneous organization process, that is, self-organization of triple complexes on the surfaces of modified silicon are driven by the balance between intermolecular and molecule-surface interactions, which may be the consequence of hydrogen bonds and the other non-covalent interactions.[16] The formation of more stable supramolecular structures in the case of complexes $\{Ni(acac)_2 \cdot MP \cdot PhOH\}$ (Fig. 8.8a) as compared with complexes $\{Ni(acac)_2 \cdot HMPA \cdot PhOH\}$ (Fig. 8.8b) is possibly related to the greater ability of $\{Ni(acac)_2 \cdot MP \cdot PhOH\}$ to form H-bonds. The association $\{Ni(acac)_2 \cdot MP \cdot PhOH\}$ systems due to H-bonding in more stable supramolecular structure in the real systems of catalytic oxidations may explain higher parameter C as compared with catalysis with $\{Ni(acac)_2 \cdot HMPA \cdot PhOH\}$ system.

Also, we observed the formation of supramolecular macrostructures (with $h \geq 10$ nm) on a surface of modified silicone on the basis of binary complexes $\{Ni(acac)_2 \cdot MP\}$ (Fig. 8.8c). In the case of the use of binary system $\{Ni(acac)_2 \cdot HMPA\}$, we did not observe the formation of supramolecular macrostructures (with $h \geq 10$ nm) on a surface of modified silicone in analogical conditions.

The received AFM data (Fig. 8.8a–c) points to the very probable stable supramolecular nanostructures' appearance based on hetero-ligand triple complexes $\{Ni(acac)_2 \cdot HMPA(or\ MP) \cdot PhOH\}_n$ due to intermolecular (phenol–carboxylate) H-bonds[12-14] and, possibly, the other non-covalent interactions, also in the real catalytic ethylbenzene oxidation with dioxygen, catalyzed by triple systems $\{Ni(acac)_2 + HMPA(MP) + PhOH\}$.

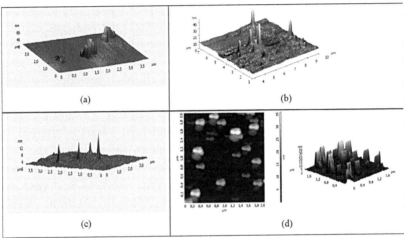

FIGURE 8.8 (a) The Atomic force microscopy (AFM) three-dimensional image $(4.0 \times 4.0\,\mu m)$ of the structures (h~80-100 nm) formed on a surface of modified silicone based on triple complexes $Ni(acac)_2 \cdot MP \cdot PhOH$ (b) The AFM three-dimensional image $(10.0 \times 10.0\,\mu m)$ of the structures (h~40 nm) formed on a surface of modified silicone based on triple complexes $\{Ni(acac)_2 \cdot HMPA \cdot PhOH\}$ (c) The AFM of three-dimensional image $(5.0 \times 5.0\,\mu m)$ of nanoparticles (h~10–12 nm) based on $\{Ni(acac)_2 \cdot MP\}$ formed on the surface of modified silicone (d) The AFM two- and three-dimensional image of nanoparticles based on $Ni_2(AcO)_3(acac) \cdot L^2 \cdot 2H_2O$ formed on the hydrophobic surface of modified silicone (h~35 nm).

8.3.3   POSSIBLE EFFECT OF TYR-FRAGMENT, BEING IN THE SECOND COORDINATION SPHERE OF METAL COMPLEX, ON MODEL SYSTEMS (AFM)

In recent years, studies in the field of homogeneous catalytic oxidation of hydrocarbons with molecular oxygen have been developed in two directions, namely, the free-radical chain oxidation catalyzed by transition metal complexes and the catalysis by metal complexes that mimic enzymes.[1,2] The findings on the mechanism of action of enzymes, and, in particular, dioxygenases and their models are very useful in the treatment of mechanism of catalysis by metal complexes in the processes of hydrocarbon oxidation with molecular oxygen. Moreover, as one will see below, the investigations of the mechanisms of catalysis by metal complexes that model the actions of enzymes can give the necessary material for the study of the mechanism of action of enzymes.

The methionine salvage pathway (MSP) plays a critical role in regulating a number of important metabolites in prokaryotes and eukaryotes. Acireductone dioxygenases (ARDs) Ni(Fe)-ARD are enzymes involved in the methionine recycle pathway, which regulates different aspects of the cell cycle.[17] Both enzymes Ni(Fe)-ARD are members of the structural superfamily, known as cupins, which also includes Fe-acetyl acetone dioxygenase (Dke1) and cysteine dioxygenase.[18,19] Regioselectivity of the oxidative carbon-carbon bond cleavage of substrate acireductone (1,2-dihydroxy-3-keto-5-(thiomethyl)pent-1-ene) is unusual and its catalysis with Ni(Fe)-ARD is determined by the nature of the metal ion. Fe-ARD catalyzes the 1,2-oxidative decomposition of acireductone to formate and 2-keto-4-(thiomethyl) butyrate, the keto-acid precursor of methionine. Ni-ARD catalyzes a 1,3-oxygenolytic reaction, yielding formate, carbon monoxide, and 3-methylthiopropionate, an off-pathway transformation of the acireductone.[20,21]

In our previous works, we have shown that formation of multidimensional forms based on nickel complexes can be one of the ways of regulating the activity of Ni-enzyme. The association of complexes $Ni_2(AcO)_3(acac)\cdot MP\cdot 2H_2O$, which is functional and structure model of Ni-ARD,[6] to supramolecular nanostructure due to intermolecular H-bonds (H_2O-MP, $H_2O-(OAc^-)$ (or $(acac^-)$)) is an evidence in favor of this assumption (see Fig. 8.8d).

We assumed that it may be necessary to take into account the role of the second coordination sphere, including tyr-fragment.[17] We first suggested the participation of tyrosine moiety in mechanisms of action of Ni(Fe)-ARD enzymes as regulatory factor (in press, *Applied Chemistry and Chemical Engineering, volume 4, chapter 18*).

Tyrosine can participate in different enzymatic reactions. It is assumed, for example, that tyr-fragment may be involved in substrate H-binding in step of O_2-activation by iron catalyst, and this can decrease the oxygenation rate of substrate in the case of homoprotocatechuate 2,3-dioxygenase action.[22] In the case of Ni-ARD, Tyr-fragment, involved in mechanism, can reduce the Ni-ARD activity.

As mentioned above, we have found[2,7] that the inclusion of PhOH in complex $Ni(acac)_2\cdot L^2$ ($L^2=MP$), which is the primary model of Ni-ARD, leads to the stabilization of formed triple complex $Ni(acac)_2 \cdot L^2 \cdot PhOH$. In this case, as we have emphasized above, ligand (acac) is not oxygenated with molecular O_2. Also, the stability of triple complexes $Ni(acac)_2 \cdot L^2 \cdot PhOH$ seem to be due to the formation of supramolecular macrostructures that are stable to oxidation with dioxygen. Formation of supramolecular macrostructures due to intermolecular (phenol–carboxylate) H-bonds and, possibly, the other non-covalent interactions,[12–14] based on the triple complexes $Ni(acac)_2 \cdot L^2 \cdot PhOH$, that we have established with the AFM-method in the case of $L^2 = MP$, HMPA, is in favor of this hypothesis (Fig. 8.8a–c).

Conclusive evidence in favor of the participation of tyrosine fragment in stabilizing primary Ni-complexes as one of the regulatory factors in mechanism of action of Ni-ARD has been obtained by AFM. We observed for the first time, the formation of nanostructures based on Ni-systems using L-Tyrosine (Tyr) as an extra-ligand[23] (in press: Mechanism of Ni(Fe) ARD Action in Methionine Salvage Pathway, in Biosynthesis of Ethylene, and Role of Tyr-Fragment as Regulatory Factor; Ludmila I.). The growth of self-assembly of supramolecular macrostructures due to intermolecular (phenol–carboxylate) H-bonds and, possibly, the other non-covalent interactions,[12–14] based on the triple systems $\{Ni(acac)_2 + MP + Tyr\}$, we observed at the apartment of a uterine H_2O solution of triple system $\{Ni(acac)_2 + MP + Tyr\}$ on surfaces of modified silicon (Fig. 8.9). Spontaneous organization process, that is, self-organization, is driven by the balance between intermolecular and molecule-surface interactions, which may be the consequence of hydrogen bonds and the other non-covalent interactions.[16]

Histogram of volumes of the particles based on systems $\{Ni^{II}(acac)_2 + MP + Tyr\}$, the empirical and theoretical cumulative normal probability distribution of volumes, and the empirical and theoretical cumulative log normal distribution of volumes of the particles based on systems $\{Ni^{II}(acac)_2 + MP + Tyr\}$, formed on the surfaces of modified silicone, are presented on Figure 8.10. As can be seen, distribution of volumes of the particles in this case is well described by a log-normal law.

In the case of binary systems $\{Ni(acac)_2 + Tyr\}$, we also observed the formation of nanostructures due to H-bonds. But these nanoparticles as well as particle based on $\{Ni(acac)_2 \cdot MP\}$ complexes (Fig. 8.8b) differ on form and high (h = 10–12 nm) from the nanostructures on the basis of triple systems $\{Ni(acac)_2 + MP + Tyr\}$ (Fig. 8.9).

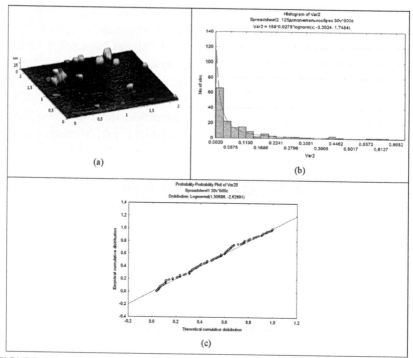

FIGURE 8.9   The AFM three-dimensional image (2.0×2.0 µm) of the structures (h~25 nm) (a), histogram of volumes of the particles based on systems {Ni^{II}(acac)$_2$ acMPPcTyr} (b), formed on a surface of modified silicone based on triple systems {Ni^{II}(acac)$_2$ acMPPcTyr}, log-normal distribution of volumes of the particles based on systems {Ni^{II}(acac)$_2$ acMPPaTyr} (c).

Earlier, we confirmed the formation of complexes {Ni(acac)$_2$·MP·Tyr} by electron absorption spectra (method of UV-spectroscopy)[23] (in press: Mechanism of Ni(Fe)ARD Action in Methionine Salvage Pathway, in Biosynthesis of Ethylene, and Role of Tyr-Fragment as Regulatory Factor, Ludmila I.).

Active site of Ni-ARD, as a member of cupins, includes histidine ligands.[18,19] The formation of stable supramolecular structures due to intermolecular H-bonds based on triple systems {Ni(acac)$_2$ + His + Tyr}, (His = L-Histidine) that we found recently by AFM (Fig. 8.10), is another fact in favor of our proposed model of Ni-ARD action.

In the case of triple systems {Ni(acac)$_2$ + His + Tyr}, we also observed growth of self-assembly of supramolecular macrostructures due to intermolecular (phenol–carboxylate) H-bonds and, possibly, the other non-covalent interactions[12–14,16] (Fig. 8.10a). The volumes of particles based on binary {Ni(acac)$_2$+His} and triple systems {Ni(acac)$_2$+His+Tyr} differ significantly (Fig. 8.10b). Besides this, the particles based on binary systems {Ni(acac)$_2$+His} are unstable compared to stable triple particles. As can be seen, the distribution of volume of the particles based on triple system {Ni(acac)$_2$+His+Tyr} is well described by a log-normal law, as in the case of triple systems {Ni(acac)$_2$+MP+Tyr} (Fig. 8.10c). This data (Fig. 8.10) are presented for the first time.

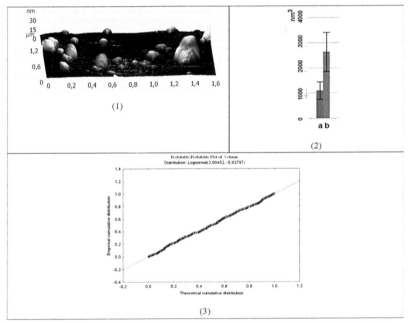

FIGURE 8.10   The AFM three-dimensional image (1.2 × 1.6 µm) of the nanostructures (h~30 nm) based on triple systems {Ni(acac)$_2$ i(His isTyr} formed on a surface of modified silicone (a). Diagram of the mean values of volumes of the particles based on binary- {Ni(acac)$_2$ caHis}(a) and triple systems {Ni(acac)$_2$ caHis isTyr}(b) (the diagram marked 95% confidence interval) (b). The empirical and theoretical cumulative log-normal distribution of volumes of the particles based on system {Ni(acac)$_2$ caHis isTyr} (c).

8.4   CONCLUSION

1. So inclusion of phenol in coordination sphere of a complex {Ni^{II}(acac)$_2$·L^2} (L^2=HMPA, MP) leads to formation of triple

complexes $\{Ni(acac)_2 \cdot L^2 \cdot PhOH\}$ with different catalytic activity. In the presence of $\{Ni(acac)_2 + L^2 + PhOH\}$, the rates of accommodation of oxidation products PEH, AP, MPC remain practically the same during the oxidation process (≤ 20–30 h), unlike the catalysis with binary system $\{Ni(acac)_2 + L^2\}$. The mechanism of by-products formation during ethylbenzene oxidation in the presence of triple $\{Ni(acac)_2 + L^2 + PhOH\}$ system does not change practically: AP and MPC form in parallel with PEH rather than as a result of PEH decomposition. The increase in selectivity $S_{PEH,max}$ ~90% at catalysis with $\{Ni(acac)_2 \cdot L^2 \cdot PhOH\}$ in comparison with no catalyzed oxidation ($S_{PEH,max} \leq 80\%$) is due to change in an order, in which products PEH, AP, and MPC form. Catalysis with $\{Ni(acac)_2 \cdot L^2 \cdot PhOH\}$ is largely associated with the involvement of these systems in the stages of chain initiation (activation of O_2) and also chain propagation (Cat + $RO_2\bullet \rightarrow$).

2. We applied AFM method at first for analytical purpose to research the possibility of the formation of supramolecular structures on the basis of hetero-ligand triple complexes $\{Ni(acac)_2 \cdot L^2 \cdot PhOH\}$ ($L^2 = HMPA$, MP) with the assistance of intermolecular H-bonds. We have shown that the self-assembly-driven growth of structures on the basis of $\{Ni(acac)_2 \cdot L^2 \cdot PhOH\}$ ($L^2 = HMPA$, MP) seems to be due to the connection of complexes with a surface of modified silicone, and further formation of supramolecular nanostructures $\{Ni(acac)_2 \cdot L^2 \cdot PhOH\}_n$ due to directional intermolecular (phenol–carboxylate) H-bonds, and possibly, other non-covalent interactions (van Der Waals attractions and π-bonding). The formation is more stable for supramolecular structures in the case of complexes $\{Ni(acac)_2 \cdot MP \cdot PhOH\}$ as compared with $\{Ni(acac)_2 \cdot HMPA \cdot PhOH\}$ complexes possibly due to the greater ability of $\{Ni(acac)_2 \cdot MP \cdot PhOH\}$ systems to form H-bonds. The received AFM data points to the very probable stable supramolecular nanostructures appearance based on hetero-ligand triple complexes. $\{Ni(acac)_2 \cdot HMPA(or\ MP) \cdot PhOH\}_n$ in the real ethylbenzene oxidation with dioxygen, is catalyzed by triple systems $\{Ni(acac)_2 + HMPA(MP) + PhOH\}$. Moreover, this can be one of the explanations of the stability of triple systems ($w \cong const$) during the selective oxidation process of the ethylbenzene oxidation into PEH with preservation of the high values of S_{PEH} and C.

3. We at first, assumed the participation of Tyr-fragment which is in the second coordination sphere in mechanism of Ni(Fe)-ARD operation, as one of possible mechanisms of decrease in Ni(Fe)-ARD enzymes activity, the role of Tyrosine residue as regulatory factor, and received experimental facts in favor of this assumption. So with AFM method, we observed the formation of supramolecular macrostructures based on triple systems {Ni(acac)$_2$+MP+Tyr} that included L-Tyrosine as an extra ligand, formed on a surface of modified silicone due to intermolecular (phenol–carboxylate) H-bonds and, possibly, the other non-covalent interactions. As an active site of Ni-ARD being a member of cupins, includes histidine ligands, the formation of stable supramolecular structures due to intermolecular H-bonds based on triple systems {Ni(acac)$_2$+His+Tyr}, (His=L-Histidine) that we found recently by AFM, is another fact in favor of our proposed model of Ni-ARD action.

GLOSSARY

Histidine (abbreviated as *His* or *H*) is an α-amino acid that is used in the biosynthesis of proteins. It contains an α-amino group (which is in the protonated $-NH_3^+$ form under biological conditions), a carboxylic acid group (which is in the deprotonated $-COO^-$ form under biological conditions), and a side chain imidazole, classifying it as a positively charged amino acid at physiological pH. Initially thought essential only for infants, longer-term studies have shown it is essential for adults also. Here *His*— imidazole is residue of Histidine as ligand. The histidine residue is part of the active centers of a variety of enzymes.

L-Histidine: (S)-2-Amino-3-(1H-imidazol-4-yl) propionic acid is one of the 22 essential amino acids.

Homogeneous catalysis: is catalysis in a solution.

DN: donor number. To assess the electron donating activity of HMPA, MP we used the values of the donor number (DN), measured by the value of the enthalpy (-ΔH, kkal/mol) of interaction of the donor (HMPA, MP) with SbCl$_5$ in dichloroethane. Unlike pK$_a$, DN number characterizes not only π-, but σ- electron donating properties of ligands. HMPA, MP coordinated with the metal ion by the oxygen atom, and are located in MP (DN=27.30)<HMPA (DN=37.07 kkal/mol).

Ligands are ions or neutral molecules that bond to a central metal atom or ion of metal complex compound. The term *"monodentate ligand"* can be translated as "one tooth", referring to the ligand binding to the center through only one atom.

L-Tyrosine: Tyrosine (*Tyr* or *Y*) or *4-hydroxyphenylalanine* is one of the 22 amino acids that are used by cells to synthesize proteins. Tyrosine exists in two optically isomeric forms of L and D. A tyrosine residue also plays an important role in biochemical processes.

Selectivity S_{PEH}, *conversion C of ethylbenzene (RH) oxidation into PEH* are determined using the formulas:

$$S_{PEH} = [PEH]/\Delta[RH] \cdot 100\%$$

$$C = \Delta[RH]/[RH]_0 \cdot 100\%.$$

KEYWORDS

- catalytic triple systems based on Ni-compounds
- ethyl benzene oxidations
- models of Ni-ARD
- nanostructures
- AFM
- L-Tyrosine
- L-Histidine

REFERENCES

1. Matienko, L. I. Solution of the Problem of Selective Oxidation of Alkylarenes by Molecular Oxygen to Corresponding Hydro Peroxides. Catalysis Initiated by Ni(II), Co(II), and Fe(III) Complexes Activated by Additives of Electron-Donor Mono- or Multidentate Extra-Ligands. In *Reactions and Properties of Monomers and Polymers;* D'Amore, A., Zaikov, G., Eds.; Nova Science Publ. Inc.: New York, USA, 2007; pp 21–41.

2. Matienko, L. I.; Mosolova, L. A.; Zaikov, G. E. *Selective Catalytic Hydrocarbons Oxidation: New Perspectives;* Nova Science Publ. Inc.: New York USA, 2010; p 150.

3. Suresh, A. K.; Sharma, M. M.; Sridhar, T. Industrial Hydrocarbon Oxidation. *Ind. Eng. Chem. Res.* **2000**, *39*, 3958.

4. Weissermel, K.; Arpe, H.-J.; *Industrial Organic Chemistry,* 3rd ed.; transl. by Lindley C. R.; VCH: New York, 1997.

5. Matienko, L. I.; Mosolova, L. A. Mechanism of Selective Catalysis with Triple System {bis(acetylacetonate)Ni(II)+metalloligand+phenol} in Ethylbenzene Oxidation with Dioxygen. Role of H-bonding Interactions. *Oxid. Commun.* **2014**, *37*, 20–31.

6. Matienko, L. I.; Binyukov, V. I.; Mosolova, L. A.; Mil, E. M.; Zaikov, G. E. *The New Approach to Research of Mechanism Catalysis with Nickel Complexes in Alkylarens Oxidation" "Polymer Yearbook 2011";* Nova Science Publ. Inc.: New York, 2012; pp 221–230.

7. Matienko, L. I.; Binyukov, V. I.; Mosolova, L. A.; Mil, E. M.; Zaikov, G. E. Supramolecular Nanostructures on the Basis of Catalytic Active Heteroligand Nickel Complexes and their Possible Roles in Chemical and Biological Systems. *J. Biol. Res.* (Nova Science Publ. Inc., USA) **2012**, *1*, 37–44.

8. Leninger, St.; Olenyuk, B.; Stang, P. J. Self-Assembly of Discrete Cyclic Nanostructures Mediated by Transition Metals. *Chem. Rev.* **2000**, *100*, 853.

9. Stang, P. J.; Olenyuk, B. Self-Assembly, Symmetry, and Molecular Architecture: Coordination as the Motif in the Rational Design of Supramolecular Metallacyclic Polygons and Polyhedra. *Acc. Chem. Res.* **1997**, *30*, 502.

10. Drain, C. M.; Varotto Radivojevic, A. I. Self-Organized Porphyrinic Materials. *Chem. Rev.* **2009**, *109*, 1630.

11. Beletskaya, I.; Tyurin, V. S.; Tsivadze, A. Yu.; Guilard, R. R.; Stem Ch. Supramolecular Chemistry of Metalloporphyrins. *Chem. Rev.* **2009**, *109*, 1659–1713.

12. Dubey, M.; Koner, R. R.; Ray, M. Sodium and Potassium Ion Directed Self-assembled Multinuclear Assembly of Divalent Nickel or Copper and L-Leucine Derived Ligand. *Inorg. Chem.* **2009**, *48*, 9294–9302.

13. Basiuk, E. V.; Basiuk, V. V.; Gomez-Lara, J.; Toscano, R. A. A Bridged High-Spin Complex bis-[Ni(II)(rac-5,5,7,12,12,14-hexamethyl-1,4,8,11-tetraazacyclotetradecane)]-2,5-pyridinedicaboxylate Diperchlorate Monohydrate. *J. Incl. Phenom. Macrocycl. Chem.* **2000**, *38*, 45–56.

14. Mukherjee, P.; Drew, M. G. B.; Gómez-Garcia, C. J.; Ghosh, A. (Ni$_2$), (Ni$_3$), and (Ni$_2$+Ni$_3$): A Unique Example of Isolated and Cocrystallized Ni$_2$ and Ni$_3$ Complexes. *Inorg. Chem.* **2009**, *48*, 4817–4825.

15. Matienko, L. I.; Binyukov, V. I.; Mosolova, L. A.; Mil, E. M.; Zaikov, G. E. The Supramolecular Nanostructures as Effective Catalysts for the Oxidation of Hydrocarbons and Functional Models of Dioxygenases. *Polym. Res. J.* (Nova Science Publ. Inc., USA), **2014**, *8*, 91–116.

16. Gentili, D.; Valle, F.; Albonetti, C.; Liscio, F.; Cavallini, M. Self-Organization of Functional Materials in Confinement. *Acc. Chem. Res.* **2014**, *47*, 2692–2699.

17. Dai, Y.; Pochapsky, Th. C.; Abeles, R. H. Mechanistic Studies of Two Dioxygenases in the Methionine Salvage Pathway of *Klebsiella Pneumoniae. Biochemistry* **2001**, *40*, 6379–6387.

18. Leitgeb, St.; Straganz, G. D.; Nidetzky, B. Functional Characterization of an Orphan Cupin Protein from Burkholderia Xenovorans Reveals a Mononuclear Nonheme

Fe^{2+}-Dependent Oxygenase that Cleaves β-Diketones. *The FEBS J.* **2009**, *276*, 5983–5997.

19. Straganz, G. D.; Nidetzky, B. Reaction Coordinate Analysis for β-Diketone Cleavage by the Non-Heme Fe^{2+}-Dependent Dioxygenase Dke 1, *J. Am. Chem. Soc.* **2005**, *127*, 12306–12314.

20. Chai, S. C.; Ju, T.; Dang, M.; Goldsmith, R. B.; Maroney, M. J.; Pochapsky, Th. C. Characterization of Metal Binding in the Active Sites of Acireductone Dioxygenase Isoforms from Klebsiella ATCC 8724. *Biochemistry* **2008**, *47*, 2428–2435.

21. Deshpande, A. R.; Wagenpfail, K.; Pochapsky, Th. C.; Petsko, G. A.; Ringe, D. Metal-Dependent Function of a Mammalian Acireductone Dioxygenase. *Biochemistry* **2016**, *55*, 1398–1407.

22. Mbughuni, M. M.; Meier, K. K.; Münck, E.; Lipscomb, J. D. Substrate-Mediated Oxygen Activation by Homoprotocatechuate 2,3-Dioxygenase: Intermediates Formed by a Tyrosine 257 Variant. *Biochemistry* **2012**, *51*, 8743–8754.

23. Matienko, Larisa A. Mosolova, Vladimir I. Binyukov, Elena M. Mil, and Gennady E. Zaikov, *Applied Chemistry and Chemical Engineering*, Appl. Acad. Press: Toronto, New Jersey, 2017; vol. 4, chap.18.

PART IV

Industrial Chemistry and Engineering Technology

CHAPTER 9

ANTIOXIDATIVE ACTIVITY OF HYALURONAN: EVALUATION AND MECHANISM

MAYSA MOHAMED SABET, TAMER MAHMOUD TAMER*, AND AHMED MOHAMED OMER

Polymer Materials Research Department, Advanced Technologies and New Materials Research Institute (ATNMRI), City of Scientific Research and Technological Applications (SRTA-City), New Borg El-Arab City, Alexandria, Egypt

*Corresponding author. E-Mail: ttamer85@gmail.com

CONTENTS

ABSTRACT

Hyaluronan (HA) is the main component of the synovial joint, eyes, extra-cellular matrix of the skin and many other tissues and organs in the body. The unique structure and properties of HA enable its usage in several bioapplications. This review describes physical–chemical properties of HA, its mechanism of action, and oxidative degradation.

9.1 INTRODUCTION

Hyaluronan (HA) is an acidic polysaccharide with repeating units of β-(1-3)-N-acetyl-D-glucosamine and β-(1–4)-D-glucuronic acid (Fig. 9.1). Both sugars are spatially linked to glucose, which in β-configuration involve all of its bulky groups (i.e., the hydroxyls, the carboxylate moiety, and the anomeric carbon of the adjacent sugar) to be in sterically promising equatorial positions. On the other hand, all of the small hydrogen atoms occupy the less sterically favorable axial positions. Therefore, the configuration of the disaccharide is energetically highly stable. Each glucuronate unit holds an anionic charge at physiological pH, and there are negative charges associated with its carboxylate group coordinated by a cation, which is generally Na^+. The molecular weight of HA is unique, which can reach up to ten million daltons. HA is formed at the cytoplasmatic membrane instead of the Golgi apparatus.[32,34,101]

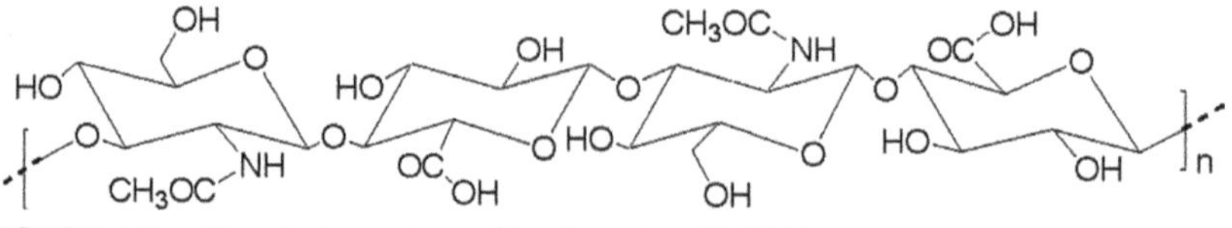

FIGURE 9.1 Chemical structure of hyaluronic acid (HA).

Meyer and Palmer discovered and isolated HA from bovine vitreous humor.[41] It exists in vivo as a polyanion and not in the protonated acid form. Survey on the Scopus database was done for the studies reporting about HA. Criteria for inclusion were articles addressed to HA and publication years from 2005 to 2016. Dramatic increases in publications referring to HA indicate a growing interest in HA over the recent years (Fig. 9.2). Figure 9.3 represents the most frequented areas in Scopus database with keywords hyaluronan or hyaluronic acid (HA). A list of top authors contributed in a research related to HA cited in Scopus database is shown in Figure 9.4.

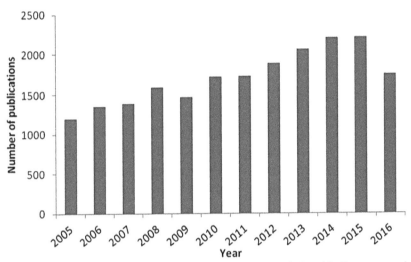

FIGURE 9.2 The annual number of articles referring to HA indexed in Scopus over the 2005–2016 period.

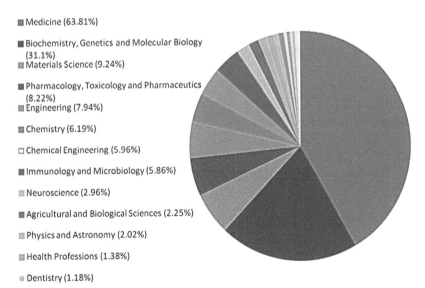

Medicine (63.81%)

Biochemistry, Genetics and Molecular Biology (31.1%)
Materials Science (9.24%)

Pharmacology, Toxicology and Pharmaceutics (8.22%)
Engineering (7.94%)

Chemistry (6.19%)

Chemical Engineering (5.96%)

Immunology and Microbiology (5.86%)

Neuroscience (2.96%)

Agricultural and Biological Sciences (2.25%)

Physics and Astronomy (2.02%)

Health Professions (1.38%)

Dentistry (1.18%)

FIGURE 9.3 The top fields involving the research of HA according to Scopus over the 2005–2016 period with the keywords hyaluronan or hyaluronic acid.

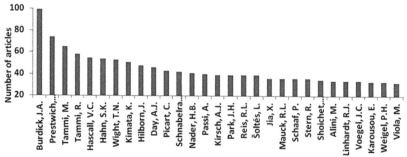

FIGURE 9.4   The top indexed scientific contributions on HA publications according to Scopus over the 2005–2016 period with the keywords hyaluronan and hyaluronic acid.

HA is employed in several medical applications for its unique physicochemical properties such as: (a) it is very hydrophilic; its viscous solutions have the most unusual rheological properties and are exceedingly lubricious; (b) biodegradability, biocompatibility, and resorbability; (c) HA is a major component of connective tissues, where it plays a significant role in lubrication, such as cell differentiation and cell growth; and (d) HA contains functional groups (carboxylic acids and alcohols) along its backbone that can be utilized to chemical modifications or structure cross-linking.[16,45]

9.2   HYALURONAN (HA) NETWORKS

The physicochemical characteristics of HA were studied in detail from 1950 onwards.[17] The molecules perform in the solution highly hydrated randomly kinked coils. The complexity point can be seen both by sedimentation analysis and viscosity.[37,42] Scott et al.[56] have given evidence that the chains, when entangling, also interact with each other and form stretches of double helices so that the network becomes mechanically more firm.

9.3   RHEOLOGICAL PROPERTIES

HA gives a highly viscous solution, viscoelastic, and the viscosity is particularly shearing dependent. Beyond the entanglement point, the viscosity rises rapidly and exponentially with strength ($\sim c^{3.3}$; where c denotes the concentration)[42] and the HA solution of 10 g/l may have a viscosity at a low shear, which is $\sim 10^6$ times the viscosity of the solvent. At a high shear, the viscosity may decline as much as $\sim 10^3$ times. The elasticity of the system improves with expanding molecular weight. The rheological properties of HA have been combined with lubrication of joints and tissues, and HA is regularly found in the body separating surfaces that move each other, for example, cartilage surfaces and muscle bundles.[13]

9.4 WATER HOMEOSTASIS

In the solution, the HA polymer takes up a stiffened helical configuration attributed to hydrogen bonding between the hydroxyl groups along the chain, forming a coiled structure that traps approximately 1000 times its weight in water. The fixed polysaccharide network offers a great defense to bulk flow of the solvent.[17] This was illustrated by Day,[19] who reported that hyaluronidase treatment removes the high interference to water flow through a fascia. Thus, HA limits excessive fluid fluxes through tissue compartments. Moreover, the osmotic pressure of the HA solution is nonideal and grows exponentially with the strength. In spite of the high molecular weight of the polymer, the osmotic pressure of a 10 g/l HA solution is of the similar order as the 10 g/l albumin solution. The exponential link performs HA as an excellent osmotic buffering substance—moderate changes in concentration lead to significant changes in osmotic pressure. Both flow resistance and osmotic buffering make HA an ideal regulator of the water homeostasis in the body.

9.5 ANTIOXIDATIVE ACTIVITY OF HA

Protection strategies from free radicals involve three levels of protection: prevention, interception, and repair. Antioxidants are materials or molecules, which have the ability to deactivate or stabilize free radicals before they attack cells. Antioxidants are critical for maintaining systemic health and optimal cellular and well-being. The most popular mechanisms of antioxidants to overcome harmful effects of free radicals are their conversion to a less harmful structure. Naturally, HA has a fundamental role in body functions according to organ type in which it is distributed.[39] The ability of HA to scavenge free radicals was reported in many studies.[36,59,85] Chemical and configuration structures allow HA to quench some reactive oxygen species (ROS), for example, OH radical by donating an H atom. HA was stabilized via series of bond rearrangements, which started fragmentation of macromolecules.

HA has many roles in the modulation and activation of the inflammatory response derived from polymorphonuclear leukocytes, which scavenge ROS, and other inflammatory cells stimulated by Nicotinamide adenine dinucleotide phosphate (NADPH) oxidase through the respiratory burst.[3,11,22,23,35,43,50,73] Oxidants and ROS include hydrogen peroxide (H_2O_2), superoxide anion radical ($O_2^{\cdot -}$), and the hydroxyl radical ($^{\cdot}OH$) species, which are now strongly associated with the pathogenesis of chronic wounds.[57,100,103] Such a characteristic is due to the wound healing process. Despite the beneficial role of ROS in killing invading microbial pathogens, excessive overproduction of ROS can be detrimental to the host tissues by inactivating enzymatic antioxidants and significantly depleting nonenzymic antioxidants levels.[53,58,70] Excess of ROS causes random damage to cellular constituents, such as proteins, lipids, and DNA.[25] ROS cause damage to components of extracellular matrix (ECM), such as proteoglycans, HA, and collagen,[100] and further alter the metabolism of cells responsible for dermal ECM synthesis.[2,33,47,78,99] In such cases, HA is chemically changed and depolymerized by ROS, whose mechanisms have now been fully explained.[100]

As a consequence of many functions attributed to HA during wound healing, researchers have proved the beneficial influences of the application of exogenous HA in the healing of chronic dermal wounds.[40] Moreover, there have been modern improvements in the development and application of HA as biomaterials in the treatment of various conditions, such as chronic wounds.[10,15,24] HA as a biomaterial employs properties of the natural biomolecules involved in the wound healing process, which are enhanced by derivatization or chemical modification. Studies have shown the benefit of applying these biomaterials regarding prevention of tissue damage associated with their resorptive properties and biocompatibility on promoting the healing process.[12,15,18] HA forms a hydrophilic gel in contact with serum or a wound exudate and provides a HA-rich scaffold in a moist microenvironment, which is thought to be conducive to granulation and wound healing.[20,26] HA has also been designated to the role as a scavenger in the joints. It has been known since the 1940s that irradiation and oxidizing systems degrade HA, and it is evident that the general denominator is a chain division induced by free radicals, typically OH radicals.[44] HA functions as a highly active scavenger of free radicals. The rapid turnover of HA in the joints has led to the suggestion that HA also functions as a scavenger of cellular debris.[38] The cellular material could be distinguished in the HA network and reduced at the same rate as HA biopolymer.[51]

The suggested roles of HA are based on its individual interactions with hyaladherins. An interesting perspective is the fact that HA controls angiogenesis but the impact is varied, which depends on its molecular weight and concentration.[55] High molecular weight and high concentrations of HA inhibit the production of capillaries, while oligosaccharides can generate angiogenesis. There are also reports of HA receptors on vascular endothelial cells, whereby HA could work on the cells. The vascularity of the joint cavity could be a result of inhibition of angiogenesis by using HA. Another interaction in joints is the adhesion of HA to cell surface proteins. Other cells such as lymphocytes may get to joints by this interaction. Injection of high doses of HA intra-articularly could attract cells expressing these proteins. Cells can also change their expression in HA-binding proteins in states of disease, whereby HA may affect immunological reactions and cellular traffic in the path of physiological processes in cells.[21] The research often describes that intra-articular injections of HA alleviate pain in joint diseases and may show a direct or indirect interaction with pain receptors.[1]

9.6   OXIDATIVE DEGRADATION OF HA

The sensitivity of HA to free radicals allows HA to be a suitable sensor for evaluation of antioxidants and inflammatory drugs. Several years ago, Soltes and his group established a novel methodology concerning to degradation of HA under oxidative stress to evaluate anti-inflammatory drugs.[7–9,61,67,74–77,79,80,90–96]

A more rapid degradation of high-molecular-weight HA occurring under inflammation or oxidative stress is accompanied by impairment and loss of its viscoelastic properties (Fig. 9.5).[27,29,49,60,69,86–88,97] Low-molecular-weight HA was found to exert different biological activities compared to the native high-molecular-weight biopolymer. HA chains of 25–50 disaccharide units are inflammatory, immune-stimulatory, and highly angiogenic. HA fragments of this size appear to function as endogenous danger signals, reflecting tissues under stress.[46,64,65,71,102] Scheme 9.1 illustrates the degradation mechanism of HA under free-radical stress.

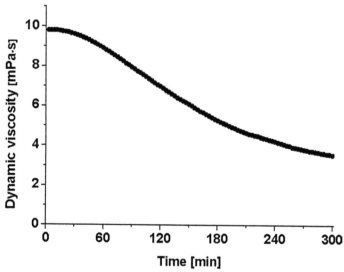

FIGURE 9.5 Degradation of high-molecular-weight HA in vitro induced by Weissberger's biogenic oxidative system (WBOS).

a) Initiation phase: The attacking of HA macromolecule by HO˙ radical generated from the Fenton-like reaction:

$$Cu^+ + H_2O_2 \rightarrow Cu^{2+} + HO^\bullet + OH^-$$

H_2O_2 has its origin in the oxidative action of WBOS. The use of systems containing copper ions requires a special attention as these may affect radical scavenging kinetics of some antioxidants.[98]

b) Generation of an alkyl radical (C-centered HA macroradical) as a result of HO˙ radical attack.

c) Propagation phase: Generation of a peroxy-type C-macroradical of HA by trapping a molecule of O_2.

d) Production of an HA-derived hydroperoxide through the reaction with extra HA macromolecule.

e) Formation of an extremely low stable alkoxy-type C-macroradical of HA on undergoing a redox reaction with a transition metal ion in a reduced state.

f) Rapid formation of alkoxy-type C-fragments and the fragments with a terminal C=O group due to the glycosidic bond scission of

HA. Alkoxy-type C fragment may continue the propagation phase of the free-radical degradation of HA. Both fragments are represented by reduced molecular weight.[6,31,54,66,72,85]

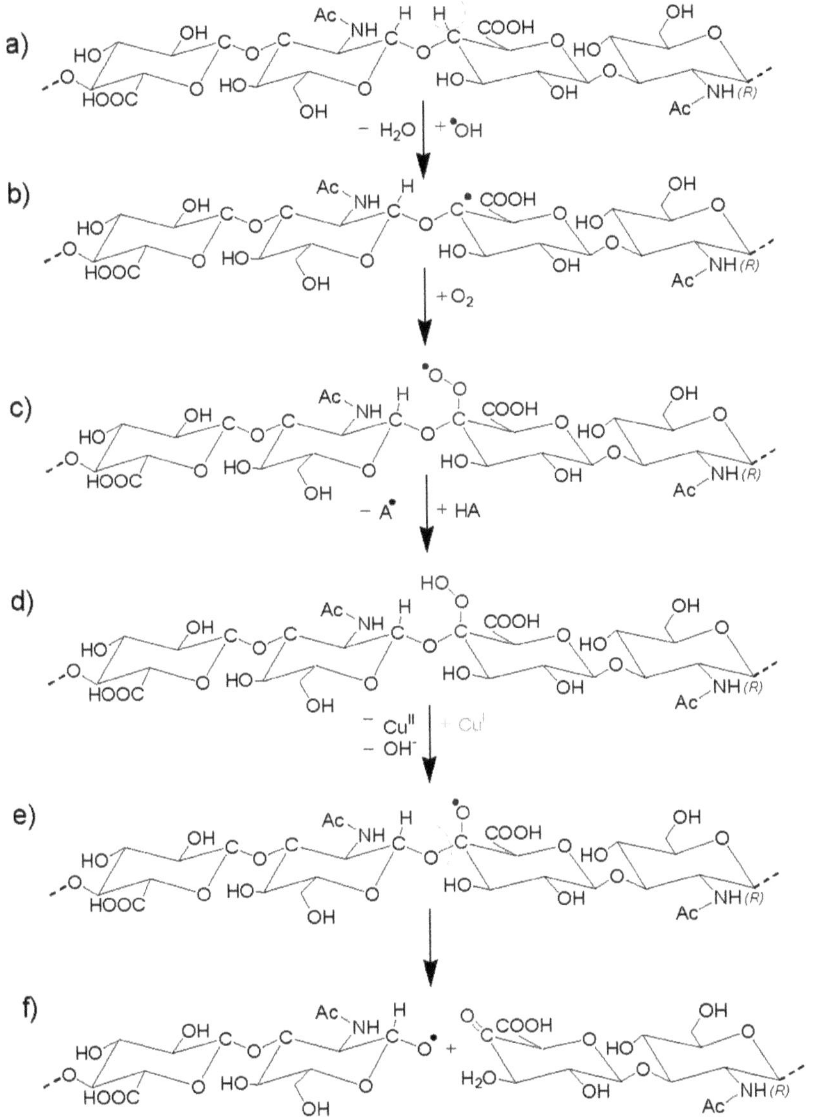

SCHEME 9.1   Schematic degradation of hyaluronan (HA) under free-radical stress.

Soltes and coworkers evaluated the influence of several natural extracts and drugs on inhibition of the degradation kinetics of a high-molecular-weight HA in vitro. High-molecular-weight HA samples were exposed to free-radical chain degradation reactions generated from ascorbate in the presence of Cu(II) ions (Scheme 9.2). The concentrations of both reactants [ascorbate and Cu(II) ions] were equivalent to those that may happen during the initial stage of the acute phase of joint inflammation.[4,5,28,30,31,51,52,62,63,68,75,81–84]

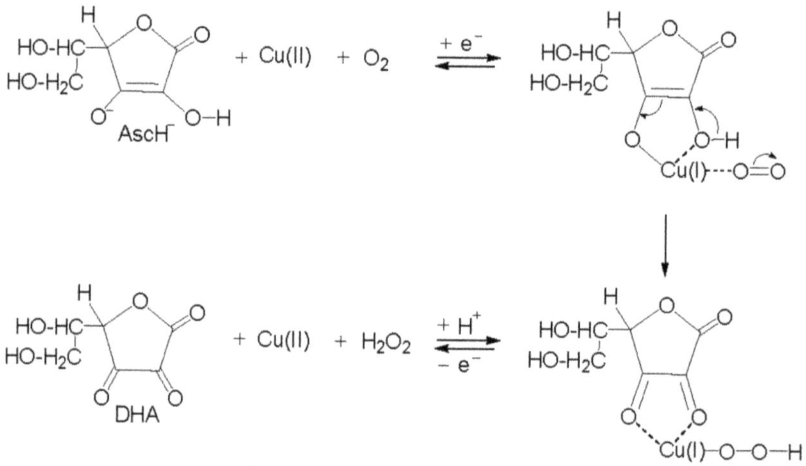

$$H_2O_2 + Cu(I)\text{---}complex \rightarrow {}^\bullet OH + Cu(II) + OH^-$$

SCHEME 9.2 Generation of H_2O_2 by Weissberger's biogenic oxidative system (WBOS) from ascorbate and Cu (II) ions under aerobic conditions.

9.7 ACKNOWLEDGMENT

The authors Maysa M. Sabet and Tamer M. Tamer would like to thank the Institute of Experimental Pharmacology and Toxicology of Slovak Academy of Sciences for inviting and orienting them in the field of medical research. They would also like to thank Slovak Academic Information Agency (SAIA) for funding them during their work at the institute.

KEYWORDS

- glycosaminoglycans
- oxidative stress
- reactive oxygen species
- rotational viscometry

REFERENCES

1. Adams, M. E. Viscosupplementation: A Treatment for Osteoarthritis. J. Rheumatol. **1993,** *20*(Suppl 39), 1–24.
2. Agren, U. M.; Tammi, R. H.; Tammi M. I. Reactive Oxygen Species Contribute to Epidermal Hyaluronan Catabolism in Human Skin Organ Culture. *Free Radical Biol. Med.* **1997,** *23*, 996–1001.
3. Artola, A; Alio, J. L., Bellot, J. L.; Ruiz, J. M. Protective Properties of Viscoelastic Substances (Sodium Hyaluronate and 2% Hydroxymethylcellulose) Against Experimental Free Radical Damage to the Corneal Endothelium. *Cornea* **1993,** *12*, 109–114.
4. Baňasová, M.; Valachová, K.; Hrabárová, E.; Priesolová, E.; Nagy, M.; Juránek, I.; Šoltés L. Early Stage of the Acute Phase of Joint Inflammation. *In vitro* Testing of Bucillamine and Its Oxidized Metabolite SA981 in the Function of Antioxidants, 16th Interdisciplinary Czech-Slovak Toxicological Conference in Prague. *Interdiscip. Toxicol.* **2011a,** *4*(2), 22.
5. Baňasová, M.; Valachová, K.; Rychlý, J.; Priesolová, E.; Nagy, M.; Juránek, I.; Šoltés, L. Scavenging and Chain Breaking Activity of Bucillamine on Free-Radical-Ediated Degradation of High Molar Mass Hyaluronan. *ChemZi* **2011b,** *7*, 205–206.
6. Baňasová, M.; Valachová, K.; Juránek, I.; Šoltés, L. Effect of Thiol Compounds on Oxidative Degradation of High Molar Hyaluronan *In Vitro. Interdiscip. Toxicol.* **2012,** *5* (Suppl. 1), 25–26.
7. Baňasová, M.; Valachová, K.; Juránek, I.; Šoltés, L. Aloe Vera and Methylsulfonylmethane as Dietary Supplements: Their Potential Benefit to Arthritic Patients with Diabetic Complications. *J. Inf. Intell. Knowl.* **2013,** *5*, 51–68.
8. Baňasová, M.; Valachová, K.; Juránek, I.; Šoltés, L. Dithiols as More Effective than Monothiols in Protecting Biomacromolecules from Free-Radical-Mediated Damage: *In Vitro* Oxidative Degradation of High-Molar-Mass Hyaluronan. *Chem. Pap.* **2014a,** *68*, 428–1434.
9. Baňasová, M.; Valachová, K.; Rychlý, J.; Janigová, I.; Csomorová, K.; Mendichi, R.; Mislovičová, D.; Juránek, I.; Šoltés, L. Free-Radical-Mediated Degradation of High-Molar-Mass Hyaluronan. Action of an Antiimmflamatory Drug. *Polymers* **2014b,** *6*, 2625–2644.

10. Band, P. A. Hyaluronan Derivatives: Chemistry and Clinical Applications. In *The Chemistry, Biology and Medical Applications of Hyaluronan and Its Derivatives (Wenner–Gren International Series);* Laurent, T. C., Ed.; Portland Press: London, 1998; pp 33–42.

11. Bellavite, P. The Superoxide-Forming Enzymatic System of Phagocytes. *Free Radical Biol. Med.* **1988,** *4,* 225–261.

12. Benedetti, L.; Cortivo, R.; Berti, A.; Pea, F.; Mazzo, M.; Moras, M.; Abatangelo, G. Biocompatibility and Biodegradation of Different Hyaluronan Derivatives (HYAFF) Implanted in Rats. *Biomaterials* **1993,** *14,* 1154–1160.

13. Bothner, H.; Wik, O. Rheology of Hyaluronate. Acta Otolaryngol. Suppl. **1987,** *442,* 25–30.

14. Camber, O.; Edman, P. Sodium Hyaluronate as an Ophthalmic Vehicle—Some Factors Governing Its Effect on the Ocular Absorption of Pilocarpine. *Curr. Eye Res.* **1989,** *8,* 563–567.

15. Campoccia, D.; Doherty, P.; Radice, M.; Brun, P.; Abatangelo, G.; Williams, D. F. Semisynthetic Resorbable Materials from Hyaluronan Esterification. *Biomaterials* **1998,** *19,* 2101–227.

16. Collins, M. N.; Birkinshaw, C. Hyaluronic Acid Based Scaffolds for Tissue Engineering: A Review. *Carbohydr. Polym.* **2003,** *92,* 1262–1279.

17. Comper, W. D.; Laurent, T. C. Physiological Functions of Connective Tissue Polysaccharides. *Physiol. Rev.* **1978,** *58,* 255–315.

18. Cortivo, R.; Brun, P.; Rastrelli, A.; Abatangelo, G. *In vitro* Studies on Biocompatibility of Hyaluronic Acid Esters. *Biomaterials* **1991,** *12,* 727–730.

19. Day, T. D. Connective Tissue Permeability and the Mode of Action of Hyaluronidase. *Nature* **1950,** *166,* 785–786.

20. Edmonds, M.; Foster, A. Hyalofill: A New Product for Chronic Wound Management. *Diabetic Foot* **2000,** *3,* 29–30.

21. Edwards, J. C. W. Consensus Statement. Second International Meeting on Synovium. Cell Biology, Physiology and Pathology. Ann. Rheum. Dis. **1995,** *54,* 389–391.

22. Foschi, D.; Castoldi, L.; Radaelli, E.; Abelli, P.; Calderini, G.; Rastrelli, A.; Mariscotti, C.; Marazzi, M.; Trabucchi, E. Hyaluronic Acid Prevents Oxygen Free-Radical Damage to Granulation Tissue: A Study in Rats. *Int. J. Tissues. React.* **1990,** *12,* 333–339.

23. Fukuda, K.; Takayama, M.; Ueno, M.; Oh, M.; Asada, S.; Kumano, F.; Tanaka, S. Hyaluronic Acid Inhibits Interleukin-1-Induced Superoxide Anion in Bovine Chondrocytes. *Inflamm. Res.* **1997,** *46,* 114–117.

24. Galassi, G.; Brun, P.; Radice, M.; Cortivo, R.; Zanon, G. F.; Genovese, P.; Abatangelo, G. *In vitro* Reconstructed Dermis Implanted in Human Wounds: Degradation Studies of the HA-Based Supporting Scaffold. *Biomaterials* **2000,** *21,* 2183–2191.

25. Halliwell, B.; Gutteridge, J. M. C.; Cross, C. E. Free Radicals, Antioxidants and Human Disease: Where Are We Now? *J. Clin. Lab. Med.* **1992,** *119,* 598–620.

26. Hollander, D. A.; Schmandra, T.; Windolf, J. A New Approach to the Treatment of Recalcitrant Wounds: A Case Report Demonstrating the Use of a Hyaluronan Ester Fleece. *Wounds* **2000,** *12,* 111–117.

27. Hrabárová, E.; Gemeiner, P.; Šoltés, L. Peroxynitrite: *In vivo* and *In Vitro* Synthesis and Oxidant Degradative Action on Biological Systems Regarding Biomolecular Injury and Inflammatory Processes. *Chem. Pap.* **2007**, *61*, 417–437.

28. Hrabárová, E.; Valachová, K.; Rychlý, J.; Rapta, P.; Sasinková, V.; Gemeiner, P.; Šoltés, L. High-Molar-Mass Hyaluronan Degradation by the Weissberger's System: Pro- and Antioxidative Effects of Some Thiol Compounds. *Polym. Degrad. Stab.* **2009**, *94*, 1867–1875.

29. Hrabárová, E.; Valachová, K.; Rapta, P.; Šoltés, L. An Alternative Standard for Trolox-Equivalent Antioxidant-Capacity Estimation Based on Thiol Antioxidants. Comparative 2,2′-Azinobis[3-ethylbenzothiazoline-6-sulfonic Acid] Decolorization and Rotational Viscometry Study Regarding Hyaluronan Degradation. *Chem. Biodiversity* **2010**, *7*(9), 2191–2200.

30. Hrabárová, E.; Valachová, K.; Juránek, I.; Šoltés, L. Free-Radical Degradation of High-Molar-Mass Hyaluronan Induced by Ascorbate Plus Cupric Ions. Antioxidative Properties of the Piešťany-Spa Curative Waters from Healing Peloidand Maturation Pool. In *Kinetics, Catalysis and Mechanism of Chemical Reactions;* Islamova, G. M., Kolesov, S. V., Zaikov, G. E., Eds.; Nova Science Publishers: New York, 2011; pp 29–36.

31. Hrabárová, E.; Valachová, K.; Juránek, I.; Šoltés, L. Free-Radical Degradation of High-Molar-Mass Hyaluronan Induced by Ascorbate Plus Cupric Ions: Evaluation of Antioxidative Effect of Cysteine-Derived Compounds. *Chem. Biodiversity* **2012**, *9*, 309–317.

32. Itano, N.; Kimata, K. Mammalian Hyaluronan Synthases. *IUBMB Life* **2002**, *54*, 195–199.

33. Kawaguchi, Y.; Tanaka, H.; Okada, T.; Konishi, H.; Takahashi, M.; Ito, M.; Asai, J. Effect of Reactive Oxygen Species on the Elastin mRNA Expression in Cultured Human Dermal Fibroblasts. *Free Radic. Biol. Med.* **1997**, *23*, 162–165.

34. Kogan, G.; Soltes, L.; Stern, R.; Gemeiner, P. Hyaluronic Acid: A Natural Biopolymer with a Broad Range of Biomedical and Industrial Applications. Biotechnol. Lett. **2007**, *29*, 17–25.

35. Kvam, B. J.; Fragonas, E.; Degrassi, A.; Kvam, C.; Matulova, M.; Pollesello, P.; Zanetti, F.; Vittur, F. Oxygen-Derived Free Radical (ODFR) Action on Hyaluronan (HA), on Two HA Ester Derivatives, and on the Metabolism of Articular Chondrocytes. *Exp. Cell Res.* **1995**, *218*, 79–86.

36. Lath, D.; Csomorova, K.; Kollarikova, G.; Stankovska, M.; Soltes, L. Molar Mass-Intrinsic Viscosity Relationship of High-Molar-Mass Hyaluronans: Involvement of Shear Rate. Chem. Pap. **2005**, *59*, 291–293.

37. Laurent, T. C.; Ryan, M.; Pictruszkiewicz, A. Fractionation of Hyaluronic Acid. The Polydispersity of Hyaluronic Acid from the Vitreous Body. *Biochim. Biophys. Acta* **1960**, *42*, 476–485.

38. Laurent, T. C.; Laurent, U. B. G.; Fraser, J. R. E. Functions of Hyaluronan. Ann. Rheum. Dis. **1995**, *54*, 429–432.

39. Laurent, T. C.; Laurent, U. B. G.; Fraser, J. R. E. The Structure and Function of Hyaluronan: An Overview. *Immunol. Cell Biol.* **1996**, *74*, 1–7.

40. Manuskiatti, W.; Maibach, HI. Hyaluronic Acid and Skin: Wound Healing and Aging. *Int. J. Dermatol.* **1996,** *35,* 539–544.

41. Meyer, K.; Palmer, J. W. The Polysaccharide of the Vitreous Humor. J. Biol. Chem. **1934,** *107,* 629–634.

42. Morris, E. R.; Rees, D. A.; Welsh, E. J. Conformation and Dynamic Interactions in Hyaluronate Solutions. *J. Mol. Biol.* **1980,** *138,* 383–400.

43. Moseley, R.; Leaver, M.; Walker, M.; Waddington, R. J.; Parsons, D.; Chen, W. Y. J.; Embery, G. Comparison of the Antioxidant Properties of HYAFFs-11p75, AQUA-CELs and Hyaluronan Towards Reactive Oxygen Species *In Vitro. Biomaterials* **2002,** *23,* 2255–2264.

44. Myint, P. The Reactivity of Various Free Radicals with Hyaluronic Acid Steady-State and Pulse Radiolysis Studies. Biochim. Biophys. Acta **1987,** *925,* 194–202.

45. Necas, J.; Bartosikov, L.; Brauner, P.; Kolar, J. Hyaluronic Acid (Hyaluronan): A Review. *Vet. Med.* **2008,** *53,* 397–411.

46. Noble, P. W. Hyaluronan and Its Catabolic Products in Tissue Injury and Repair. *Matrix Biol.* **2002,** *1,* 25–29.

47. O'Toole, E. A.; Goel, M.; Woodley, D. T. Hydrogen Peroxide Inhibits Human Keratinocyte Migration. *Dermatol. Surg.* **1996,** *22,* 525–529.

48. Omer, A. M.; Tamer, T. M.; Mohyeldin, M. S. High-Molecular Weight of Biopolymer. In *Analysis and Performance of Engineering Materials Key Research and Development;* Zaikov, G. E., Ed.; Apple Academic Press: Waretown, Oakville, 2014; pp 56–70.

49. Parsons, B. J.; Al-Assaf, S.; Navaratnam, S.; Phillips, G. O. Comparison of the Reactivity of Different Oxidative Species (ROS) Towards Hyaluronan. In *Hyaluronan: Chemical, Biochemical and Biological Aspects;* Kennedy, J. F., Phillips, G. O., Williams, P. A., Hascall, V. C., Eds.; Woodhead Publishing Ltd: Cambridge, 2002; pp 141–150.

50. Presti, D.; Scott, J. E. Hyaluronan-Mediated Protective Effect Against Cell Damage Caused by Enzymatically Produced Hydroxyl (OH) Radicals is Dependent on Hyaluronan Molecular Mass. *Cell Biochem. Funct.* **1994,** *12,* 281–288.

51. Rapta, P.; Valachová, K.; Gemeiner, P.; Šoltés, L. High-Molar-Mass Hyaluronan Behavior During Testing Its Radical Scavenging Capacity in Organic and Aqueous Media: Effects of the Presence of Manganese (II) Ions. *Chem. Biodivers.* **2009,** *6,* 162–169.

52. Rapta, P.; Valachová, K.; Zalibera, M.; Šnirc, V.; Šoltés, L. Hyaluronan Degradation by Reactive Oxygen Species: Scavenging Effect of the Hexapyridoindole Stobadine and Two of Its Derivatives. In *Monomers, Oligomers, Polymers, Composites, and Nanocomposite;* Pethrick, R. A., Petkov, P., Zlatarov, A., Zaikov, G. E., Rakovsky, S. K., Eds.; Nova Science Publishers: New York, 2010; pp 113–126.

53. Rasik, A. M.; Shukla, A. Antioxidant Status in Delayed Healing Type of Wounds. *Int. J. Exp. Pathol.* **2000,** *81,* 257–263.

54. Rychlý, J.; Šoltés, L.; Stankovská, M.; Janigová, I.; Csomorová, K.; Sasinková, V.; Kogan, G.; Gemeiner, P. Unexplored Capabilities of Chemiluminescence and Thermoanalytical Methods in Characterization of Intact and Degraded Hyaluronans. *Polym. Degrad. Stab.* **2006,** *91*(12), 3174–3184.

55. Sattar, A.; Kumar, S.; West, D. C. Does Hyaluronan Have a Role in Endothelial Cell Proliferation of the Synovium. Semin. *Arthritis Rheum.* **1992**, *22*, 37–43.

56. Scott, J. E.; Cummings, C.; Brass, A.; Chen, Y. Secondary and Tertiary Structures of Hyaluronan in Aqueous Solution, Investigated by Rotary Shadowing-Electron Microscopy and Computer Simulation. *Biochem. J.* **1991**, *274*, 600–705.

57. Senel, O.; Cetinkale, O.; Ozbay, G.; Ahc-ioglu, F.; Bulan, R. Oxygen Free Radicals Impair Wound Healing in Ischaemic Rat Skin. *Ann. Plast. Surg.* **1997**, *39*, 516–523.

58. Shukla, A.; Rasik, A. M.; Patnaik, G. K. Depletion of Reduced Glutathione, Ascorbic Acid, Vitamin E and Antioxidant Defence Enzymes in a Healing Cutaneous Wound. *Free Radic. Res.* **1997**, *26*, 93–101.

59. Šoltés, L.; Mendichi, R.; Kogan, G.; Mach, M. Associating Hyaluronan Derivatives: A Novel Horizon in Viscosupplementation of Osteoarthritic Joints. *Chem. Biodiversity* **2004**, *1*, 468–472.

60. Šoltés L.; Stankovská M.; Kogan G.; Germeiner P.; Stern R. Contribution of Oxidative Reductive Reactions to High Molecular Weight Hyaluronan Catabolism. *Chem. Biodiversity* **2005**, *2*, 1242–1245.

61. Šoltés, L.; Brezová, V.; Stankovská, M.; Kogan, G.; Gemeiner, P. Degradation of High-Molecular-Weight Hyaluronan by Hydrogen Peroxide in the Presence of Cupric Ions. *Carbohydr. Res.* **2006a**, *341*, 639–644.

62. Šoltés, L.; Mendichi, R.; Kogan, G.; Schiller, J.; Stankovska, M.; Arnhold, J. Degradative Action of Reactive Oxygen Species on Hyaluronan. *Biomacromolecules* **2006b**, *7*, 659–668.

63. Šoltés, L.; Stankovská, M.; Brezová, V.; Schiller, J.; Arnhold, J.; Kogan, G.; Gemeiner, P. Hyaluronan Degradation by Copper (II) Chloride and Ascorbate: Rotational Viscometric, EPR Spin-Trapping, and MALDI-TOF Mass Spectrometric Investigations. *Carbohydr. Res.* **2006c**, *341*, 2826–2834.

64. Šoltés, L.; Valachová, K.; Mendichi, R.; Kogan, G.; Arnhold, J.; Gemeiner, P. Solution Properties of High-Molar-Mass Hyaluronans: The Biopolymer Degradation by Ascorbate. *Carbohydr. Res.* **2007**, *342*, 1071–1077.

65. Šoltés, L.; Kogan, G. Impact of Transition Metals in the Free-Radical Degradation of Hyaluronan Biopolymer. In *Kinetics and Thermodynamics for Chemistry and Biochemistry;* Pearce, E. M., Zaikov, G. E., Kirshenbaum, G., Eds.; Nova Science Publishers: New York, 2009; pp 181–199.

66. Šoltés, L. Hyaluronan – A High-Molar-Mass Messenger Reporting on the Status of Synovial Joints: Part II. Pathophysiological Status. In *New Steps in Chemical and Biochemical Physics: Pure and Applied Science;* Pearce, E. M., Kirshenbaum, G., Zaikov, G. E., Eds.; Nova Science Publishers: New York, 2010; pp. 137–152.

67. Šoltés, L.; Kogan, G. Hyaluronan: A Harbinger of the Status and Functionality of the Joint. In *Engineering of Polymers and Chemicals Complexity Volume II: New Approaches, Limitations and Control;* Focke, W., Radusch, H. J., Zaikov, G. E., Haghi, A. K., Eds.; Apple Academic Press, Taylor & Francis Group: Oakville, Waretown, 2014; pp 259–286.

68. Stankovská, M.; Šoltés, L.; Vikartovská, A.; Mendichi, R.; Lath, D.; Molnárová, M.; Gemeiner, P. Study of Hyaluronan Degradation by Means of Rotational Viscometry: Contribution of the Material of Viscometer. *Chem. Pap.* **2004**, *58*, 348–352.

69. Stankovská, M.; Šoltés, L.; Vikartovska, A.; Gemeiner, P.; Kogan, G.; Bakos, D. Degradation of High-Molecular-Weight Hyaluronan: A Rotational Viscometry Study. *Biologia* **2005**, *60*(Suppl 17), 149–152.
70. Steiling, H.; Munz, B.; Werner, S.; Brauchle, M. Different Types of ROS-Scavenging Enzymes Are Expressed During Cutaneous Wound Repair. *Exp. Cell Res.* **1999**, *247*, 484–494.
71. Stern, R.; Kogan, G.; Jedrzejas, M. J.; Šoltés, L. The Many Ways to Cleave Hyaluronan. *Biotechnol. Adv.* **2007**, *25*, 537–557.
72. Surovčíková, L.; Valachová, K.; Baňasová, M.; Šnirc, V.; Priesolova, E.; Nagy, M.; Juránek, I.; Šoltés, L. Free-Radical Degradation of High-Molar-Mass Hyaluronan Induced by Ascorbate Plus Cupric Ions: Testing of Stobadine and Its Two Derivatives in Function as Antioxidants. *Gen. Physiol. Biophys.* **2012**, *31*, 57–64.
73. Suzuki, Y.; Yamaguchi, T. Effects of Hyaluronic Acid on Macrophage Phagocytosis and Active Oxygen Release. *Agents Actions* **1993**, *38*, 32–37.
74. Tamer, T. M. Hyaluronan and Synovial Joint: Function, Distribution and Healing. *Interdiscip. Toxicol.* **2013**, *6*, 111–125.
75. Tamer, T. M.; Valachová, K.; Šoltés, L. Inhibition of Free Radical Degradation in Medical Grade Hyaluronic Acid. In *Hyaluronic Acid for Biomedical and Pharmaceutical Applications;* Collins, M. N., Ed.; Smithers Rapra Technology Ltd.: United Kingdom, 2014; pp 103–117.
76. Tamer, M. T.; Valachová, K.; Mohamed, S. M.; Šoltés, L. Free Radical Scavenger Activity of Cinnamyl Chitosan Schiff Base. *J. Appl. Pharm. Sci.* **2016a**, *6*, 130–136.
77. Tamer, M. T.; Valachová, K.; Mohyeldin, M. S.; Šoltés, L. Free Radical Scavenger Activity of Chitosan and Its Aminated Derivative. *J. App. Pharm. Sci.* **2016b**, *6*, 195–201.
78. Tanaka, H.; Okada, T.; Konishi, H.; Tsuji, T. The Effect of Reactive Oxygen Species on the Biosynthesis of Collagen and Glycosaminoglycans in Cultured Human Dermal Fibroblasts. *Arch. Derm. Res.* **1993**, *285*, 352–355.
79. Topoľská, D.; Valachová, K.; Nagy, M.; Šoltés, L. Determination of Antioxidative Properties of Herbal Extracts: *Agrimonia herba, Cynare folium,* and *Ligustri folium. Neuroendocrinol. Lett.* **2014**, *35*, 192–196.
80. Topoľská, D.; Valachová, K.; Rapta, P.; Šilhár, S.; Panghyová, E.; Horváth, A.; Šoltés, L. Antioxidative Properties of *Sambucus nigra* Extracts. *Chem. Pap.* **2015**, *69*, 1202–1210.
81. Valachová, K.; Hrabárová, E.; Gemeiner, P.; Šoltés, L. Study of Pro- and Anti-oxidative Properties of D-Penicillamine in a System Comprising High-Molar-Mass Hyaluronan, Ascorbate, and Cupric Ions. *Neuroendocrinol. Lett.* **2008a**, *29*, 697–701.
82. Valachová, K.; Kogan, G.; Gemeiner, P.; Šoltés, L. Hyaluronan Degradation by Ascorbate: Protective Effects of Manganese (II). *Cellul. Chem. Technol.* **2008b**, *42*(9–10), 473–483.
83. Valachová, K.; Rapta, P.; Kogan, G.; Hrabárová, E.; Gemeiner, P.; Šoltés, L. Degradation of High-Molar-Mass Hyaluronan by Ascorbate Plus Cupric Ions: Effects of D-Penicillamine Addition. *Chem Biodivers.* **2009a**, *6*, 389–395.
84. Valachová, K.; Kogan, G.; Gemeiner, P.; Šoltés, L. Hyaluronan Degradation by Ascorbate: Protective Effects of Manganese (II) Chloride. In *Progress in Chemistry*

and Biochemistry. Kinetics, Thermodynamics, Synthesis, Properties and Application; Pearce, E. M., Kirshenbaum, G., Zaikov, G. E., Eds.; Nova Science Publishers: New York, 2009b; pp 201–215.

85. Valachová, K.; Mendichi, R.; Šoltés, L. Effect of L-Glutathione on High-Molar-Mass Hyaluronan Degradation by Oxidative System Cu(II) Plus Ascorbate. In *Monomers, Oligomers, Polymers, Composites, and Nanocomposites;* Pethrick, R. A., Petkov, P., Zlatarov, A., Zaikov, G. E., Rakovsky, S. K., Eds.; Nova Science Publishers: New York, 2010; pp 101–111.

86. Valachová, K.; Hrabárová, E.; Priesolová, E.; Nagy, M.; Baňasová, M.; Juránek, I.; Šoltés, L. Free-Radical Degradation of High-Molecular-Weight Hyaluronan Induced by Ascorbate Plus Cupric Ions. Testing of Bucillamine and Its SA981-Metabolite as Antioxidants. *J. Pharm. Biomed. Anal.* **2011a,** *56,* 664–670.

87. Valachová, K.; Vargová, A.; Rapta, P.; Hrabárová, E.; Dráfi, F.; Bauerová, K.; Juránek, I.; Šoltés, L. Aurothiomalate in Function of Preventive and Chain-Breaking Antioxidant at Radical Degradation of High-Molar-Mass Hyaluronan. *Chem. Biodivers.* **2011b,** *8,* 1274–1283.

88. Valachová, K.; Hrabárová, E.; Juránek, I.; Šoltés, L. Radical Degradation of High-Molar-Mass Hyaluronan Induced by Weissberger Oxidative System. Testing of Thiol Compounds in the Function of Antioxidants. 16th Interdisciplinary Slovak-Czech Toxicological Conference in Prague. *Interdiscip. Toxicol.* **2011c,** *4*(2), 65.

89. Valachová, K.; Rapta, P.; Slováková, M.; Priesolová, E.; Nagy, M.; Mislovičová, D.; Dráfi, F.; Bauerová, K.; Šoltés, L. Radical Degradation of High-Molar-Mass Hyaluronan Induced by Ascorbate Plus Cupric Ions. Testing of Arbutin in the Function of Antioxidant. In *Advances in Kinetics and Mechanism of Chemical Reactions;* Zaikov, G. E., Valente, A. J. M., Iordanskii, A. L., Eds.; Apple Academic Press: Oakville, Waretown, 2012; pp 1–19.

90. Valachová, K.; Baňasová, M.; Machová, Ľ.; Juránek, I.; Bezek, Š.; Šoltés, L. Testing Various Hexahydropyridoindoles to Act as Antioxidants. In *Pharmaceutical and Medical Biotechnology: New Perspectives;* Orlicki, R., Cienciala, C., Krylova, L. P., Pielichowski, J., Zaikov, G. E., Eds.; Nova Science Publishers: New York, 2013a; pp 93–110.

91. Valachová, K.; Baňasová, M.; Machová, Ľ.; Juránek, I.; Bezek, Š.; Šoltés, L. Practical Hints on Testing Various Hexahydropyridoindoles to Act as Antioxidants. In *Chemistry and Physics of Modern Materials;* Aneli, J. N., Jimenez, A., Kubica, S., Eds.; Apple Academic Press, Taylor & Francis Group: Toronto, 2013b; pp 137–158.

92. Valachová, K.; Topoľská, D.; Nagy, M.; Gaidau, C.; Niculescu, M.; Matyašovský, J.; Jurkovič P.; Šoltés L. Radical Scavenging Activity of *Caesalpinia spinosa. Neuroendocrinol. Lett.* **2014,** *35,* 197–200.

93. Valachová, K.; Tamer, M. T.; Šoltés, L. Comparison of Free-Radical Scavenging Properties of Glutathione Under Neutral and Acidic Conditions. In *Chemistry and Chemical Biology, Methodologies and Applications;* Joswik, R., Dalinkevich, A. A., Eds; Apple Academic Press, Taylor & Francis Group: New Jersey, 2015a; pp 227–245.

94. Valachová, K.; Tamer, M. T.; Šoltés, L. Comparison of Free-Radical Scavenging Properties of Glutathione Under Neutral and Acetic Conditions. In *Chemistry and*

Chemical Biology Methodologies and Applications; Joswik, R., Dalinkevich, A. A., Zaikov, G. E., Haghi, A. K., Eds.; Apple Academic Press Inc. Apple Academic Press Inc.: Oakville, Waretown, 2015b; pp 228–245.

95. Valachová, K.; Baňasová, M.; Topoľská, D.; Sasinková, V.; Juránek, I.; Collins, M. N; Šoltés, L. Influence of Tiopronin, Captopril and Levamisole Therapeutics on the Oxidative Degradation of Hyaluronan. *Carbohydr. Polym.* **2014,** *134,* 516–523.

96. Valachová, K.; Tamer, M. T.; MohyEldin, M.; Šoltés, L. Radical-Scavenging Activity of Glutathione, Chitin Derivatives and Their Combination. *Chem. Pap.* **2016a,** *70,* 820–827.

97. Valachová, K.; Topoľská, D.; Mendichi, R.; Collins, M. N.; Sasinková, V.; Šoltés, L. Hydrogen Peroxide Generation by the Weissberger Biogenic Oxidative System During Hyaluronan Degradation. *Carbohydr. Polym.* **2016b,** *148,* 189–193.

98. Valent, I.; Topoľská, D.; Valachová, K.; Bujdák, J.; Šoltés, L. Kinetics of ABTS Derived Radical Cation Scavenging by Bucillamine, Cysteine, and Glutathione. Catalytic Effect of Cu^{2+} Ions. *Biophys. Chem.* **2016,** *212,* 9–16.

99. Vassey, D. A.; Lee, K. H.; Blacker, K. L. Characterization of the Oxidative Stress Initiated in Cultured Human Keratinocytes by Treatment with Peroxides. *J. Invest. Derm.* **1992,** *99,* 859–863.

100. Waddington, R. J.; Moseley, R.; Embery, G. Reactive Oxygen Species: A Potential Role in the Pathogenesis of Periodontal Diseases. *Oral. Dis.* **2000,** *6,* 138–151.

101. Weigel, P. H.; Hascall, V. C.; Tammi, M. Hyaluronan Synthases. *J. Biol. Chem.* **1997,** *272,* 13997–14000.

102. West, D. C.; Hampson, I. N.; Arnold, F.; Kumar, S. Angiogenesis Induced by Degradation Products of Hyaluronic Acid. *Science* **1985,** *228,* 1324–1326.

103. White, M. J.; Heckler, F. R. Oxygen Free Radicals and Wound Healing. *Clin. Plast. Surg.* **1990,** *17,* 473–484.

CHAPTER 10

PREPARATION OF TWO QUERCETIN DERIVATIVES AND EVALUATION OF THEIR ANTIOXIDATIVE ACTIVITY

KATARÍNA VALACHOVÁ[1,*], TAMER MAHMOUD TAMER[1,2], MIROSLAV VEVERKA[3], AND LADISLAV ŠOLTÉS[1]

[1]Laboratory of Bioorganic Chemistry of Drugs, Institute of Experimental Pharmacology and Toxicology, SK 81404, Bratislava, Slovakia

[2]Polymer Materials Research Department, Advanced Technologies and New Materials Research Institute (ATNMRI), City of Scientific Research and Technological Applications (SRTA-City), New Borg El-Arab City, Alexandria 21934, Egypt

[3]EUROFINS Bel/Novamann Ltd., SK 94002, NovéZámky, Slovakia

*Corresponding author. E-mail: katarina.valachova@savba.sk

CONTENTS

ABSTRACT

In this study, we synthesized phosphatidylcholine quercetin (PCQ) and phosphatidylcholine dihydro quercetin (PCDQ) and assessed their anti-oxidative activities under acidic conditions, which occurs in vivo during inflammation. We determined IC_{50} values of both derivatives using two standard spectrophotometric methods—the 2,2'-azino-bis(3-ethylbenzo-thiazoline-6-sulphonic acid) (ABTS) and 2,2-Diphenyl-1-picrylhydrazyl (DPPH) assays. Further, these methods were used to examine the PCQ and PCDQ electron donor properties. An original rotational viscometric method was used to evaluate the ability of both quercetin derivatives to act as a donor of H atom.

As observed, the quercetin derivative PCQ was not effective in scavenging the ˙OH radicals; on the other hand, PCDQ, at the highest applied concentration, was effective against the forming hydroxyl radicals. PCQ showed concentration-dependent inhibition of the alkoxy-/peroxy-type radicals formed during the propagation phase of the degradation of hyaluronan—the biopolymer applied as a sensitive probe of free radical processes. Similarly, PCDQ in a concentration-dependent manner was effective in retarding hyaluronan degradation.

On using the ABTS assay, the kinetics of reduction of the $ABTS^{˙+}$ cation radical by PCQ at the lower concentration demonstrated a continual reduction process. In contrast, PCDQ reduced the $ABTS^{˙+}$ cation radical quickly by a total radical quenching within seconds. Both quercetin derivatives showed concentration-dependent reduction of the DPPH˙ radicals.

10.1 INTRODUCTION

Inflammation is the body's effort of self-protection, whereas the goal is to eliminate harmful stimuli, including irritants, damaged cells or pathogens, and begin the healing process. When something harmful or irritating affects a part of the body, there is a biological response to minimize or remove it. The signs and symptoms of inflammation, specifically during the acute inflammations, show that the body is trying to heal itself. Chronic inflammation is a pathological situation identified by continued active inflammation response followed usually by an extensive destruction of body tissues. During the stage of the chronic inflammation many of the immune cells, including neutrophils, eosinophils, and macrophages, are

directly involved and/or indirectly participate in responses to inflammation by production of inflammatory cytokines.[11,16]

Extensive experimental and clinical researches, during the past two
decades have established the mechanism by which continual oxidative
stress can result in chronic inflammation, which in turn could mediate
diseases including cancer, diabetes, as well as cardiovascular, neurological,
and pulmonary diseases.[20,26] Clinically, severe physicochemical changes
were observed in the inflammatory area rather than in normal tissues. For
example, acidosis has often been reported during inflammation and thus
lowering of pH has been claimed in inflamed tissues.[36] Release and elimination of acid–base relevant ions take place at different sites of the body,
whose processes are interconnected by the circulating blood as the main
transport medium in vertebrates. The inflammatory exudate shows a high
vascular permeability, which leads to further protein clearance.[10] Therefore, the acidosis observed in the inflammatory exudate may affect the
vascular compartment through a compensatory exchange with blood of
acid–base relevant ions such as strong anions or weak acids (i.e., proteins
and inorganic phosphate). The relationship between chronic inflammation and tissue pH should be taken into account to maximize the effect
of drugs, which are applied to treat primarily the fever and those drugs
introduced to patients with the aim to diminish (as much as it is possible)
the time of healing.

Administration of the drug to patients means that a disease already
developed. Currently, in population, there is an increased interest in prevention of diseases. Population becomes aware of many overlooked indicators
of unhealthy regimen and lack of movement. The idea to consume fewer
calories or to increase movement activity along with the consumption of
healthy foods becomes widespread in rationally thinking families. Apart
from advices to consume sufficient amount of fresh vegetables and fruits,
the market offers hundreds or maybe thousands of preparations declared as
a natural supplements. Many of them contain vitamins and trace elements
as well as antioxidants, which are either isolated from natural sources or
are chemically synthesized. One group of antioxidative preparations is
polyphenols, which it is necessary to introduce quercetin.

Quercetin (Fig. 10.1; IUPAC name 2-(3,4-dihydroxyphenyl)-3,5,7-tri-
hydroxy-4H-chromen-4-one; $C_{15}H_{10}O_7$), a flavonol found in many fruits,
vegetables, leaves, red wine, tea, and grains, is believed to be a supplement
promoting a wide spectrum of diseases. Quercetin induces mitochondrial-
derived apoptosis via reactive oxygen species-mediated ERK activation

in HL-60 leukemia cells, xenograft, and hepatocyte cells.[5,18] Quercetin dietary consumption is 25–50 mg per day.[7]

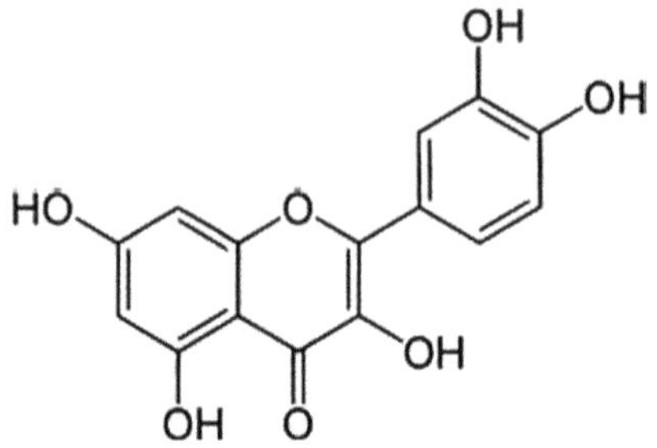

FIGURE 10.1   Structural formulae of quercetin.

In the solid state, quercetin forms a yellow crystalline powder, which is practically insoluble in water[22] and partly soluble in aqueous alkaline solutions. Although the molar mass of the quercetin is 302.24 g/mol, its oral bioavailability in humans is low. The half-life 1–2 h (postquercetin intravenous application) indicates either its rapid metabolism or a quick elimination from the blood circulation. As known, following dietary ingestion, quercetin undergoes rapid and extensive metabolism which makes the biological effects presumed in in vitro studies unlikely to be applied in vivo. The major quercetin metabolites are quercetin-3-glucuronide, 3'-methylquercetin-3-glucuronide, and quercetin-3'-sulfate.[4]

The parent quercetin molecules have been reported to efficiently inhibit the oxidation of other molecules in vitro and hence, quercetin is classified as an antioxidant. However, the European Food Safety Authority with consumption of quercetin declared that no cause and effect relationship has been established for any physiological effect in human health or diseases.[8,18] Although quercetin is under basic and early-stage clinical research for a variety of disease conditions, the US FDA has been emphasized that at present quercetin cannot be assigned as a drug for treating a concrete human disease.[1]

Quercetin has a wide range of biological effects, including antioxidant, antihypertension, anti-inflammatory, antifibrotic, anticoagulative, antibacterial, antiatherogenic, antihypertensive, chemopreventive, and antiproliferative properties.[2,6,28,15,19,27] Antioxidative effects of quercetin can be explained by scavenging reactive oxygen species and chelating metal cations.[29] Quercetin becomes oxidized into various oxidation products such as quinone, which has four tautomeric forms, that is, orthoquinone and three quinine methides. It is known that oxidation products such as semiquinone radicals and quinones have toxic effects for their ability to arylate protein thiols. Moreover, quinone is very reactive toward thiols and can form adduct with glutathione as the most abundant endogenous thiol.[2] Quercetin can interfere with reactive oxygen species and thus causes apoptosis. On the one hand, quercetin markedly increases intracellular levels of reactive oxygen species, as quercetin radicals (Qu-O') can be formed after peroxidase-catalyzed oxidation to scavenge reactive peroxy-type radicals.[8]

However, regarding the chemical structure of quercetin molecule, there exists an extensive attempt of investigators to exploit existing conjugated quercetin π electrons for donation of, for example, a single electron to a radical –R' yielding an anionic structure (–R⁻). Quercetin—a flavonol—(cf. Fig. 10.1) with its four or five free aromatic –OH groups implies a presumption that a modification of quercetin molecules is not a complicated task. Such structural modification can result remarkable changes in biochemical properties of new derivatives such as their higher water solubility, prolongation of the half-life of their elimination from organism, and especially changes in reaching the targeted regions inside and/or outside of the cell.

An ideal antioxidant is classified as a small molecule, which has the ability to stabilize and scavenge free radicals before they attack cells. It can be stated, that antioxidants are crucial for maintaining optimal cellular and systemic health and well-being. Quercetin is supposed to be a powerful antioxidant not only due to its capacity to scavenge free radicals but also to its property to bind transition metal ions.

It is generally known that flavonoid glycosides are only slightly absorbed in the small intestine due to the presence of sugar moieties, which can elevate their hydrophilicity.[23] In vivo bioavailability of quercetin is elevated when added into oils.[27] To enhance quercetin bioavailability, liposomes composed of quercetin and phosphatidylcholine were prepared. Nassibullin and Sharafutdinova[21] have investigated the interaction of the quercetin with the cell membrane phosphatidylcholin by methods such as the quantum chemistry, ^{1}H, and ^{13}C NMR. The rings of the quercetin and the choline group of the phosphatidylcholine molecule by means of π-system of the heterocycle electrons were established.

Soloviev et al.[32] showed that quercetin-filled phosphatidylcholine liposomes have antioxidative and membrane repairing properties and ability to inhibit protein kinase C activity. Moreover, this compound is shown to be a potential medication in the case of ionizing irradiation accident and for the patients with neoplasm who have to receive external radiotherapy as well.

Shaji et al.[30] prepared and optimized quercetin liposomes with phosphatidylcholine and also Cai et al.[3] studied how to increase the dissolution rate and oral bioavailability of quercetin enriched in soybean phosphatidylcholine. Tikhonov et al.[39] invented a complex composed of phosphatidylcholine dihydro quercetin (PCDQ) with quercetin for cosmetic industry. Grigor'ev et al.[9] invented a stable phospholipid emulsion based on dihydroquercetin, its compounds and derivatives. The disclosed emulsion is characterized by elevated bioavailability and selectivity of the effect of active compounds and by the increased stability and prolonged storage terms.

The aim of our present study was to synthesize quercetin analogs such as phosphatidylcholine quercetin (PCQ) and PCDQ (cf. Fig. 10.2) to characterize primarily their electron donating properties as well as the ability to donate atom(s) of hydrogen, for example, from the molecules classified as polar connecting bridge.

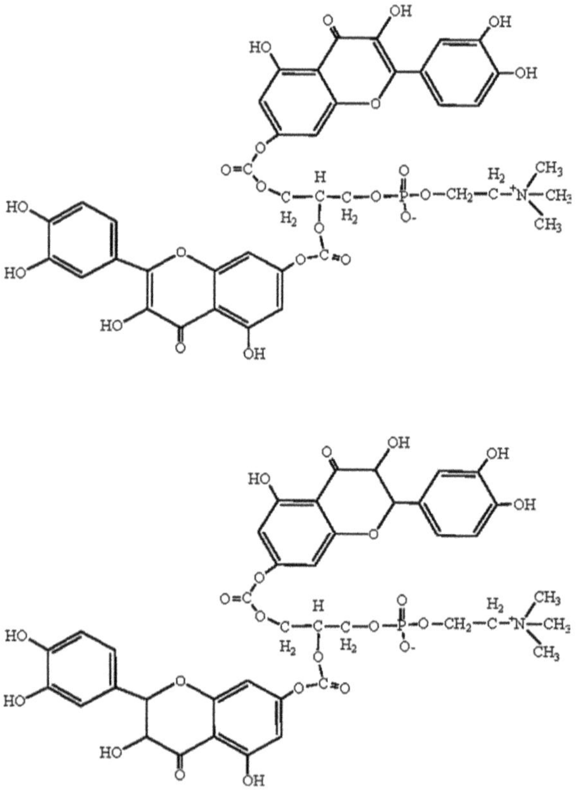

FIGURE 10.2   Chemical structures of phosphatidylcholine quercetin (PCQ) (top) and phosphatidylcholine dihydro quercetin (PCDQ) (bottom).

10.2 MATERIALS AND METHODS

10.2.1 MATERIALS

The high molar mass hyaluronan sample Lifecore P9710-2A, kindly donated by Lifecore Biomedical Inc., Chaska, MN, USA ($M_w = 808.7$ kDa), was used on experiments. The analytical purity grade NaCl and $CuCl_2·2H_2O$ (Slavus Ltd., Slovakia); $K_2S_2O_8$ (p.a. purity, max 0.001% nitrogen; Merck, Germany); 2,2'-azinobis [3-ethylbenzothiazoline-6-sulfonic acid] diammonium salt ((2,2'-azino-bis(3-ethylbenzothiazoline-6-sulphonic acid) (ABTS); purum, >99%; Fluka, Germany) were used. 2,2-Diphenyl-L-picrylhydrazyl (DPPH; 95%), methanol, L-ascorbic acid, quercetin (95%), acetic acid, dimethylsulfoxide (DMSO), dimethylformamide, dichloromethane, and cyclohexane were purchased from Sigma-Aldrich, Germany. Soybean lecithin phosphatidylcholine (Phospholipon 90 G, 96.9% purity) was obtained as a gift from Phospholipid GmbH, Germany. Dihydroquercetine (95% purity) was purchased from the Favorsky Irkutsk Institute of Chemistry (Irkutsk, Russia). Acidified DMSO used was prepared by diluting 10 ml DMSO with 90 ml of 0.5% acetic acid. Deionized high purity grade H_2O, with conductivity of ≤ 0.055 µS/cm, was produced by using the TKA water purification system (Water Purification Systems GmbH, Niederelbert, Germany).

10.2.2 METHODS

10.2.2.1 SYNTHESIS OF PHOSPHATIDYLCHOLINE QUERCETINE (PCQ) AND PHOSPHATIDYLCHOLINE DIHYDRO QUERCETIN (PCDQ)

The PCQ or PCDQ complexes were prepared by stirring appropriate flavonoid (100 mg) and phosphatidylcholine in a weight ratio 1:1.[31] Both reactants were placed in 10 ml round bottom flask containing 7 ml of solvent (Method A–dimethylformamide and Method B–dichloromethane). Reaction mixture was vigorously stirred at temperate 45°C for 3 days under nitrogen (Method A) or by gentle refluxing the mixtures until all flavonoids dissolved (Method B). Thereafter, the solvents were removed by evaporation under reduced pressure. Cyclohexane (3 ml) was added to the residue and then stirred for 6 h. The complex was isolated by centrifugation

(6000 g, 20 min) and washed three times with 2 ml cyclohexane. The amorphous material was dried under vacuum at 25°C and stored in a desiccator. The infrared (IR) spectra, nuclear magnetic resonance (NMR), and differential scanning calorimetric thermograms confirmed apparent identity of corresponding products obtained by methods A and B.

10.2.2.2   2,2'-AZINO-BIS(3-ETHYLBENZOTHIAZOLINE-6-SULPHONIC ACID) (ABTS) ASSAY

The ABTS$^{·+}$ cation radical was formed by the reaction of $K_2S_2O_8$ (3.3 mg) in H_2O (5 ml) with ABTS (17.2 mg) and stored overnight in the dark below 0°C. The ABTS$^{·+}$ solution (1 ml) was diluted in deionized water to a final volume, that is, 60 ml. PCQ and PCDQ (1.0 and 0.1 mmol/l) stock solutions were prepared in acidified DMSO. A modified ABTS assay[24] was used to assess the radical scavenging activity of quercetin derivatives using a UV-1800 spectrophotometer (SHIMADZU, Japan). The UV/VIS spectra were recorded in time interval 1–20 min in 1-cm quartz cuvette after admixing of the substance solution (50 µl) to the ABTS$^{·+}$ cation radical solution (2 ml).

10.2.2.3   2,2-DIPHENYL-1-PICRYLHYDRAZYL (DPPH) ASSAY

The DPPH$^·$ was prepared by dissolving 1.1 mg DPPH in distilled methanol (50 ml). The UV/VIS spectra using UV-1800 spectrophotometer (SHIMADZU, Japan) were recorded in time interval 1–10 min in 1-cm quartz cuvette after admixing of the substance solution (50 µl) to the DPPH$^·$ radical solution (2 ml).

102.2.4   DPPH ASSAY—DETERMINATION OF IC_{50} VALUES

In the DPPH assay, the DPPH$^·$ solution (55 µmol/l) was prepared by dissolving DPPH in methanol. Then 225 µl of the above mentioned DPPH$^·$ radical solution was added to 25 µl of methanolic solutions of PCQ or PCDQ, which concentrations ranged from 25 to 1.6 mmol/l. The absorbance of the sample mixtures was measured at 517 nm for 30 min after mixing the reactants. The measurements were performed in quadruplicate in a 96-well Greiner UV-Star microplate (Greiner-Bio-One GmbH, Germany) with Tecan Infinite M 200 reader (Tecan AG, Austria). The IC_{50} values were calculated using a standard PC program.

10.3 ROTATIONAL VISCOMETRY

10.3.1 PREPARATION OF STOCK AND WORKING SOLUTIONS

The hyaluronan samples (20 mg) were dissolved in 0.15 mol/l of aqueous NaCl solution for 24 h in the dark. Hyaluronan working solutions were prepared in two steps: first, 4.0 ml and after 6 h, 3.90 or 3.85 ml of 0.15 mol/l NaCl were added when working in the absence or presence of the samples, respectively. Solutions of ascorbate (16 mmol/l) and cupric chloride (160 μmol/l) were prepared in deionized water. Quercetin derivatives were dissolved in acidified DMSO.

10.3.2 UNINHIBITED HYALURONAN DEGRADATION

At first, hyaluronan degradation was induced by the oxidative system comprising $CuCl_2$ (1.0 μmol/l) and ascorbic acid (100 μmol/l). The procedure was as follows: a volume of 50 μl of $CuCl_2$ solution (160 μmol/l) was added to the hyaluronan solution (7.90 ml), and the mixture was left to stand for 7 min 30 s at room temperature after a 30-s stirring. Then, 50 μl of ascorbic acid solution (16 mmol/l) was added to the solution and stirred for 30 s. The solution was then immediately transferred into the viscometer Teflon® cup reservoir.

10.3.3 INHIBITED HYALURONAN DEGRADATION

The procedures to investigate effectiveness of PCQ and PCDQ were as follows:

I) A volume of 50 μl of 160 μmol/l $CuCl_2$ solution was added to the hyaluronan solution (7.85 ml), and the mixture after a 30-s stirring was left to stand for 7 min 30 s at room temperature. Then, 50 μl of acidified DMSO or quercetin derivatives (100 μmol/l or 1 mmol/l) was added to the hyaluronan mixture, followed by stirring again for 30 s. Finally, 50 μl of ascorbic acid solution (16 mmol/l) was added to the solution, and the mixture was stirred for 30 s. The solution was then immediately transferred into the viscometer Teflon® cup reservoir.

II)   In the second experimental regime, a procedure similar to that
       described in (i) was applied; however, after standing for 7 min 30 s
       at room temperature, 50 µl of ascorbic acid solution (16 mmol/l) was
       added to the mixture and a 30-s stirring followed. One hour later,
       finally 50 µl of acidified DMSO or quercetin derivatives was added
       to the reaction mixture, followed by stirring for 30 s after addition of
       each component. The reaction mixture was then immediately trans-
       ferred into the viscometer Teflon® cup reservoir.[33,35,42]

10.3.4   VISCOSITY MEASUREMENTS

Dynamic viscosity of the reaction mixture (8 ml) containing hyaluronan
(2.5 mg/ml), ascorbate (100 µmol/l) and Cu(II) ions (1 µmol/l) in the
absence and presence of samples started to be monitored by a Brookfield
LVDV-II+PRO digital rotational viscometer for 2 min after the addition of
all reactants (Brookfield Engineering Labs., Inc., Middleboro, MA, U.S.A.)
at $25.0 \pm 0.1°C$ and at a shear rate of $237.6 \ s^{-1}$ for 5 h in the Teflon® cup
reservoir.[40,44]

10.3.5   RESULTS AND DISCUSSION

As seen in Figure 10.3, left panel, hyaluronan was subjected to degradation by the Weissberger biogenic oxidative system (WBOS; ascorbate
+ Cu(II) under aerobic conditions in the presence of acidified DMSO)
reaching the decrease in value of dynamic viscosity (η) of the hyaluronan
solution within 5 h by 4.30 mPa·s (red curve). This curve was marked as
a reference. PCQ at concentration 0.625 µmol/l had a mild prooxidative
effect (green curve). The result of PCQ addition at a higher concentration
(6.25 µmol/l, blue curve) was almost identical to the reference. Neither PCQ
at the highest concentration (100 µmol/l, cyan curve) showed a protective
effect against forming ˙OH radicals, which was evidenced within the time
interval 0–180 min. On the other hand, PCDQ (right panel) was shown to
be significantly triggered by hyaluronan degradation at lower concentrations (green and blue curves). Only the addition of the PCDQ substance at
concentration 100 µmol/l inhibited the degradation of hyaluronan in part
(cyan curve), whereas the decrease in dynamic viscosity of the hyaluronan
solution was 2.31 mPa·s.

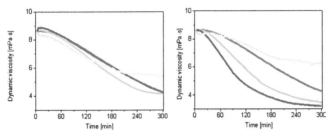

FIGURE 10.3 Effects of PCQ (left panel) and PCDQ (right panel) on hyaluronan degradation induced by Weissberger biogenic oxidative system (WBOS) (red curve) when added to the reaction system before initiating hyaluronan degradation. Concentrations of both quercetin derivatives in the reaction system were 0.625 (green curve), 6.25 (blue curve), and 100 μmol/l (cyan curve).

Results of the addition of PCQ into the hyaluronan reaction mixture showed a slight dose-dependent inhibition of alkoxy/peroxy-type radicals during the propagation phase of hyaluron degradation (Fig. 10.4, left panel). Within 5 h, the dynamic viscosity values of the hyaluronan solutions decreased by 3.9, 3.47, and 2.62 mPa·s at PCQ concentrations 0.625, 6.25, and 100 μmol/l, respectively. Results in Figure 10.4, right panel showed that PCDQ was not effective in inhibiting alkoxy-/peroxy-type radicals during the propagation phase of hyaluronan degradation (green and blue curve); however, PCDQ at the highest concentration resulted in a moderate prevention of hyaluronan degradation (cyan curve) reaching the decrease in the dynamic viscosity of the hyaluronan solution 2.34 mPa·s.

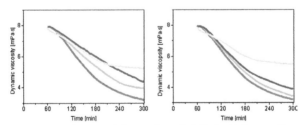

FIGURE 10.4 Effects of PCQ (left panel) and PCDQ (right panel) on hyaluronan degradation induced by WBOS (red curve) when added to the reaction system 1 h after initiating hyaluronan degradation. Concentrations of both quercetin derivatives in the reaction system were 0.625 (green curve), 6.25 (blue curve), and 100 μmol/l (cyan curve).

During our studies when working with rotation viscometric methods, the WBOS was used to form H_2O_2 as a result of catalytic oxidation of ascorbic acid by Cu(II) ions under aerobic condition, followed by production of ·OH radicals (Scheme 10.1).[12]

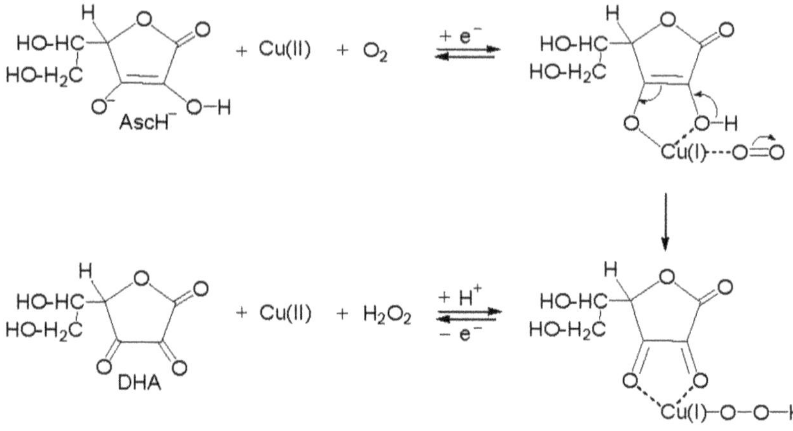

$$H_2O_2 + Cu(I)\text{---}complex \rightarrow {}^{\bullet}OH + Cu(II) + OH^-$$

SCHEME 10.1   Weissberger's biogenic oxidative system (WBOS).

High molar mass hyaluronan was used as a model for studying its oxidative degradation. Attacking of hyaluronan by hydroxyl radicals, which are generated by WBOS system, caused a multistep oxidation of hyaluronan macromolecules, which resulted in the formation of hyaluronan fragments (Scheme 10.2).

Consequently, during the initiation phase of hyaluronan degradation, the concentration of ·OH radicals are the actors, however in a short time along with ·OH radicals the dominated radicals will be alkoxy- and peroxy-type.[34,13,14,41,43,45,14,37,38]

10.3.5.1   ABTS ASSAY

A spectrophotometric decolorization assay of ABTS·+ cation radicals, originally working with colorless 2,2'-azinobis[3-ethylbenzthiazoline sulfonate], is applicable for both lipophilic and hydrophilic supposed antioxidant compounds. ABTS·+ cation radicals display a bluish green color with maximum absorbance values at 815, 734, and 645 nm. The reduction in absorbance was followed to observe the consumption of colored ABTS·+.

This assay is usually employed for the screening of scavenging ability of various antioxidants such as flavonoids, carotenoids, hydroxycinnamates, and plasma antioxidants.[25]

$$ABTS^{\bullet +} + e^- \rightarrow ABTS$$

bluish-green   colorless

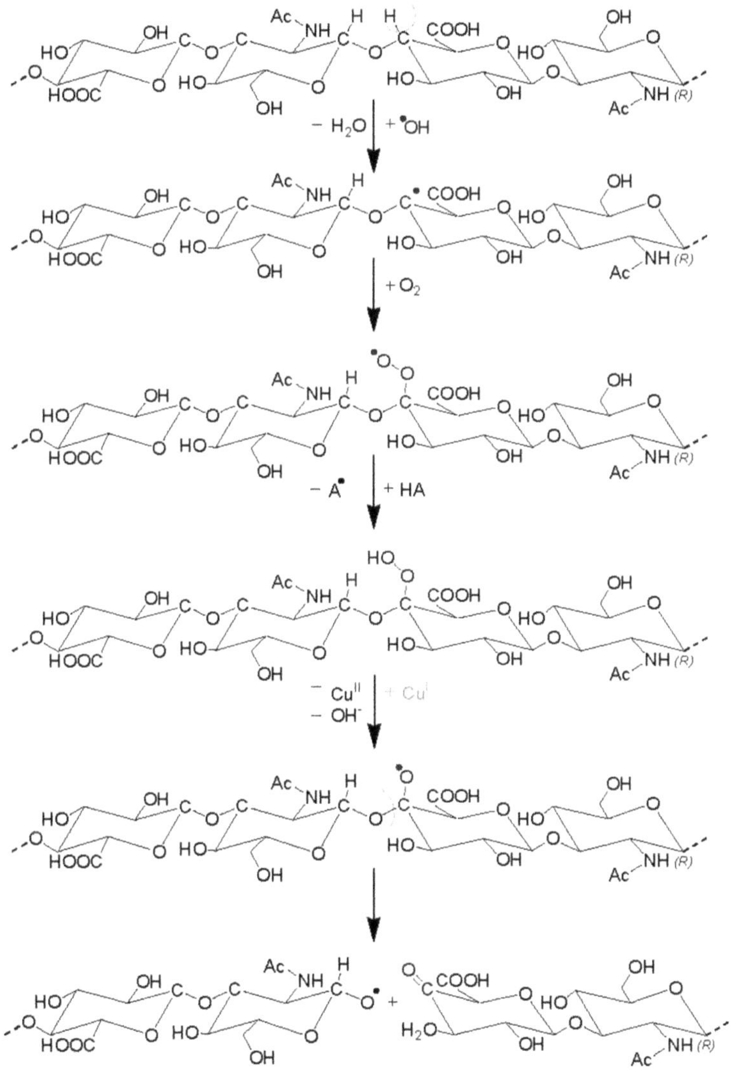

SCHEME 10.2   Schematic degradation of hyaluronan under free radical stress.

Figure 10.5 displays the ability of quercetin derivatives, namely PCQ and PCDQ, to reduce ABTS$^{\cdot+}$ cation radicals. Concerning the results of the quercetin PCQ derivative (left top panel), this substance scavenges ABTS$^{\cdot+}$ cation radicals continually within 20 min. The kinetics measurement showed 26.2 and 28.3% not reduced ABTS$^{\cdot+}$ cation radicals after addition of PCQ at concentrations 2.5 and 25 μmol/l within 20 min (left bottom panel). As seen in right top panel, PCDQ at concentration of 2.5 μmol/l scavenged ABTS$^{\cdot+}$ cation radicals in part within 1 min. Results of kinetics showed that PCDQ dose dependently reduced ABTS$^{\cdot+}$ cation radicals. The percentages of the unreduced ABTS$^{\cdot+}$ cation radicals were 69.4 and 32.0% at the applied PCDQ concentrations 2.5 and 25 μmol/l, respectively (right bottom panel).

Efficacy in the reduction of PCDQ and especially of PCQ can be explained by its phenolic structure that works as a suitable electron donor.

Time dependence of absorbance at 734 nm measured after addition of the tested compounds at concentrations 2.5 (violet) and 25 μmol/l (dark yellow) into the ABTS$^{\cdot+}$ solution under the same experimental conditions (bottom panels).

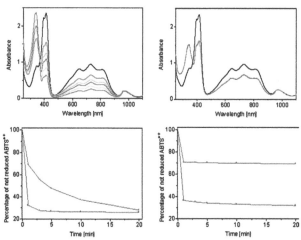

FIGURE 10.5   Effect of PCQ (top left panel) and PCDQ (top right panel) at concentration 2.5 μmol/l on the reduction of 2,2'-azino-bis(3-ethylbenzothiazoline-6-sulphonic acid) (ABTS$^{\cdot+}$) cation radicals (black) in the time: 1 (red), 3 (green), 5 (blue), 10 (cyan), and 20 min(magenta) (top panels).

Results in Figure 10.6 show that both quercetin derivatives effectively reduced DPPH· radical in a dose-dependent manner. The amounts of not scavenged DPPH· were 15, 51 and 89% for PCQ and 15, 44 and 83% for PCDQ.

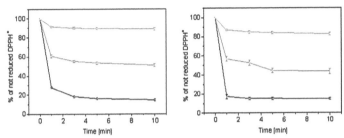

FIGURE 10.6 Effect of PCQ (left panel) and PCDQ (right panel) at concentrations 2.5 (green), 12.5 (red), and 25 µmol/l (black) on the reduction of 2,2-Diphenyl-1-picrylhydrazyl (DPPH·) radical in time: 1, 3, 5, and 10 min. The measurements were performed in triplicate.

Results in Table 10.1 show that both quercetin derivatives effectively reduced ABTS·+; however, these values were higher compared with quercetin. In DPPH assay, the ability of PCQ to reduce DPPH· was similar to quercetin. On the other hand, PCDQ was shown to be a poor antioxidant in reducing DPPH·.

TABLE 10.1 IC_{50} Values of Quercetin and its Derivatives.

Sample	ABTS IC_{50} (µM)	DPPH IC_{50} (µM)
Quercetin	9.8±0.6	4.36*
PCQ	35.4±0.6	9.4±0.78
PCDQ	21.2±0.7	269±2.4

*The value for quercetin by DPPH assay was adopted from Kurin et al.[17]

10.4 CONCLUSION REMARKS

The investigated quercetin derivatives PCQ and PCDQ underwent a rotational viscometric measurement, where a high molar mass hyaluronan probe is degraded by a two-component oxidative system, WBOS under aerobic conditions. According to Scheme 10.1, when applied 100 µmol/l of ascorbate, 100 µmol/l hydroxyl radicals should be generated. To scavenge such concentration of hydroxyl radicals, it is evident that 100 μmol/l of H radicals could be effective. However, as evident from Figure 10.3, either PCQ or PCDQ at the highest applied concentration did not result in inhibiting the degradation of hyaluronan. Thus, these results led us not to classify the two quercetin preparations as preventive antioxidants. By comparing the data recorded in Figure 10.3, the efficiency of PCDQ derivative was a bit higher than that of PCQ when applied at the highest concentration (100 μmol/l).

Figure 10.4 illustrates that by applying one or another quercetin derivative (PCQ or PCDQ), chain-breaking antioxidant activity was indicated in both derivatives. Moreover, the results proved that this action was classifiable as dose-dependent. However, on applying the derivatives at the highest concentration (100 μmol/l), the scavenging of alkoxy-/peroxy-type radicals reached values 56.0 and 50.2% for PCQ and PCDQ, respectively.

In the ABTS assay, PCQ rather than PCDQ was a more efficient donor of electrons. While the PCQ derivative gradually donated electrons even at concentration 2.5 μmol/l (Fig. 10.5, bottom left panel), such a concentration on applying the PCDQ derivative resulted in fast ABTS$^{•+}$ cation radical decolorization of within 1 min (Fig. 10.5, bottom right panel). In the DPPH$^{•}$ assay both quercetin derivatives reduced DPPH$^{•}$ in a similar manner.

In conclusion, the experimental part when rotational viscometry was used on investigating the scavenging properties of the two original quercetin derivatives, neither one can be classified as a proper donor of H$^{•}$ radicals nor a sequester of Cu(II) ions.

10.5 ACKNOWLEDGMENTS

The study was financed by the grant VEGA 2/0065/15 and APVV-15-0308. Dr. Tamer thanks to the agency SAIA for providing postdoctoral scholarship at the Institute of Experimental Pharmacology and Toxicology, Slovak Academy of Sciences.

KEYWORDS

- ABTS assay
- antioxidants
- DPPH assay
- glycosaminoglycans
- hyaluronan

- inflammation
- phosphatidylcholine quercetin (PCQ)
- phosphatidylcholine dihydro quercetin (PCDQ)
- rotational viscometry

REFERENCES

1. Adams, A. M. River Hills Harvest dba Elderberrylife. Inspections, Compliance, Enforcement, and Criminal Investigations, US FDA, 2014.
2. Boots, A. W.; Haenen, G. R. M. M.; Bast, A. Health Effects of Quercetin: From Antioxidant to Nutraceutical. *Eur. J. Pharmacol.* **2008,** *585,* 325–337.
3. Cai, X.; Jiaxing, S.; Deqing, S.; Lisi, Q.; Yuanyan, G.; Yuxia, L. Quercetin-Phospholipid Complex Loaded Nanosuspension with Potential for Oral Delivery. *J. Nanosci. Nanotechnol.* **2016,** *16,* 7266–7272.
4. Day, A. J.; Rothwell, J. A.; Morgan, R. Characterization of Polyphenol Metabolites. In *Phytochemicals in Health and Disease;* Bao, Y., Fenwick, R., Eds.; Dekker: New York, NY, 2004; pp50–67.
5. European Food Safety Agency (EFSA) NDA Panel, 2014.
6. Fonseca-Silva, F.; Inacio, J. D. F.; Canto-Cavalheiro, M. M.; Almeida-Amaral, E. E. Reactive Oxygen Species Production and Mitochondrial Dysfunction Contribute to Quercetin Induced Death in *Leishmania Amazonensis. Plos One* **2011,** *6*(2), e14666.
7. Formica, J. V.; Regelson, W. Review of the Biology of Quercetin and Related Bioflavonoids. *Food Chem. Toxicol.* **1995,** *33*(12), 1061–1080.
8. Gibellini, L.; Pinti, M.; Nasi, M.; De Biasi, S.; Roat, E.; Bertoncelli, L.; Cossarizza, A. Interfering with ROS Metabolism in Cancer Cells: The Potential Role of Quercetin. *Cancers* **2010,** *2,* 1288–1311.
9. Grigor'ev, A. M.; Evteev, A. V.; Smyslov, A. P.; Smyslov, P. A.; Cvetkov, M. V. Phospholipid Emulsion Containing Dihydroquercetin, and Method of Producing Thereof. U.S. Patent 20,110,097,391 A1, 2011.
10. Hambleton, P.; Miller, P. Studies on Carrageenan Air Pouch Inflammation in the Rat. *Br. J. Exp. Pathol.* **1989,** *70,* 425–433.
11. Hollman, P. C.; Katan, M. B. Absorption, Metabolism and Health Effects of Dietary Flavonoids in Man. *Biomed. Pharmacother.* **1997,** *51*(8), 305–310.
12. Hrabarova, E. Free Radical Hyaluronan Degradation by Reactive Oxygen Species. Evaluation of Antioxidative Properties of Endogenous and Exogenous Substances Containing Thiol Group. Ph.D. Thesis, Slovak, 2012.

13. Hrabárová, E.; Valachová, K.; Rapta, P.; Soltés, L. An Alternative Standard for Trolox-Equivalent Antioxidant-Capacity Estimation Based on Thiol Antioxidants. Comparative 2,2'-azinobis[3-ethylbenzothiazoline-6-sulfonic acid] Decolorization and Rotational Viscometry Study Regarding Hyaluronan Degradation. *Chem. Biodivers.* **2010,** *7*(9), 2191–2200.

14. Hrabarova, E.; Valachova, K.; Juranek, I.; Soltes, L. Free-Radical Degradation of High-Molar-Mass Hyaluronan Induced by Ascorbate Plus Cupric Ions: Evaluation of Antioxidative Effect of Cysteine-Derived Compounds. *Chem. Biodivers.* **2012,** *9*, 309–317.

15. Jeong, C. H.; Joo, S. H. Down Regulation of Reactive Oxygen Species in Apoptosis. *J. Cancer Prev.* **2016,** *21*(1), 13–20.

16. Khansari, N.; Shakiba, Y.; Mahmoudi, M. Chronic Inflammation and Oxidative Stress as a Major Cause of Age-Related Diseases and Cancer. *Recent Pat. Inflamm. Allergy Drug Discov.* **2009,** *3*(1), 73–80.

17. Kurin, E.; Mucaji, P.; Nagy, M. In vitro Antioxidant Activities of Three Red Wine Polyphenols and Their Mixtures: An Interaction Study. *Molecules* **2012,** *17*(12), 14336–14348.

18. Lee, W. J.; Hsiao, M.; Chang, J. L.; Yang, S. F.; Tseng, T. H.; Cheng, C. W.; Chow, J. M.; Lin, K. H.; Lin, Y. W.; Liu, C. C.; Lee, L. M.; Chien, M. H. Quercetin Induces Mitochondrial-Derived Apoptosis via Reactive Oxygen Species-Mediated ERK Activation in HL-60 Leukemia Cells and Xenograft. *Arch. Toxicol.* **2015,** *89*, 1103–1117.

19. Li, Y.; Yao, J.; Han, C.; Yang, J.; Chaudhry, M. T.; Wang, S.; Liu, H.; Yin, Y. Quercetin, Inflammation and Immunity. *Nutrients* **2016,** *8*, 167.

20. Machlin, L. J.; Bendich, A. Free Radical Tissue Damage: Protective Role of Antioxidant Nutrients. *FASEB J.* **1987,** *1*(6), 441–445.

21. Nassibullin, R. S.; Sharafutdinova, R. R. Study of the Quercetin-Phosphatidylcholine Complex. *Izvestiya Eng. Sci.* **2008,** *5*, 213–216.

22. Quercetin dihydrate safety sheet. September 16, 2011, http://www.pvp.com.br (English).

23. Ramadan, M. F. Phenolipids: New Generation of Antioxidants with Higher Bioavailability. *Austin J. Nutr. Food Sci.* **2014,** *1*(1), 1–2.

24. Rapta, P.; Valachova, K.; Gemeiner, P.; Soltes, L. High-Molar-Mass Hyaluronan Behavior During Testing its Radical Scavenging Capacity in Organic and Aqueous Media: Effects of the Presence of Manganese (II) Ions. *Chem. Biodivers.* **2009,** *6*, 162–169.

25. Re, R.; Pellegrini, N.; Proteggente, A.; Pannala, A.; Yang, M.; Rice-Evans, C. Antioxidant Activity Applying an Improved ABTS Radical Cation Decolorization Assay. *Free Radical Biol. Med.* **1999,** *26*(9–10), 1231–1237.

26. Reuter, S.; Gupta, S. C.; Chaturvedi, M. M.; Aggarwal, B. B. Oxidative Stress, Inflammation, and Cancer: How are they Linked? *Free Radical Biol. Med.* **2010,** *49*(11), 1603–1616.

27. Rich, G. T.; Buchweitz, M.; Winterbone, M. S.; Krovn, P. A.; Wilde, P. J. Towards on Understanding of the Low Bioavailability of Quercetin: A Study of its Interaction with Intestinal Lipids. *Nutrients* **2017,** *9*, 111.

28. Russo, G. L.; Russo, M.; Spagnuolo, C.; Tedesco, I.; Bilotto, S.; Iannitti, R.; Palumbo, R. Quercetin: APleiotropic Kinase Inhibitor Against Cancer. *Cancer Treat. Res.* **2014,** *159,* 185–205.

29. Sakanashi, Y.; Oyama, K.; Matsui, H.; Oyama, T. B.; Oyama, T. M.; Nishimura, Y.; Sakai, H.; Oyama, Y. Possible Use of Quercetin, an Antioxidant, for Protection of Cells Suffering from Overload of Intracellular Ca^{2+}: A Model Experiment. *Life Sci.* **2008,** *83,* 164–169.

30. Shaji, J.; Iyer, S. Double-Loaded Liposomes Encapsulating Quercetin and Quercetin Beta-Cyclodextrin Complexes: Preparation, Characterization and Evaluation. *Asian J. Pharm.* **2012,** *July–Sept,* 218–226.

31. Sing, D.; Rawat, M. S. M.; Semalty, A.; Semalty, M. Quercetin-Phospholipid Complex: An Amorphous Pharmaceutical System in Herbal Drug Delivery. *Curr. Drug Discov. Technol.* **2012,** *9,* 17–24.

32. Soloviev, A.; Tishkin, S.; Kyrychenko, S. Quercetin-Filled Phosphatidylcholine Liposomes Restore Abnormalities in Rat Thoracic Aorta BK_{Ca} Channel Function Following Ionizing Irradiation. *Acta Physiol. Sin.* **2009,** *61*(3), 201–210.

33. Soltes, L.; Stankovska, M.; Kogan, G.; Gemeiner, P.; Stern, R. Contribution of Oxidative-Reductive Reactions to High-Molecular-Weight Hyaluronan Catabolism. *Chem. Biodivers.* **2005,** *2,* 1242–1246.

34. Šoltés, L.; Stankovská, M.; Brezová, V.; Schiller, J.; Arnhold, J.; Kogan, G.; Gemeiner, P. Hyaluronan Degradation by Copper(II) Chloride and Ascorbate: Rotational Ciscometric, EPR Spin-Trapping, and MALDI–TOF Mass Spectrometric Investigations. *Carbohydr. Res.* **2006,** *341,* 2826–2834.

35. Soltes, L.; Kogan, G.; Stankovska, M.; Mendichi, R.; Rychly, J.; Schiller, J.; Gemeiner, P. Degradation of High-Molar-Mass Hyaluronan and Characterization of Fragments. *Biomacromolecules* **2007,** *8*(9), 2697–2705.

36. Steen, K. H.; Steen, A. E.; Reeh, P. W. A Dominant Role of Acid pH in Inflammatory Excitation and Sensitization of Nociceptors in Rat Skin in vitro. *J. Neurosci.* **1995,** *15*(5), 3982–3989.

37. Tamer, T. M.; Valachová, K.; Mohyeldin, M. S.; Šoltés, L. Free Radical Scavenger Activity of Chitosan and its Aminated Derivative. *J. Appl. Pharm. Sci.* **2016a,** *6*(04), 195–201.

38. Tamer, T. M.; Valachová, K.; MohyEldin, M. S.; Šoltés, L. Free Radical Scavenger Activity of Cinnamyl Chitosan Schiff Base. *J. Appl. Pharm. Sci.* **2016b,** *6*(01), 130–136.

39. Tikhonov, V. P.; Sidljarov, D. P.; Zaveshchevskaja, T. L. Antioxidant and Antihypoxant Dihydro quercetin for Cosmetic Products, 2009, RU2367409C1.

40. Topol'ská, D.; Valachová, K.; Rapta, P.; Šilhár, S.; Panghyová, E.; Horváth, A.; Šoltés, L. Antioxidative Properties of Sambacusnigra Extracts. *Chem. Pap.* **2015,** *69,* 1202–1210.

41. Valachova, K.; Hrabarova, E.; Drafi, F.; Juranek, I.; Bauerova, K.; Priesolova, E.; Nagy, M.; Soltes, L. Ascorbate and Cu(II)-Induced Oxidative Degradation of High-Molar-Mass Hyaluronan. Pro- and Antioxidative Effects of Some Thiols. *Neuro. Lett.* **2010,** *31*(2), 101–104.

42. Valachova, K.; Vargova, A.; Rapta, P.; Hrabarova, E.; Drafi, F.; Bauerova, K.; Juranek, I.; Soltes, L. Aurothiomalate as Preventive and Chain-Breaking Antioxidant in Radical Degradation of High-Molar-Mass Hyaluronan. *Chem. Biodivers.* **2011,** *8,* 1274–1283.

43. Valachova, K.; Tamer, T. M.; Soltes, L. Comparison of Free-Radical Scavenging Properties of Glutathione Under Neutral and Acidic Conditions. *J. Nat. Sci. Sustainable Technol.* **2014,** *8*(4), 645–660.

44. Valachova, K.; Banasova, M.; Topolska, D.; Sasinkova, V.; Juranek, I.; Collins, M. N.; Soltes, L. Influence of Tiopronin, Captopril and Levamisole Therapeutics on the Oxidative Degradation of Hyaluronan. *Carbohydr. Polym.* **2015,** *134,* 516–523.

45. Valachova, K.; Tamer, T. M.; Mohyeldin, M. S.; Soltes, L. Radical-Scavenging Activity of Glutathione, Chitin Derivatives and their Combination. *Chem. Pap.* **2016,** *70*(6), 820–827.

CHAPTER 11

PEROXOCOMPLEXES OF SOME TRANSITION METAL IONS WITH d^0 ELECTRONIC CONFIGURATIONS: STRUCTURAL AND REACTIVITY CONSIDERATIONS

M. LUÍSA RAMOS*, LICÍNIA L. G. JUSTINO, PEDRO F. CRUZ, AND HUGH D. BURROWS

Department of Chemistry, Coimbra Chemistry Centre, University of Coimbra, 3004-535 Coimbra, Portugal, Tel.: +351-239-854453, Fax:+351-239-827703

*Corresponding author. E-mail: mlramos@ci.uc.pt

CONTENTS

ABSTRACT

This chapter considers aspects of structure and reactivity of peroxocom-plexes of three transition metal ions with the d^0 configuration (V(V), Mo(VI), and W(VI)). Their importance in both chemical and biological systems, such as industrial and enzymatic catalysis, will be discussed.

11.1 INTRODUCTION

Peroxy complexes of transition metals provide a route to the activation of molecular oxygen for a variety of processes of chemical and biological interest. This is relevant because molecular oxygen is not particularly reactive, since whilst the two-electron (E^0 $O_2/H_2O_2 = +0.695$ V, pH 7) and four-electron (E^0 $O_2/H_2O = +0.815$ V) reduction potentials are favorable for oxidation, the one-electron one (E^0 $O_2/O_2^- = -0.16$ V) is not.[1] Transition metal peroxocomplexes are relevant for both metabolic processes and industrial synthesis.[2–6] A metal ion induced activity of oxygen species in biomedical and cellular systems has been reported.[7,8]

In this chapter, we will focus on the peroxocomplexes of the transition metal ions V(V), Mo(VI), and W(VI) with the d^0 configuration, as these species with a closed shell electronic structure have a particular interest and relevance for the activation of molecular oxygen, both for the development of sustainable industrial processes and for the understanding of biological activity. Although, as discussed elsewhere,[9] we will place particular emphasis on examples from our own work, we believe that these results are generally relevant to the understanding of both the structure and reactivity in these systems. Hydrogen peroxide (H_2O_2) is of great importance as a reagent in both industrial and biological processes,[10–13] and we will show how these d^0 transition metal complexes are important for its activation.

Peroxocomplexes of vanadium (V) have been intensively studied due to their biological importance in insulino-mimetic and antitumor activities,[14–18] their role as functional models for the haloperoxidase enzymes,[19–21] and their application in the oxidation of several substrates, such as benzene and other aromatics, alkenes, allylic alcohols, sulfides, halides, and primary and secondary alcohols.[22] Peroxocomplexes of molybdenum (VI) and tungsten (VI) are also highly relevant. They are known to be involved in stoichiometric oxygen transfer reactions, including epoxidation of alkenes and oxidation of organic and inorganic substrates,[2,23–31] and can be regarded as environmentally friendly alternatives to traditional oxidation reactions.[32] These complexes are also important in analytical chemistry and are involved in the chromatographic separation of vanadate, tungstate, molybdate, and chromate ions in the presence of hydrogen peroxide.[33] They have industrial applications as precursors of new materials,[34,35] and as bleaching accelerators.[36] In addition, molybdenum (VI) and tungsten (VI) are very efficient catalysts in the oxidation of bromide by hydrogen peroxide, where they may mimic vanadium peroxidases.[37] Peroxomolybdates and peroxotungstates are also potent stimulators of insulin effects in rat adipocytes.[38]

11.2 STRUCTURAL DETAILS AND SPECTROSCOPIC STUDIES

Upon reaction with hydrogen peroxide, vanadium(V), molybdenum(VI), and tungsten(VI) form peroxometalates or oxoperoxometalates,[2,39–44] where other ligands are present in peroxocomplexes. In some cases, it is possible to isolate solids and characterize them by X-ray diffraction ,[2,45–58] IR [46–54,59] and Raman[46,48,51–53]spectroscopy.

The peroxocomplexes of vanadium show a wider range of structures than those of molybdenum and tungsten; examples are given in Figures 11.1–11.3. Coordination numbers of seven or, less frequently, six are observed, with each metal center coordinating one or two peroxide groups in geometries close to pentagonal bipyramidal or octahedral. With the binuclear complexes, the metal centers are linked, either through oxo bridges or water molecules, frequently assisted by additional bridges, which can be established by ligand donor groups, or by oxygen atoms of the peroxo groups coordinated simultaneously to the two metal ions. Figure 11.3 shows binuclear V(V) complexes involved in different types of binding of the metal centers.

With molybdenum (VI) and tungsten (VI), a coordination number of seven with a structure close to pentagonal bipyramidal is normally seen; the complexes are homologous, characterized by a MO^{4+} center, with the M=O (M=Mo, W) group occupying one of the apical positions. The oxodiperoxo complexes of molybdate and tungstate are more stable than the oxomonoperoxo species, and most reported oxoperoxo complexes of molybdate and tungstate have two peroxide groups coordinated in the equatorial plane. There are exceptions, and the key to stabilization of oxomonoperoxo complexes is the inclusion of heteroligands that can effectively coordinate three sites in the pentagonal plane.[60] The metal ions in mono- or binuclear Mo(VI) and W(VI) peroxocomplexes have a coordination number of seven, and structures close to pentagonal bipyramidal, with two peroxo groups coordinated to each metal center. Examples are shown for Mo(VI) in Figure 11.4.

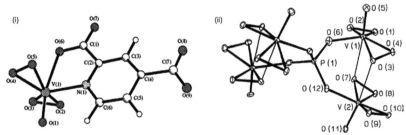

FIGURE 11.1 Diperoxocomplexes of vanadium (V): (i) mononuclear and (ii) tetra nuclear, in which two dinuclear units are connected by a phosphate group.

Source: Reproduced with permission from ref 57, © 2004, Springer Nature, and ref 46 © 1996, Elsevier, respectively.

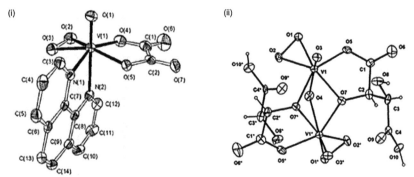

FIGURE 11.2 Monoperoxocomplexes of V(V): (i) mononuclear and (ii) binuclear with an oxo bridge and two ligand donor atoms forming bridges between the two metal atoms.

Source: Reproduced with permission from ref 48 © 2003, Royal Chemical Society, and ref 49© 1998, Elsevier, respectively.

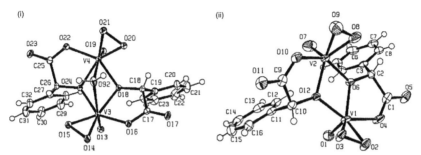

FIGURE 11.3 Monoperoxocomplexes of binuclear V(V): (i) one water molecule and two donor atoms from the ligands establish bridges between the two metal atoms and (ii) only ligand donor atoms form bridges between the two metal atoms.
Source: Reproduced with permission from ref 50, © 2003, Elsevier.

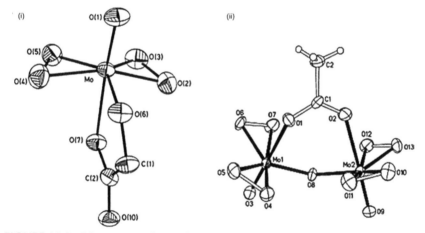

FIGURE 11.4 Diperoxocomplexes of Mo(VI): (a) mononuclear and (b) binuclear.
Source: Reproduced with permission from ref 54, © 2003, Elsevier, and ref 55 © 2004, Royal Society of Chemistry, respectively.

The novel binuclear peroxotungstate complex in Figure 11.5(i)[41] has been obtained from species and in Figure 11.5(ii) by reaction with hydrogen peroxide in nonaqueous media. Each metal center is coordinated with two peroxo groups, whereas one additional peroxo group forms a bridge with two oxygen atoms coordinated simultaneously to the two metal centers. This contrasts with the structure of the peroxotungstate species in Figure 11.5 (ii).[2]

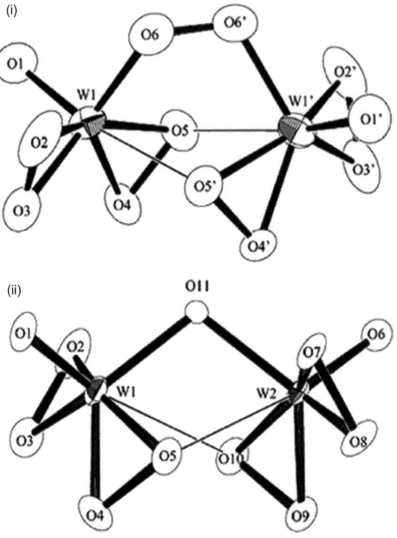

FIGURE 11.5 Binuclear diperoxocomplexes of W(VI): (i) peroxo bridge and (ii) oxo bridge; (in both the species, two atoms of the coordinated peroxo group form additional bridges between the two metal atoms).

Source: Reproduced with permission from ref 41, © 2007, American Chemical Society.

One of the most versatile spectroscopic techniques for characterizing peroxometalates, oxoperoxometalates, and peroxocomplexes is multi-nuclear nuclear magnetic resonance (NMR) spectroscopy, and this has provided insights into their coordination behavior in solution, including identification and characterization of new species, and the study of their properties and reactivity.[2,39,43,44,46,48,50–53,55,58,61–81] Our own research on oxo- and oxoperoxocomplexes of V(V), Mo(VI), and W(VI) in aqueous solutions involved studies on the complexation of vanadate, tungstate, and molybdate with α-hydroxy acids, sugar acids, phospho-sugar acids, and 8-hydroxyquinoline-5-sulfonate.[82–94] In some cases, this involves reactions with hydrogen peroxide of particular biological or synthetic relevance.[66–70] All of the NMR-active nuclei present in the complexes ^{1}H, ^{13}C, ^{17}O, ^{51}V, ^{95}Mo, and ^{183}W have been studied using one and two-dimensional NMR to obtain a wealth of information which is not available using a single technique (or nucleus) for such complex systems. ^{1}H and ^{13}C NMR allow identification of the nature and positions of binding of the ligand to the metal through changes in chemical shifts upon coordination, whereas the metal ion (^{51}V, ^{95}Mo, ^{183}W) and ^{17}O NMR spectra supply further structural information, including the type of metal center present. The value of this method can be seen by the excellent agreement between structures proposed by us on the basis of multinuclear NMR for the various peroxocomplexes of V(V), Mo(VI), and W(VI) present in aqueous solution, and those obtained by various authors on single crystals in the solid state using X-ray diffraction. The strengths of the NMR technique are that it is not necessary to prepare single crystals, and it is not affected by crystal packing effects.

^{51}V NMR spectroscopy has provided information on the metal centers in large number of vanadium (V) peroxovanadates and oxoperoxocomplexes.[46,48,50–52,61–68,71–74,81,92] Although ^{51}V NMR signals are broad due to the dominant quadrupolar relaxation (I = 7/2), the large chemical shift range allows the identification of the vanadium center, whereas the spectral line width can also provide indications of its symmetry. The relaxation of a quadrupolar nucleus is mainly due to the interaction of the quadrupole moment with electrical field gradients present at the position of the nucleus, and this allows efficient transfer of energy from the nucleus to molecular rotational motions. In the limit of fast motion, the nuclear spin quadrupolar relaxation follows the equation:[95]

$$\frac{1}{T_{1Q}} = \frac{1}{T_{2Q}} = \frac{3\pi^2}{10} \times \frac{(2I+3)}{I^2(2I-1)}\left(1 + \frac{\eta^2}{3}\right)\left(\frac{e^2 Q q_{zz}}{h}\right)^2 \times \tau_c$$

where T_{1Q} and T_{2Q} are the quadrupolar spin–lattice and spin–spin relaxation times, respectively, I is the nuclear spin, e^2qQ/h is the quadrupolar coupling constant, q is the electric field gradient at the nucleus, e is the charge of the electron, Q is the nuclear electric quadrupole moment, h is the Planck's constant, τ_c is the rotational correlation time, and η is an asymmetry parameter ($0<\eta<1$) that gives the deviation of the electric field gradient from axial symmetry, which mainly depends on the lack of spherical symmetry of the p electron density. The electric field gradient is a very important parameter in the relaxation of quadrupolar nuclei, depending on the arrangement and nature of the ligands around the central atom. Cubic, tetrahedral, octahedral, or spherical symmetry have, in principle, a zero field gradient ($q=0$), which gives rise to sharp signals. Asymmetry in the ligand field produces an increase in the NMR line width of the signals.[95] Signals of oligomeric vanadate species (deca-, mono-, di-, tetra-, and hexavanadate), oxocomplexes with octahedral and bipyramidal centers, monoperoxo-, diperoxovanadates, oxomono-, and oxodiperoxocomples of V(V) have been detected in well separated, characteristic regions of the ^{51}V spectra.[46,48,50–52, 61–68,71–74,81,82,84,93]

Although less has been reported on chemical shifts of ^{95}Mo and ^{183}W in peroxomolybdates, peroxotungstates, and oxoperoxocomplexes of Mo(VI) and W(VI), there is data on model systems.[2,41,58,69,70,75–78] With molybdenum, the dominant quadrupolar relaxation mechanism leads to broad ^{95}Mo NMR signals (I = 5/2), whereas with the dipolar ^{183}W (I = 1/2) sharp NMR signals are seen. Using the line width of the signals and the vicinal coupling constants ($^{3}J_{\text{W–H}}$) in the ^{95}Mo and ^{183}W NMR spectra, additional information can be obtained on the symmetry of the metal center and the conformation of the ligands around it, respectively. Furthermore, the well separated ^{95}Mo and ^{183}W shift ranges allow characterization of the nature of the metal M(VI) center in the complexes, such as oxoions of molybdate and tungstate, and octahedral monomeric oxocomplexes with MO_2^{2+} centers, whereas dinuclear complexes with $M_2O_5^{2+}$ centers, mixed complexes (possessing $M_2O_5^{2+}$ and MO_2^{2+} centers) as a function of the ligand configuration, monoperoxo-, diperoxo- molybdates and tungstates, and pentagonal bipyramidal oxodiperoxocomplexes have all given good characteristic ^{95}Mo and ^{183}W NMR spectra.[2,69,70,85–92,94]

For example, Figure 11.6 shows the ^{183}W, ^{95}Mo, and ^{51}V NMR spectra of diperoxo complexes of W(VI) and Mo(VI) with D-galactonic acid and the monoperoxo complex of V(V) with glycolic acid.[67,87]

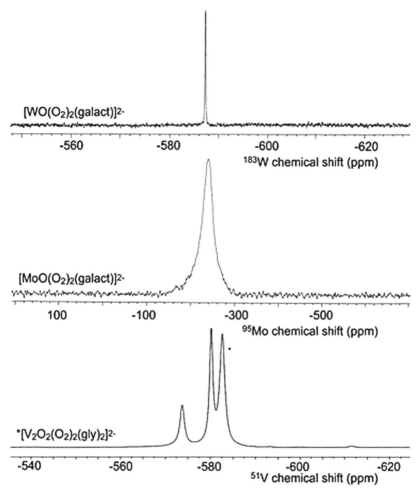

FIGURE 11.6   183W (20.825 MHz), 95Mo (32.565 MHz), and 51V (131.404 MHz) NMR spectra, respectively, of diperoxo complexes of W(VI) and Mo(VI) with D-galactonic acid and monoperoxo complex of V(V) with glycolic acid.

Source: Reproduced with permission from ref 9, © 2011, Royal Society of Chemistry.

A large number of oxo- and peroxocomplexes have been characterized, and we have noted that the characteristic ^{183}W and ^{95}Mo NMR chemical shift ranges of homologous complexes of these two metal ions show a linear correlation ($\delta W = 1.8\delta Mo - 137.3$), which parallels the ratio of nuclear charges (Fig. 11.7).[9,69]

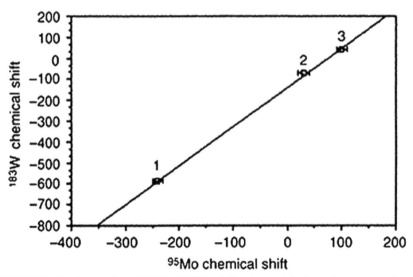

FIGURE 11.7 Correlation of ^{183}W with ^{95}Mo shifts observed for homologous complexes of (1) MO^{4+} (diperoxocomplexes) (2) $M_2O_5^{2+}$ (dinuclear oxocomplexes), and (3) MO_2^{2+} (mononuclear oxocomplexes) (M=Mo, W).

Source: Reproduced with permission from ref 9, © 2011, Royal Society of Chemistry.

   With the peroxocomplexes, the metal nuclei are more shielded than with the corresponding oxocomplexes, irrespective of the metal ion, suggesting that strong σ donors such as the peroxo group produce upfield shifts.[77] Since diperoxocomplexes should be more shielded than mono-peroxo ones, it is reasonable that the ^{183}W NMR spectra of the species are shown in Figures 11.5(i) and 11.5(ii) show signals at −587.5 and −462.2 ppm, respectively, assigned to the diperoxotungstate and monoperoxo-tungstate complexes.[41] These observations can be attributed to the domi-nant paramagnetic (σ_p) contribution for NMR chemical shifts of metal ions and the particular contributions for each case.

   ^{17}O NMR spectra can provide useful information on the structure and reactivity of these complexes.[39,43,44,79,80] Natural abundance spectra show a severe broadening of the peroxide ^{17}O resonances, due to nuclear quadrupole relaxation, and only terminal $M=O_t$ terminal groups are observed.[39,79] Figure 11.8 shows natural abundance ^{17}O NMR spectra of oxodiperoxo complexes of W(VI) and Mo(VI) with D-galactonic acid and the monoperoxo complex of V(V) with glycolic acid, respectively. These show the HDO water signal, the free H_2O_2, the free carboxylate of the ligand (CO_2^-), the terminal MO (M=V, Mo, and W) oxo groups of the peroxocomplexes. Shielding of the oxygen nuclei is seen to decrease in the order W>Mo>V.

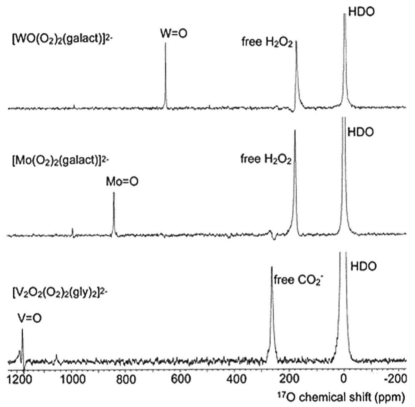

FIGURE 11.8 ^{17}O NMR spectra (67.792 MHz) of oxodiperoxo complexes of W(VI) and Mo(VI) with D-galactonic acid and monoperoxo complex of V(V) with glycolic acid, respectively, in natural abundance.

Source: Reproduced with permission from ref 9, © 2011, Royal Society of Chemistry.

Using hydrogen peroxide enriched with ^{17}O, the resonances of the bound peroxo groups can also be detected (Fig. 11.9). The ^{17}O chemical shifts of coordinated peroxo groups follow the same trend of W>Mo>V as the terminal oxo groups shown in Figure 11.8.

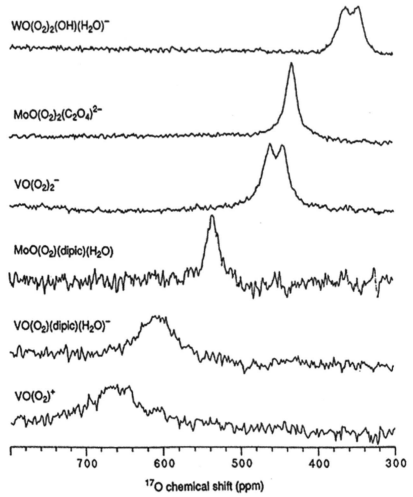

FIGURE 11.9 ^{17}O NMR spectra of coordinated peroxo groups in a series of peroxocomplexes prepared from hydrogen peroxide enriched with ^{17}O.

Source: Reproduced with permission from ref 39 © 1996, American Chemical Society.

Chemical shift depends on the electron density in molecules, and there is a relationship between the ^{17}O chemical shift of the coordinated peroxo groups and the reciprocal of the energy of the ligand-to-metal charge transfer band in the UV–visible absorption spectrum (Fig. 11.10). ^{17}O chemical shifts can be analyzed using the Ramsey equation:

$$\sigma_p = -\frac{2e^2h^2}{3m^2c^2}\langle r^{-3}\rangle P_u\left(1/\Delta E\right)$$

where $\langle r^{-3}\rangle$, P_u, and ΔE correspond, respectively, to the inverse of the average volume, the noncompensated charge of a valence orbital and the average of the energy difference between the ground state and all possible excited states. The value of ΔE is commonly estimated on the basis of the electronic spectrum of the molecule. Figure 11.10 suggests that, as δ depends linearly upon $1/\Delta E$, either the properties of the orbitals represented by the factors $\langle r^{-3}\rangle$ and P_u vary relatively little along the series or they vary in a way such that their effects cancel out.[39]

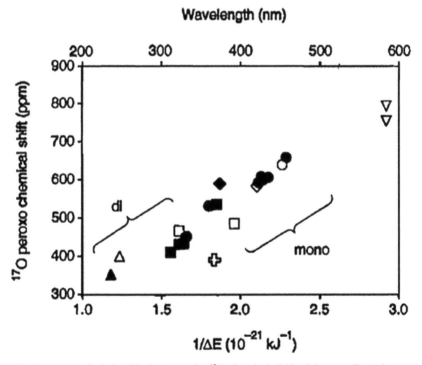

FIGURE 11.10   Relationship between the ^{17}O chemical shift of the coordinated peroxo groups and the reciprocal of the energy of the ligand-to-metal charge transfer band. Filled and open symbols correspond to the following central metal ions: chromium (∇) molybdenum ($\square$) rhenium ($+$), titanium ($\lozenge$), tungsten (Δ) vanadium ($\circ$).

Source: Reproduced with permission from ref 39 © 1996, American Chemical Society.

Computational studies at various levels suggest that with the monoperoxocomplexes the metal–peroxo bonds maybe described through σ interactions between the empty d_{xy} orbital of the metal and the filled π^* orbital of the peroxo group situated on the same plane, whereas with the diperoxocomplex, the d_{x2-y2} orbital of the metal interacts in a similar fashion with the π^* orbital of the second peroxo ligand. Therefore, as the overlap of the metal orbitals with the π^* orbitals of the peroxo groups increases, this will normally lead to a decrease in the electron density in the antibonding orbital. This has implications on the use of these complexes in catalysis as the activation of hydrogen peroxide by d^0 transition metals is often explained in terms of the increasing electrophilic character of the peroxo group resulting from its complexation with a highly charged metal ion. Thus, the reactivity of the peroxo group results from a balance between an increase in the electrophilic character and an increase in the strength of the O–O bond accompanying the decrease in the electron density in the π^* orbital.[39]

Density functional theory (DFT) has helped identify the structures and simulate the solution NMR spectra of these complexes. Comparison between the experimental and theoretical NMR chemical shifts is particularly useful for structural evaluation, while also allowing validation of the computational methods used.[81,96,97]

11.3   REACTIVITY

Coordination to a metal center can activate peroxide toward the oxidation of a variety of substrates, making peroxometal complexes important intermediates in both enzymatic and synthetic catalysis. Peroxocomplexes of metals with a d^0 configuration, such as V(V), Mo(VI), and W(VI), are relatively selective efficient oxidants with low toxicity for a variety of substrates,[41,45,61,98–102] including alkenes, alcohols, aromatic and aliphatic hydrocarbons, and thioethers.[41,45,61,98–102]

Based on experiments using ^{183}W NMR, infrared, UV–visible, Raman spectroscopy, and X-ray diffraction, Kamata et al.41 have proposed, that the peroxotungstate **II**, $[\{WO(O_2)_2\}_2(\mu\text{-}O_2)]^{2-}$, obtained from the previously characterized peroxotungstate **I**, $[\{WO(O_2)_2\}_2(\mu\text{-}O)]^{2-2}$ by addition of hydrogen peroxide in nonaqueous solvents, promoted the catalytic epoxidation of terminal, cyclic, and internal olefins (Fig. 11.11).[2]

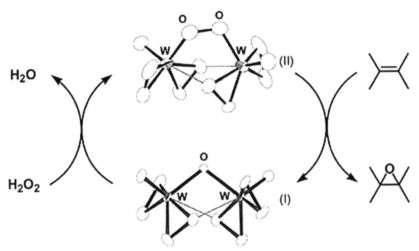

FIGURE 11.11   Reaction scheme proposed for catalytic epoxidation of alkenes by the peroxotungstate II $[(n\text{-}C_4H_9)_4N]_2$ $[\{WO(O_2)_2\}_2(\mu\text{-}O_2)]$.
Source: Reproduced with permission from ref 41, © 2007, American Chemical Society.

In a green chemistry perspective, peroxotungstates are less polluting rather than the reagents used in many conventional industrial oxidation processes, while still being efficient oxidants.[42,103,104] For example, current industrial routes to the important adipic acid (used extensively in the manufacture of nylon and other synthetic fibers) involve oxidation of cyclohexene with nitric acid.[103] A process which is both more efficient and environmentally friendly has been developed by substituting nitric acid in this reaction with tungsten (VI) complexes and hydrogen peroxide.[104]

With transition metal peroxocomplexes, the reactivity depends on the metal, its oxidation state, the number of peroxo ligands, and the nature of the other ligands in the coordination sphere of the complex. There is a direct relationship between the structure and reactivity, and generally, the reactivity decreases in the order $W(VI) > Mo(VI) > (V)$. In addition, increasing the donor character of the ligand is suggested to decrease the reactivity of the complex.[39,61]

Peroxocomplexes of transition metal with d^0 configuration are also involved in many biological processes.[105] The haloperoxidases (chloroperoxidases, bromoperoxidases, and iodoperoxidases) are enzymes containing peroxocomplexes of vanadium (V),[106] and have been isolated from marine algae and lichens. They catalyze the oxidation of the halide ions (Cl^-, Br^-,

I⁻) by hydrogen peroxide, and maybe involved in the biosynthesis of various marine natural products. Owing to their strong antifungal, antibacterial, anti-viral or anti-inflammatory properties,[107] and potential activity against human immunodeficiency virus (HIV) haloperoxidases are important at the pharmacological level. The haloperoxidase enzymes are named, based on the most electronegative halide that they can oxidize, that is, the chloroperoxidases can oxidize Cl⁻, Br,⁻ and I⁻, whereas the bromoperoxidases can only oxidize Br⁻ and I⁻.[105] The structure of the active site of the vanadium chloroperoxidase of the fungus *Curcuvalaria inaequalis* is shown in Figure 11.12, and has a central vanadium (V) with a trigonal bipyramidal geometry.[106]

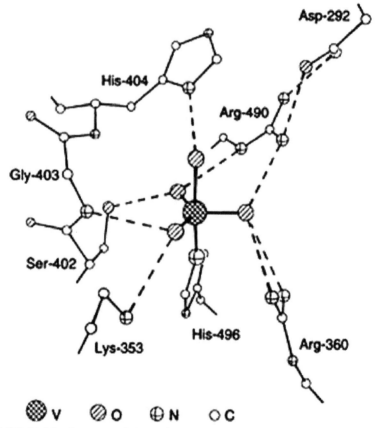

FIGURE 11.12 Structure of the active site of the chloroperoxidase of the fungus *Curcuvalaria inaequalis*.

Source: Reproduced with permission from ref 106, © 2004. American Chemical Society.

The mechanism proposed for the oxidation of Cl⁻ by this enzyme is shown in Figure 11.13.[106]

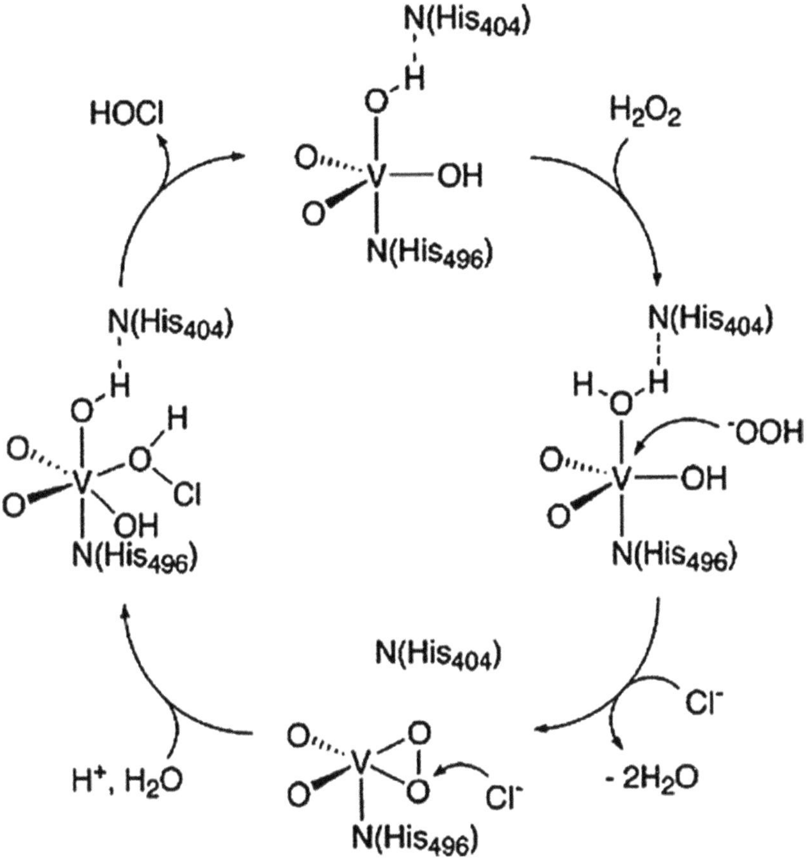

FIGURE 11.13   Proposed mechanism for the oxidation of chloride ion catalyzed by the vanadium enzyme chloroperoxidase.

Source: Reproduced with permission from ref 106, © 2004. American Chemical Society.

Peroxocomplexes of vanadium also present properties which mimic those of systems such as insulin, diet regulators while showing other important biological properties (e.g., antibacterial, antiviral, and antineoplastic, etc.).[106]

11.4   CONCLUSION

This chapter has discussed structural aspects of peroxocomplexes of V(V), Mo(VI), and W(VI); transition metal ions with a d^0 configuration. These are relevant for both biological processes and catalytic activity. We have paid particular emphasis to the wealth of information obtained on these complexes from multinuclear NMR spectroscopy, considering all the NMR active nuclei present in the complexes: ^1H, ^{13}C, ^{17}O, ^{51}V, ^{95}Mo, and ^{183}W. Chemical shifts in ^1H and ^{13}C NMR spectroscopy allow the characterization of the primary coordination spheres of the metal ions through the changes in the chemical shifts resulting from coordination. Further, information on the metal center is obtained from chemical shifts in the metal ion (^{51}V, ^{95}Mo, and ^{183}W) and ^{17}O NMR spectra. These facets are interrelated, and that the metal plays a role in determining both the structure and reactivity of the complex, with geometry, oxidation states, and accessibility of all the orbitals involved. A combination of structural studies using multinuclear NMR and X-ray diffraction with DFT calculations and other theoretical methodologies will be needed to fully understand this important group of compounds. This will extend the existing knowledge on the link between structure and chemical or biological catalytic activity.

11.5   ACKNOWLEDGMENT

The authors are grateful for funding from "The Coimbra Chemistry Centre" which is supported by the Fundação para a Ciência e a Tecnologia (FCT), Portuguese Agency for Scientific Research, through the programmes UID/QUI/UI0313/2013 and COMPETE. NMR data was obtained at the UC-NMR facility which is supported in part by FEDER—European Regional Development Fund through the COMPETE Programme (Operational Programme for Competitiveness) and by National Funds through FCT with the grants REEQ/481/QUI/2006, RECI/QEQ-QFI/0168/2012, CENTRO-07-CT62-FEDER-002012, and the Rede Nacional de Ressonância Magnética Nuclear (RNRMN). The authors also thank the Laboratory for Advanced Computing at University of Coimbra for providing computing resources that have contributed to the research results reported within this paper (URL http://www.lca.uc.pt). LLGJ thanks FCT for the postdoctoral grant (SFRH/BPD/97026/2013).

KEYWORDS

- peroxocomplexes
- vanadium (V)
- molybdenum (VI) and tungsten (VI)
- NMR
- reactivity

REFERENCES

1. Wood, P. M. The Potential Diagram for Oxygen at pH 7. *Biochem. J.* **1988**, *253*, 287–289.
2. Dickman, M. H.; Pope, M. T. Peroxo and Superoxo Complexes of Chromium, Molybdenum, and Tungsten. *Chem. Rev.* **1994**, *94*, 569–584.
3. Cramer, C. J.; Tolman, W. B.; Theopold, K. H.; Rheingold, A. L. Variable Character of O–O and M–O Bonding in Side-on (η^2) 1:1 Metal Complexes of O_2. *Proc. Natl. Acad. Sci. U. S. A.* **2003**, *100*, 3635–3640.
4. Deubel, D. V.; Frenking, G.; Gisdakis, P.; Herrmann, W. A.; Rösch, N.; Sundermayer, J. Olefin Epoxidation with Inorganic Peroxides. Solutions to Four Long-Standing Controversies on the Mechanism of Oxygen Transfer. *Acc. Chem. Res.* **2004**, *37*, 645–652.
5. Burke, A. J. Chiral Oxoperoxomolybdenum(VI) Complexes for Enantioselective Olefin Epoxidation: Some Mechanistic and Stereochemical Reflections. *Coord. Chem. Rev.* **2008**, *252*, 170–175.
6. Conte, V.; Floris, B. Vanadium and Molybdenum Peroxides: Synthesis and Catalytic Activity in Oxidation Reactions. *Dalton Trans.* **2011**, *40*, 1419–1436.
7. Imlay, J. A. Cellular Defenses Against Superoxide and Hydrogen Peroxide. *Annu. Rev. Biochem.* **2008**, *77*, 755–776.
8. Halliwell, B.; Gutteridge, J. M. C. Role of Free Radicals and Catalytic Metal Ions in Human Disease: An Overview. *Methods Enzymol.* **1990**, *186*, 1–85.
9. Ramos, M. L.; Justino, L. L. G.; Burrows, H. D. Structural Considerations and Reactivity of Peroxocomplexes of V(V), Mo(VI) and W(VI)). *Dalton Trans.* **2011**, *40*, 4374–4383.
10. Jones, C. W. *Applications of Hydrogen Peroxide and Derivatives;* Royal Society of Chemistry Monograph: Cambridge, 1997.
11. Bach, R. D. Transition Metal Peroxides, Synthesis and Role in Oxidation Reactions. In *The Chemistry of Peroxides;* Rappoport, Z., Ed.; John Wiley, 2006; Vol. 2, Part 1.
12. Ando W. *Peroxide Chemistry. Mechanistic and Preparative Aspects of Oxygen Transfer;* Adam, W., Ed.; Wiley-VCH Verlag GmbH: Weinheim, 2000.

13. Sheldon R. A. *Organic Peroxides*, Ed. W. Ando, John Wiley & Sons Ltd: Germany, 1992.

14. Thompson, K. H.; McNeill, J. H.; Orvig, C. Vanadium Compounds as Insulin Mimics. *Chem. Rev.* **1999**, *99*, 2561–2572.

15. Shaver, A.; Ng, J. B.; Hall, D. A.; Lum, B. S.; Posner, B. I. Insulin-Mimetic Peroxovanadium Complexes–Preparation and Structure of Potassium Oxodiperoxo(pyridine-2-carboxylato)Vanadate(v), $K_2[VO(O_2)_2(C_5H_4NCOO)].2H_2O$, and Potassium Oxodiperoxo(3-hydroxypridine-2-carboxylato)Vanadate(v), $K_2[VO(O_2)2(OHC_5H_3NCOO)].3H_2O$, and their Reactions with Cysteine. *Inorg. Chem.* **1993**, *32*, 3109–3113.

16. Posner, B. I.; Faure, R.; Burgess, J. W.; Bevan, A. P.; Lachance, D.; Zhang-Sun, G.; Fantus, I. G.; Ng, J. B.; Hall, D. A.; Lum, B. S.; Shaver, A. Peroxovanadium Compounds. A New Class of Potent Phosphotyrosine Phosphatase Inhibitors Which are Insulin Mimetics. *J. Biol. Chem.* **1994**, *269*, 4596–4604.

17. Bevan, A. P.; Burgess, J. W.; Yale, J. F.; Drake, P. G.; Lachance, D.; Baquiran, G.; Shaver, A.; Posner, B. I. In vivo Insulin Mimetic Effects of pV Compounds: Role for Tissue Targeting in Determining Potency. *Am. J. Physiol.* **1995**, *268*, E60–66.

18. Djordjevic, C.; Lee, M.; Sinn, E. Oxoperoxo(citrato)- and dioxo(citrato)vanadates(V): Synthesis, Spectra, and Structure of a Hydroxyl Oxygen Bridged Dimer, $K_2[VO(O_2)(C_6H_6O_7)]_2 \cdot 2H_2O$. *Inorg. Chem.* **1989**, *28*, 719–723.

19. Colpas, G. J.; Hamstra, B. J.; Kampf, J. W.; Pecoraro, V. L. A Functional Model for Vanadium Haloperoxidase. *J. Am. Chem. Soc.* **1994**, *116*, 3627–3628.

20. Colpas, G. J.; Hamstra, B. J.; Kampf, J. W.; Pecoraro, V. L. Functional Models for Vanadium Haloperoxidase. *J. Am. Chem. Soc.* **1996**, *118*, 3469–3478.

21. Kanamori, K.; Nishida, K.; Miyata, N.; Okamoto, K. Synthesis and Characterization of a Peroxovanadium (V) Complex Containing N-Carboxymethylhistidine as a Model for the Vanadium Haloperoxidase Enzymes. *Chem. Lett.* **1998**, *27*, 1267–1268.

22. Butler, A.; Clague, M. J.; Meister, G. E. Vanadium Peroxide Complexes. *Chem. Rev.* **1994**, *94*, 625–638.

23. Bonchio, M.; Conte, V.; Di Furia, F.; Carofiglio, T.; Magno, F.; Pastore, P. Co(II) Induced Radical Oxidation by Peroxomolybdenum Complexes. *J. Chem. Soc. Perkin Trans. II*, **1993**, *2*, 1923–1926.

24. Nardello, V.; Marko, J.; Vermeersch, G.; Aubry, J. M. ^{95}Mo NMR and Kinetic Studies of Peroxomolybdic Intermediates Involved in the Catalytic Disproportionation of Hydrogen Peroxide by Molybdate Ions. *Inorg. Chem.* **1995**, *34*, 4950–4957.

25. Won, T.-J.; Sudam, B. M.; Thompson, R. C. Characterization of Oxoperoxo(2,6-Pyridinedicarboxylato)Molybdenum(VI) and Oxoperoxo(Nitrilotriacetato) Molybdate(VI) in Aqueous-Solution and a Kinetic-Study of Their Reduction by a (Thiolato)Cobalt(III) Complex, Dimethyl-Sulfoxide, and Iron(II). *Inorg. Chem.* **1994**, *33*, 3804–3810.

26. Reynolds, M. S.; Morandi, S. J.; Raebiger, J. W.; Melican, S. P.; Smith, S. P. E. Kinetics of Bromide Oxidation by (oxalato)oxodiperoxomolybdate(VI). *Inorg. Chem.* **1994**, *33*, 4977–4984.

27. Arzoumanian, H.; Sanchez, J.; Strukul, G.; Zennaro, R. Dioxygen Ligand Transfer from Platinum to Molybdenum. Isolation of a Highly Reactive Molybdenum(VI) Oxoperoxo Dimer. *Bull. Soc. Chim. Fr.* **1995**, *132*, 1119–1122.

28. Salles, L.; Robert, F.; Semmer, V.; Jeannin, Y.; Brégeault, J.-M. Novel di- and Trinuclear Oxoperoxosulfato Species in Molybdenum(VI) and Tungsten(VI) Chemistry: The Key Role of Pairs of Bridging Peroxo Groups. *Bull. Soc. Chim. Fr.* **1996**, *133*, 319–328.

29. Tachibana, J.; Inamura, T.; Sasaki, Y. A Reversible Metalloporphyrin Oxygen Carrier Both in the Solid State and in Solution: Preparation, Characterization, and Kinetics of Formation of a Molybdenum(VI) 5,10,15,20-Tetramesitylporphyrin Dioxygen Complex. *Bull. Chem. Soc. Jpn.* **1998**, *71*, 363–369.

30. Glas, H.; Spiegler, M.; Thiel, W. R. Spectroscopic Evidence for Intramolecular $MoVI(O_2)$ $HO-C$ Hydrogen Bonding in Solution. *Eur. J. Inorg. Chem.* **1998**, *1998*(2), 275–281.

31. Nardello, V.; Marko, J.; Vermeersch, G.; Aubry, J. M. ^{183}W NMR Study of Peroxotungstates Involved in the Disproportionation of Hydrogen Peroxide into Singlet Oxygen ($^1O_2$, $^1\Delta_g$) Catalyzed by Sodium Tungstate in Neutral and Alkaline Water. *Inorg. Chem.* **1998**, *37*, 5418–5423.

32. Bolm, C.; Beckmann, O.; Dabard, O. A. G. The Search for New Environmentally Friendly Chemical Processes. *Angew. Chem.* **1999**, *38*, 907–909.

33. Sato, H.; Yaeda, T.; Hishioka, Y. Separation and Detection of Vanadate, Tungstate, Molybdate and Chromate Ions. *J. Chromatogr. A* **1997**, *789*, 267–271.

34. Itoh, K.; Yamagishi, K.; Nagasono, M.; Murabayashi, M. Photochromic Reaction of Peroxopolytungstic Acid Thin Films. *Ber. Bunsenges. Phys. Chem.* **1994**, *98*, 1250–1255.

35. Nakajima, H.; Kudo, T.; Mizuno, M. Reaction of Tungsten Metal Powder with Hydrogen Peroxide to Form Peroxo Tungstates, An Useful Precursor of Proton Conductor. *Chem. Lett.* **1997**, *8*, 693–694.

36. Thompson, K. M.; Spiro, M.; Griffith, W. P. Mechanism of Bleaching by Peroxides. 4. Kinetics of Bleaching of Malvin Chloride by Hydrogen Peroxide at Low pH and its Catalysis by Transition-Metal Salts. *J. Chem. Soc. Faraday Trans.* **1996**, *92*(14), 2535–2540.

37. Meister, G. E.; Butler, A. Molybdenum(VI)-Mediated and Tungsten(VI)-Mediated Biomimetic Chemistry of Vanadium Bromoperoxidase. *Inorg. Chem.* **1994**, *33*, 3269–3275.

38. Li, J.; Elberg, G.; Gefel, D.; Shechter, Y. Permolybdate and Pertungstate—Potent Stimulators of Insulin Effects in Rat Adipocytes: Mechanism of Action. *Biochemistry* **1995**, *34*, 6218–6225.

39. Reynolds, M.; Butler, A. Oxygen-17 NMR, Electronic, and Vibrational Spectroscopy of Transition Metal Peroxo Complexes: Correlation with Reactivity. *Inorg. Chem.* **1996**, *35*, 2378–2383.

40. Bortolini, O.; Conte, V. Vanadium(V) Peroxocomplexes: Structure, Chemistry and Biological Implications. *J. Inorg. Biochem.* **2005**, *99*, 1549–1557.

41. Kamata, K.; Kuzuya, S.; Uehara, K.; Yamaguchi, S.; Mizuno, N. μ-η^1:η^1-Peroxo-Bridged Dinuclear Peroxotungstate Catalytically Active for Epoxidation of Olefins. *Inorg. Chem.* **2007**, *46*, 3768–3774.

42. Barrio, L.; Campos-Martin, J. M.; Fierro, J. L. G. Spectroscopic and DFT Study of Tungstic Acid Peroxocomplexes. *J. Phys. Chem. A* **2007**, *111*, 2166–2171.

43. Taube, F.; Andersson, I.; Toth, I.; Bodor, A.; Howarth, O.; Pettersson, L. Equilibria and Dynamics of some Aqueous Peroxomolybdate Catalysts: A O-17 NMR Spectroscopic Study. *J. Chem. Soc. Dalton Trans.* **2002**, *23*, 4451–4456.

44. Howarth, O. W. Oxygen-17 NMR Study of Aqueous Peroxotungstates. *Dalton Trans.* **2004**, 476–481.

45. Maiti, S. K.; Banerjee, S.; Mukherjee, A. K.; Malik, K. M. A.; Bhattacharyya, R. Oxoperoxo Molybdenum(VI) and Tungsten(VI) and Oxodiperoxo Molybdate(VI) and Tungstate(VI) Complexes with 8-quinolinol: Synthesis, Structure and Catalytic Activity. *New J. Chem.* **2005**, *29*, 554–563.

46. Schwendt, P.; Oravcová, A.; Tyršelová, J.; Pavelčík, F. The First Tetranuclear Vanadium(V) Peroxo Complex: Preparation, Vibrational Spectra and X-ray Crystal Structure of $K_6[V_4O_4(O_2)_8(PO_4)]\cdot9H_2O$. *Polyhedron* **1996**, *15*, 4507–4511.

47. Schwendt, P.; Švančárek, P.; Smatanová, I.; Marek, J. Stereospecific Formation of α-hydroxycarboxylato Oxo Peroxo Complexes of Vanadium(V). Crystal Structure of $(NBu_4)_2[V_2O_2(O_2)_2(l\text{-lact})_2]\cdot2H_2O$ and $(NBu_4)_2[V_2O_2(O_2)_2(D\text{-lact})(L\text{-lact})]\cdot2H_2O$. *J. Inorg. Biochem.* **2000**, *80*, 59–64.

48. Tatiersky, J.; Schwendt, P.; Marek, J.; Sivák, M. Racemic Vanadium(V) Oxo Monoperoxo Complexes with Two Achiral Bidentate Heteroligands. Synthesis, Characterization, Crystal Structure and Stereochemistry of $K[VO(O_2)(ox)(bpy)]\cdot3H_2O$ and $Pr_4N[VO(O_2)(ox)(phen)]$. *New J. Chem.* **2004**, *28*, 127–133.

49. Schwendt, P.; Švančárek, P.; Kuchta, L.; Marek, J. A New Coordination Mode for the Tartrato Ligand. Synthesis of Vanadium(V) Oxo Peroxo Tartrato Complexes and the X-ray Crystal Structure of $K_2[\{VO(O_2)(L\text{-tartH}_2)\}2(\mu\text{-}H_2O]5H_2O)$. *Polyhedron* **1998**, *17*, 2161–2166.

50. Ahmed, M.; Schwendt, P.; Marek, J.; Sivák, M. Synthesis, Solution and Crystal Structures of Dinuclear Vanadium(V) Oxo Monoperoxo Complexes with Mandelic Acid: $(NR_4)_2[V_2O_2(O_2)_2(mand)_2]\cdot xH_2O$ [R=H, Me, Et; mand=mandelato$(2^-)=C_8H_6O_3^{2-}$. *Polyhedron* **2004**, *23*, 655–663.

51. Tatiersky, J.; Schwendt, P.; Sivák, M.; Marek, J. Racemic Monoperoxovanadium(V) Complexes with Achiral OO and ON Donor Set Heteroligands: Synthesis, Crystal Structure and Stereochemistry of $[NH_3(CH_2)2NH_3][VO(O_2)(ox)(pic)]\cdot2H_2O$ and $[NH_3(CH_2)2NH_3][VO(O_2)(ox)(pca)]$. *Dalton Trans.* **2005**, (13), 2305–2311.

52. Schwendt, P.; Ahmed, M.; Marek, J. Complexation Between Vanadium (V) and Phenyllactate: Synthesis, Spectral Studies and Crystal Structure of (NEt_4) $(NH_4)_3[V_2O_2(O_2)_2(R\text{-3-phlact})_2][V_2O_2(O_2)_2(S\text{-3-phlact})_2]$ $6H_2O$, [3-phlact = 3-phenyl-lactato(2^-)]. *Inorg. Chim. Acta* **2005**, *358*, 3572–3580.

53. Chrappová, J.; Schwendt, P.; Marek, J. NMR Study of the $[VO(O_2)_2F]^{2-}$ Ion in Solution: Synthesis, Vibrational Spectra and Crystal Structure of $[NH_3CH_2CH_2NH_3]$ $[VO(O_2)_2F]$. *J. Fluorine Chem.* **2005**, *126*, 1297–1302.

54. Hou, S.-Y.; Zhou, Z.-H.; Wan, H.-L.; Ng, S.-W. A Carboxylate-Bridged μ-oxo-bis [diperoxomolybdate (VI)]. *Inorg. Chem. Commun.* **2003**, *6*, 1246–1348.

55. Zhou, Z.-H.; Hou, S.-Y.; Wan, H.-L. Peroxomolybdate(VI)-Citrate and -Malate Complex Interconversions by pH-Dependence. Synthetic, Structural and Spectroscopic Studies. *Dalton Trans.* **2004**, *9*, 1393–1399.

56. Hou, S.-Y.; Zhou, Z.-H.; Lin, T.-R.; Wan, H.-L. Peroxotungstates and Their Citrate and Tartrate Derivatives. *Eur. J. Inorg. Chem.* **2006,** *2006,* 1670–1677.

57. Sergienko, V. S. Structural Chemistry of Oxoperoxo Complexes of Vanadium (V): A Review. *Crystallogr. Rep.* **2004,** 49, 401–426; (English translation of Kristallografiya **2004,** *49,* 467–491).

58. Dengel, A. C.; Griffith, W. P.; Powell, R. D.; Skapski, A. C. Studies on Transition-Metal Peroxo Complexes. Part 7. Molybdenum(VI) and Tungsten(VI) Carboxylato Peroxo Complexes, and the X-ray Crystal Structure of $K_2[MoO(O_2)_2(glyc)]\cdot 2H_2O$. *J. Chem. Soc., Dalton Trans.* **1987,** 991–995.

59. Butler, A.; Clague, M. J.; Meister, G. E. Vanadium Peroxide Complexes. *Chem. Rev.* **1994,** *94,* 625–638.

60. Won, T.-J.; Sudam, B. M. Characterization of Oxoperoxo(2,6-Pyridinedicarboxylato) Molybdenum(VI) and Oxoperoxo(Nitrilotriacetato)Molybdate(VI) in Aqueous-Solution and a Kinetic-Study of Their Reduction by a (Thiolato)Cobalt(III) Complex, Dimethyl-Sulfoxide, and Iron(II). *Inorg. Chem.* **1994,** *33,* 3804–3810.

61. Conte, V.; Di Furia, F.; Moro, S. The Versatile Chemistry of Peroxo Complexes of Vanadium, Molybdenum and Tungsten as Oxidants of Organic Compounds. *J. Phys. Org. Chem.* **1996,** *9,* 329–336.

62. Gorzsás, A.; Andersson, I.; Pettersson, L. Speciation in the Aqueous $H^+/H_2VO_4^-/H_2O_2/L$-(+)-Lactate System. *Dalton Trans.* **2003,** 2503–2511.

63. Andersson, I.; Gorzsás, A.; Pettersson, L. Speciation in the Aqueous $H^+/H_2VO_4^-/H_2O_2/$Picolinate System Relevant to Diabetes Research. *Dalton Trans.* **2004,** 421–428.

64. Gorzsás, A.; Getty, K.; Andersson, I.; Pettersson, L. Speciation in the Aqueous $H^+/H_2VO_4^-/H_2O_2/$Citrate System of Biomedical Interest. *Dalton Trans.* **2004,** 2873–2882.

65. Andersson, I.; Gorzsás, A.; Kerezsi, C.; Tóth, I.; Pettersson, L. Speciation in the Aqueous $H^+/H_2VO_4^-/H_2O_2/$Phosphate System. *Dalton Trans.* **2005,** 3658–3666.

66. Justino, L. L. G.; Ramos, M. L.; Caldeira, M. M.; Gil, V. M. S. Peroxovanadium (V) Complexes of L-Lactic Acid as Studied by NMR Spectroscopy. *Eur. J. Inorg. Chem.* **2000,** *2000*(7), 1617–1621.

67. Justino, L. L. G.; Ramos, M. L.; Caldeira, M. M.; Gil, V. M. S. Peroxovanadium (V) Complexes of Glycolic Acid as Studied by NMR Spectroscopy. *Inorg. Chim. Acta* **2000,** *311,* 119–125.

68. Justino, L. L. G.; Ramos, M. L.; Caldeira, M. M.; Gil, V. M. S. NMR Spectroscopy Study of the Peroxovanadium (V) Complexes of L-Malic Acid. *Inorg. Chim. Acta* **2003,** *356,* 179–186.

69. Ramos, M. L.; Caldeira, M. M.; Gil, V. M. S. Peroxo Complexes of Sugar Acids with Oxoions of Mo(VI) and W(VI) as Studied by NMR Spectroscopy. *J. Chem. Soc. Dalton Trans.* **2000,** 2099–2103.

70. Ramos, M. L.; Pereira, M. M.; Beja, A. M.; Silva, M. R.; Paixão, J. A.; Gil, V. M. S. NMR and X-ray Diffraction Studies of the Complexation of D-(−) Quinic Acid with Tungsten (VI) and Molybdenum (VI). *J. Chem. Soc. Dalton Trans.* **2002,** 2126–2131.

71. Rehder, D.; Weidemann, C.; Duch, A.; Priebsch, W. ^{51}V Shielding in Vanadium (V) Complexes: A Reference Scale for Vanadium Binding Sites in Biomolecules. *Inorg. Chem.* **1988,** *27,* 584–587.

72. Andersson, I.; Angus-Dunne, S.; Howarth, O.; Pettersson, L. Speciation in Vanadium Bioinorganic Systems: 6. Speciation Study of Aqueous Peroxovanadates, Including Complexes with Imidazole. *J. Inorg. Biochem.* **2000,** *80,* 51–58.

73. Campbell, N. J.; Dengel, A. C.; Griffith, W. P. Studies on Transition Metal Peroxo Complexes—X. The Nature of Peroxovanadates in Aqueous Solution. *Polyhedron* **1989,** *8,* 1379–1386.

74. Harrison, A. T.; Howarth, O. W. High-Field Vanadium-51 and Oxygen-17 Nuclear Magnetic Resonance Study of Peroxovanadates (V). *J. Chem. Soc. Dalton Trans.* **1985,** 1173–1177.

75. Nardello, V.; Marko, J.; Vermeersch, G.; Aubry, J. M. Mo-95 NMR and Kinetic-Studies of Peroxomolybdic Intermediates Involved in the Catalytic Disproportion-ation of Hydrogen-Peroxide by Molybdate Ions. *Inorg. Chem.* **1995,** *34,* 4950–4957.

76. Campbell, N. J.; Dengel, A. C.; Edwards, C. J.; Griffith, W. P. Studies on Transition Metal Peroxo Complexes. Part 8. The Nature of Peroxomolybdates and Peroxotung-states in Aqueous Solution. *J. Chem. Soc. Dalton Trans.* **1989,** 1203–1208.

77. Nakajima, H.; Kudo, T.; Mizuno, N. Reaction of Tungsten Metal Powder with Hydrogen Peroxide to form Peroxo Tungstates, an Useful Precursor of Proton Conductor. *Chem. Lett.* **1997,** 693–694.

78. Nardello, V.; Marko, J.; Vermeersch, G.; Aubry, J. M. ^{183}W NMR Study of Peroxo-tungstates Involved in the Disproportionation of Hydrogen Peroxide into Singlet Oxygen ($^1O_2$, $^1\Delta_g$) Catalyzed by Sodium Tungstate in Neutral and Alkaline Water. *Inorg. Chem.* **1988,** *37,* 5418–5423.

79. Postel, M.; Brevard, C.; Arzoumanian, H.; Riess, J. G. O-17 NMR as a Tool for Studying Oxygenated Transition-Metal Derivatives—1st Direct O–17 NMR Obser-vations of Transition-Metal-Bonded Peroxidic Oxygen-Atoms—Evidence for the Absence of Oxo-Peroxo Oxygen-Exchange in Molybdenum(VI) Compounds. *J. Am. Chem. Soc.* **1983,** *105,* 4922–4926.

80. Conte, V.; Di Furia, F.; Modena, G.; Bortolini, O. Metal Catalysis in Oxidation by Peroxides. 29. A Oxygen-17 NMR Spectroscopic Investigation of Neutral and Anionic Molybdenum Peroxo Complexes. *J. Org. Chem.* **1988,** *53,* 4581–4582.

81. Schwendt, P.; Tatiersky, J.; Krivosudský, L.; Šimuneková, M. Peroxido Complexes of Vanadium. *Coord. Chem. Rev.* **2016,** *318,* 135–157.

82. Caldeira, M. M.; Ramos, M. L.; Oliveira, N. C.; Gil, V. M. S. Complexes of Vanadium (V) with α-hydroxycarboxylic Acids Studied by ^{1}H, ^{13}C, and ^{51}V Nuclear Magnetic Resonance Spectroscopy. *Can. J. Chem.* **1987,** *65,* 2434–2440.

83. Caldeira, M. M.; Ramos, M. L.; Gil, V. M. S. Complexes of W(VI) and Mo(VI) with Glycolic, Lactic, Chloro- and Phenyl-lactic, Mandelic, and Glyceric Acids Studied by ^{1}H and ^{13}C Nuclear Magnetic Resonance Spectroscopy. *Can. J. Chem.* **1987,** *65,* 827–832.

84. Caldeira, M. M.; Ramos, M. L.; Cavaleiro, A. M.; Gil, V. M. S. Multinuclear NMR Study of Vanadium(V) Complexation with Tartaric and Citric Acids. *J. Mol. Struct.* **1988,** *174,* 461–466.

85. Ramos, M. L.; Caldeira, M. M.; Gil, V. M. S. Multinuclear NMR Study of Complex-ation of D-galactaric and Mannaric Acids with Tungsten(VI) Oxoions. *J. Coord. Chem.* **1994,** *33,* 319–329.
86. Ramos, M. L.; Caldeira, M. M.; Gil, V. M. S. Multinuclear NMR Study of the Complexation of D-galactaric and Mannaric Acids with Molybdenum (VI). *Polyhedron* **1994,** *13,* 1825–1833.
87. Ramos, M. L.; Caldeira, M. M.; Gil, V. M. S. NMR Study of the Complexation of D-galactonic Acid with Tungsten (VI) and Molybdenum (VI). *Carbohydr. Res.* **1997,** *297,* 191–200.
88. Ramos, M. L.; Caldeira, M. M.; Gil, V. M. S. NMR Spectroscopy Study of the Complexation of L-mannonic Acid with Tungsten(VI) and Molybdenum(VI). *Carbohydr. Res.* **1997,** *299,* 209–220.
89. Ramos, M. L.; Caldeira, M. M.; Gil, V. M. S. NMR Spectroscopy Study of the Complexation of D-gluconic Acid with Tungsten(VI) and Molybdenum(VI). *Carbohydr. Res.* **1997,** *304,* 97–109.
90. Ramos, M. L.; Caldeira, M. M.; Gil, V. M. S. NMR Study of the Complexation of D-gulonic Acid with Tungsten(VI) and Molybdenum(VI). *Carbohydr. Res.* **2000,** *329,* 387–397.
91. Ramos, M. L.; Gil, V. M. S. Multinuclear NMR Study of the Complexes of 6-phospho-D-gluconic Acid with W(VI) and Mo(VI). *Carbohydr. Res.* **2004,** *339,* 2225–2232.
92. Ramos, M. L.; Justino, L. L. G.; Gil, V. M. S.; Burrows, H. D. NMR and DFT Studies of the Complexation of W (VI) and Mo (VI) with 3-phospho-D-glyceric and 2-phospho-D-glyceric Acids. *Dalton Trans.* **2009,** 9616–9624.
93. Ramos, M. L.; Justino, L. L. G.; Fonseca, S. M.; Burrows, H. D. NMR, DFT and Lminescence Studies of the Complexation of V(V) Oxoions in Solution with 8-Hydroxyquinoline-5-Sulfonate. *New J. Chem.* **2015,** *39,* 1488–1497.
94. Ramos, M. L.; Justino, L. L. G.; Abreu, P. E.; Fonseca, S. M.; Burrows, H. D. Oxocomplexes of Mo(VI) and W(VI) with 8-Hydroxyquinoline-5-Sulfonate in Solution: Structural Studies and the Effect of the Metal Ion on the Photophysical Behaviour. *Dalton Trans.* **2015,** *44,* 19076–19089.
95. Akitt, J. W.; Mason, J., Eds. *Multinuclear NMR;* Plenum Press: New York, 1987, pp 259–292.
96. Justino, L. L. G.; Ramos, M. L.; Nogueira, F.; Sobral, A. J. F. N.; Geraldes, C. F. G. C.; Kaupp, M.; Burrows, H. D.; Fiolhais, C.; Gil, V. M. S. Oxoperoxo Vanadium (V) Complexes of L-lactic Acid: Density Functional Theory Study of Structure and NMR Chemical Shifts. *Inorg. Chem.* **2008,** *47,* 7317–7326.
97. Justino, L. L. G.; Ramos, M. L.; Kaupp, M.; Burrows, H. D.; Fiolhais, C.; Gil, V. M. S. Density Functional Theory Study of the Oxoperoxo Vanadium (V) Complexes of Glycolic Acid. Structural Correlations with NMR Chemical Shifts. *Dalton Trans.* **2009,** *44,* 9735–9745.
98. Maiti, S. K.; Dinda, S.; Banerjee, S.; Mukherjee, A. K.; Bhattacharyya, R. Oxidoperoxidotungsten(VI) Complexes with Secondary Hydroxamic Acids: Synthesis, Structure and Catalytic Uses in Highly Efficient, Selective and Ecologically Benign Oxidation of Olefins, Alcohols, Sulfides and Amines with H_2O_2 as a Terminal Oxidant. *Eur. J. Inorg. Chem.* **2008,** *12,* 2038–2051.

99. Gharah, N.; Chakraborty, S.; Mukherjee, A. K.; Bhattacharyya, R. Oxoperoxo Molybdenum(VI)- and Tungsten(VI) Complexes with 1-(2 .ntVI) Complexes Ethanone Oxime: Synthesis, Structure and Catalytic Uses in the Oxidation of Olefins, Alcohols, Sulfides and Amines Using H_2O_2 as a Terminal Oxidant. *Inorg. Chim. Acta* **2009,** *362,* 1089–1100.

100. Gharah, N.; Drew, M. G. B.; Bhattacharyya, R. Synthesis and Catalytic Epoxidation Potentiality of Oxodiperoxo Molybdenum(VI) Complexes with Pyridine-2-Carboxaldoxime and Pyridine-2-Carboxylate: The Crystal Structure of $PMePh_3[MoO(O_2)_2(PyCO)]$. *Transition Met. Chem.* **2009,** *34,* 549–557.

101. Cai, S. F.; Wang, L. S.; Fan, C. L. Catalytic Epoxidation of a Technical Mixture of Methyl Oleate and Methyl Linoleate in Ionic Liquids Using $MoO(O_2)2\cdot2QOH$ (QOH = 8-quinilinol) as Catalyst and $NaHCO_3$ as co-Catalyst. *Molecules* **2009,** *14,* 2935–2946.

102. Conte, V.; Floris, B. Vanadium Catalyzed Oxidation with Hydrogen Peroxide. *Inorg. Chim. Acta* **2010,** *363,* 1935–1946.

103. Brégeault, J.-M. Transition-Metal Complexes for Liquid-Phase Catalytic Oxidation: Some Aspects of Industrial Reactions and of Emerging Technologies. *Dalton Trans.* **2003,** 3289–3302.

104. Sato, K.; Aoki, M.; Noyori, R. A "Green" Route to Adipic Acid: Direct Oxidation of Cyclohexenes with 30 Percent Hydrogen Peroxide. *Science* **1998,** *281,* 1646–1647.

105. Rehder, D. *Bioinorganic Vanadium Chemistry;* John Wiley: Chichester, 2008.

106. Crans, D. C.; Smee, J. J.; Gaidamauskas, E.; Yang, L. Q. The Chemistry and Biochemistry of Vanadium and the Biological Activities Exerted by Vanadium Compounds. *Chem. Rev.* **2004,** *104,* 849–902.

107. Butler, A.; Walker, J. V. Marine Haloperoxidases. *Chem. Rev.* **1993,** *93,* 1937–1944.

CHAPTER 12

PHYSICOCHEMICAL AND GELLING PROPERTIES OF β-GLUCAN FROM A LOW-QUALITY OAT VARIETY

N. RAMÍREZ-CHÁVEZ[2], J. SALMERON-ZAMORA[2], E. CARVAJAL-MILLAN[1,*], AND R. PÉREZ LEAL[2]

[1]CTAOA, Laboratory of Biopolymers, Research Center for Food and Development, CIAD, A. C., Hermosillo, Sonora 83000, Mexico

[2]Faculty of Agro-Technological Sciences, Autonomous University of Chihuahua, Chihuahua 31125, Mexico

*Corresponding author. E-mail: ecarvajal@ciad.mx

CONTENTS

ABSTRACT

The main component of the soluble dietary fiber of oat is β-glucan. Research on oat has been intensified in the last few years. In clinical studies, oat β-glucan was shown to reduce serum cholesterol levels and attenuate postprandial blood glucose and insulin responses in a viscosity-related fashion. Oat is extensively planted as a forage crop in Northern Mexico, where drought resulted in a lower oat seed quality crop which failed to meet the requirements of the market. In this chapter, β-glucan from a low-quality oat variety was extracted and characterized for the first time as a potential gelling agent for the food industry. β-glucan presented an intrinsic viscosity ($[\eta]$) of 315 ml/g and a viscosimetric molecular weight (M_v) of 369 kDa. Fourier-transform infrared spectroscopy spectrum of β-glucan shows the main band centered at 1035 cm^{-1} which could be assigned to C–OH bending, with shoulders at 1158, 995, and 897 cm^{-1} that were related to the antisymmetric C–O–C stretching mode of the glycosidic link and β,(1–4) linkages. The gelling capability of β-glucans at 10%·w/v was investigated by rheological measurements. G′ and G″ attained the respective plateau values of 155 and 51 Pa after an induction period. The gel set time (G′>G″) corresponded to 20 h. The results attained suggest that β-glucans from low-quality oat can be a potential source of a gelling polysaccharide for the food industry.

12.1   INTRODUCTION

Research on oat grain properties (*Avena sativa*) has been intensified in the past few years as it has been reported to reduce serum cholesterol levels and attenuate postprandial blood glucose and insulin responses, related to the presence of β-glucan.[1–4,10] On the other hand, β-glucans have a high potential application as food texturizing, fat-mimetics, and encapsulation agent.[11] β-glucan is a linear polysaccharide made up entirely of sequences of (1→4)-linked D-glucopyranosyl units separated by single (1→3)-β-linked units.[18] Oats are being extensively planted as a forage crop in Northern Mexico, where rainfall has an erratic distribution and therefore, seed yields and quality are low. These low-quality oats have been studied on the basis of forage yield and nutritional value but the extraction and characterization of purified β-glucan from these grains have not been reported elsewhere.[6–8,12–17] This chapter has been focused on the extraction, characterization, and gelling capability of β-glucan from a low-quality oat.

12.2 MATERIALS AND METHODS

12.2.1 MATERIALS

Whole oat seeds from *A. sativa* cultivar Karma harvested under drought conditions were provided by the National Institute for Investigation in Forestry, Agriculture and Animal Production in Mexico (INIFAP). The seed hull was removed by hand. Whole oat seeds were milled down to 0.84 mm particle size using a M20 Universal Mill (IKA®, Werke Staufen, Germany). The enzymes amyloglucosidase, pronase, and amylase thermo usedwere of analytical grade and purchased from Sigma Chemical Co. (St. Louis, MO, USA).

12.2.2 METHODS

12.2.2.1 β-GLUCAN EXTRACTION

β-glucan was water extracted from milled oat seeds (1 Kg/3 l) for 15 min at 25°C. The water extract was then centrifuged (12096 g, 20°C, 15 min) and the supernatant recovered. Starch was then enzymatically degraded (Termamyl®120, 100°C, 30 min, 2800 U/g of sample and, amyloglucosidase, 3 h, 50°C, pH 5, 24 U/g of sample). The extract was then treated with pronase (pH 7.5, 20°C, 16 h followed for 100°C, 10 min, 0.4 U/g of sample) to eliminate protein. The supernatant was precipitated in 65% ethanol treated for 4 h at 4°C. The precipitate was recovered and dried by solvent exchanged (80%, v/v ethanol, absolute ethanol, and acetone) to give oat gum. β-glucan was then dried by solvent exchange.

12.2.2.2 CHEMICAL ANALYSIS

The β-glucan content was determined by the method of McCleary and Gennie-Holmes[21] using Megazyme® mixed linkage β-glucan assay kit.

The protein content of the β-glucan preparation was according to the method of Bradford.[5]

12.2.2.3 INTRINSIC VISCOSITY

Specific viscosity, η_{sp}, was measured by registering β-glucan solutions flow time in an Ubbelohde capillary viscometer at $25 \pm 0.1°C$, immersed in a temperature-controlled bath. β-glucan solutions were prepared at different concentrations, dissolving dried β-glucan in water for 18 h with stirring at room temperature. β-glucan solution and the solvent were filtered using 0.45 μm membrane filters before viscosity measurements. The intrinsic viscosity ($[\eta]$) was estimated from relative viscosity measurements, η_{rel}, of β-glucan solutions by extrapolation of Kraemer and Mead and Fouss curves to "zero" concentration.[19-22]

12.2.2.4 VISCOSIMETRIC MOLECULAR WEIGHT

The viscosimetric molecular weight (M_v) was calculated from the Mark–Houwink relationship, $M_v = (\eta/k)^{1/\alpha}$, where the constants k and α are 0.013 and 1.099, respectively.

12.2.2.5 FTIR SPECTROSCOPY

Infrared spectra were collected using a Nicolet FTIR spectrometer (Nicolet Instrument Corp. Madison, USA). The samples were pressed into KBr pellets. A blank KBr disk was used as background. In order to obtain more exact band positions, Fourier self-deconvolution was applied.[23-40] Spectra were recorded between 600 and 1800 cm^{-1}.

12.2.2.6 β-GLUCAN GELATION

The gelling β-glucan solution was prepared at 10%·w/v in water for 18 h with stirring at room temperature. After preparation, the sample was transferred onto the rheometer. Rheological tests were performed by small amplitude oscillatory shear by using a strain-controlled rheometer (AR-1500ex, TA Instruments, U.S.A.) in the oscillatory mode. A plate geometry (5.0 cm in diameter) was used and exposed edges of the sample were covered with mineral oil fluid to prevent evaporation during measurements. β-glucan gelation kinetic was monitored at 25°C for 60 h by following the storage (G') and loss (G″) modulus. All measurements were carried out at a frequency of 0.25 Hz and 1% strain (in linear domain). The mechanical spectrum of β-glucan gels was obtained by frequency sweep from 0.1 to 100 Hz at 1% strain. All measurements were at 25°C.

12.2.2.7 STATISTICAL ANALYSIS

All determinations were made in triplicates and the coefficients of variation were lower than 8%. All results are expressed as mean values.

12.3 RESULTS AND DISCUSSION

12.3.1 EXTRACTION AND CHARACTERIZATION

The β-glucan yield was 1.8% (w β-glucan/w oat seeds, dry basis d.b.), which means that 35% of the β-glucan initially present in the oat seed (5.2%·w β-glucan /w oat seed d.b.) was recovered. The chemical composition of β-glucan is presented in Table 12.1. The extraction protocol adopted in the present study provided a β-glucan of high purity (β-glucan content of 92%·d.b.) and low protein residues (4.1%·d.b.) [η] of β-glucan was 3.15 dl/g. This value is in the range of the calculated values found by other researchers for oat β-glucan (2.58–9.63 dl/g) under similar conditions. Intrinsic viscosity provides a convenient measure of the space-occupancy of individual polymer coils. M_v of β-glucan was 369 kDa. The molecular size plays an important role in the solubility, conformation, and rheological properties of the molecule in solution.

TABLE 12.1 Compositional and Physicochemical Features of Oat β-Glucan.

β-glucan (%·d.b.)	92.0±1.50
Protein (%·d.b.)	4.1±0.30
Mw (kDa)	369±23
[η] (ml/g)	315±29

Values are means (± SD) of triplicate measurements. All results are obtained from triplicates

12.3.2 RHEOLOGY

The gelation capability of aqueous β-glucan dispersions was moni-tored as a function of time at 25°C. The time-dependent evolution of the storage (G′) and loss (G″) modulus was monitored periodically at 0.25 Hz frequency and strain level of 1% (Fig. 12.1). During gelation, G′ and G″ increased with time and attained a plateau value with the sample adopting gel properties. The initial rise of G″, which is sharper than the rise of G″, involves the establishment of a three-dimensional network structure. The subsequent slower rise in G′ could be attributed to further cross-linking and rearrangement of cross-links as well as lateral chain aggregation. The transition from the stage of rapidly increasing modulus to the plateau stage occurs when most of the sol fraction has been converted into the gel phase.[28] At the end of the gel curing experiment, the mechanical spectra became typical of elastic gel networks (G′≫G″ and G′ nearly indepen-dent on frequency). The condition of linear viscoelasticity is fulfilled during the gel cure experiments, as demonstrated by strain sweeps at 0.25 Hz performed before and after the time dependence experiments.

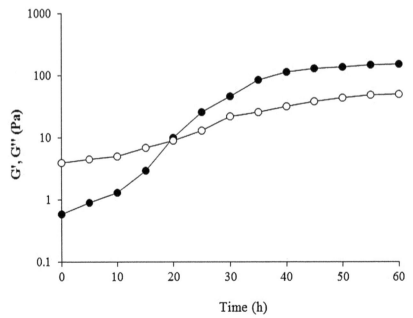

FIGURE 12.1   Time dependence of G′ (●) and G″ (○) for oat β-glucan preparation at 10% w/v. Data obtained at 25°C, 0.25 Hz and 1% strain.

12.3.3   FOURIER-TRANSFORM INFRARED SPECTROSCOPY (FTIR)

In order to confirm the identity of oat β-glucan, the sample was analyzed by Fourier-transform infrared FTIR spectroscopy (Fig. 12.2). It was found that β-glucan extract spectrum exhibited similarities in its absorption pattern to that reported by other authors. FTIR spectrum in the wave number between 950 and 1200 cm^{-1} is considered as the "fingerprint" region for carbohydrates, which allows the identification of major chemical groups in polysaccharides,[9] β-glucan showed a broad absorbance band for polysaccharides in this region. The main band centered at 1035 cm^{-1} could be assigned to C–OH bond, with shoulders at 1158, 994, and 897 cm^{-1} that were related to the antisymmetric C–O–C stretching mode of the glycosidic link and β,(1–4) linkages. A weak absorbance was observed at 1648 cm^{-1} implying a low degree of esterification with aromatic esters such as hydroxycinnamic acids. Two bands at 1648 and 1541 cm^{-1}, related to protein content together with a band at 1640 cm^{-1} attributed to adsorbed water, were also observed. The protein signal detected could be related to protein cell wall or contaminating endosperm proteins (Fig. 12.3).

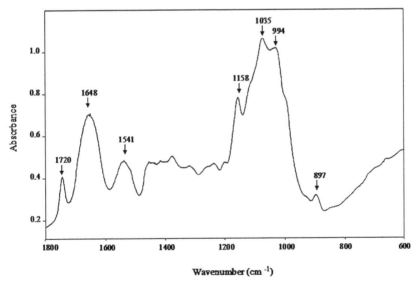

FIGURE 12.2   FTIR spectrum of β-glucan extracted from a low-quality Mexican oat seed (karma variety).

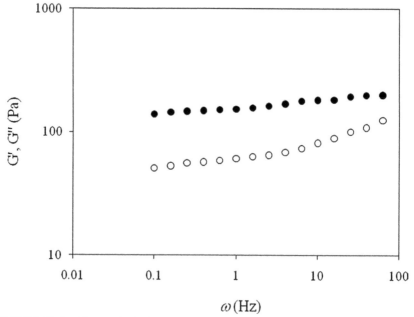

FIGURE 12.3   The mechanical spectrum of oat β-glucan gels at 10%·w/v (G' •, G" ○) after gel network development (60 h). Data obtained at 25°C and 1% strain.

12.4 CONCLUSION

The results attained suggest that low-quality oat cultivar Karma can be a potential source of gelling β-glucans for the food industry. Further research is undergoing in order to explore the additional functional properties of this polysaccharide.

12.5 ACKNOWLEDGMENTS

The authors are pleased to acknowledge Alma C. Campa-Mada and Karla G. Martínez-Robinson for their technical assistance.

KEYWORDS

- **polysaccharide gels**
- **elasticity**
- **gel structure**
- **diffusion**

REFERENCES

1. AACC. Approved Methods of the American Association of Cereal Chemists; AACC: Minnesota, USA, 1998.
2. Adapa, S.; Schmidt, K. A. l.; Toledo, R. Functional Properties of Skim Milk Processed with Continuous High Pressure Throttling. *J. Dairy Sci.* **1997,** *80*, 1941–1948.
3. Arcila-Lozano, C. C.; Loarca-Piña, G.; Lecona-Uribe, S.; González de Mejía, E. Oregano: Properties, Composition and Biological Activity. *Arch. Latinoam. Nutr.* **2004,** *54*, 100–111.
4. Arnason, J. T.; Baum, B.; Gale, J.; Lambert, J. D. H.; Bergvinson, D.; Philogène, B. J. R.; Serratos, J. A.; Mihm, J.; Jewell, D. C. Variation in Resistance of Mexican Landraces of Maize to Maize Weevil Sitophilus zeamais, in Relation to Taxonomic and Biochemical Parameters. *Euphytica* **1994,** *74*, 227–236.
5. Bradford, M. A. Rapid and Sensitive Method for the Quantification of Microgram Quantities of Protein Utilizing the Principle of Protein-Dye Binding. *Anal. Biochem.* **1976,** *72*, 248–254.
6. Carvajal-Millan, E.; Guigliarelli, B.; Belle, V.; Rouau, X.; Micard, V. Storage Stability of Arabinoxylan Gels. *Carbohydr. Polym.* **2005a,** *59*, 181–188.
7. Carvajal-Millan, E.; Landillon, V.; Morel, M. H.; Rouau, X.; Doublier, J. L.; Micard, V. Arabinoxylan Hydrogels: Impact of the Feruloylation Degree on their Structure and Properties. *Biomacromolecules* **2005b,** *6*, 309–317.
8. Carvajal-Millan, E.; Guilbert, S.; Morel, M. H.; Micard V. Impact of the Structure of Arabinoxylan Gels on their Rheological and Protein Transport Properties. *Carbohydr. Polym.* **2005c,** *60*, 431–438.
9. Cernà, M.; Barros, A. S.; Nunes, A.; Rocha, S. M.; Delgadillo, I.; Copíkovà, J.; Coimbra, M. A. Use of FT-IR Spectroscopy as a Tool for the Analysis of Polysaccharide Food Additives. *Carbohydr. Polym.* **2003,** *19*, 793–801.
10. Chanliaud, E.; Saulnier, L.; Thibault, J. F. Alkaline Extraction and Characterization of Heteroxylans from Maize Bran. *J. Cereal Sci.* **1995,** *21*, 195–203.
11. Cui, S. W. *Polysaccharide Gums from Agricultural Products. Processing, Structures and Functionality;* Technomic Publishing: Pennsylvania, 2001; pp 252–258.

12. Figueroa-Espinoza, M. C.; Rouau X. Oxidative Cross-Linking of Pentosans by a Fungal Laccase and a Horseradish Peroxidase: Mechanism of Linkage between Feruloylated Arabinoxylans. *Cereal Chem.* **1998**, *75*, 259–265.

13. García-Lara, S.; Bergvinson, D. J.; Burt, A. J.; Ramputh, A. I.; Díaz-Pontones, D. M.; Arnason, J. T. The Role of Pericarp Cell Wall Components in Maize Weevil Resistance. *Crop Sci.* **2004**, *44*, 1546–1552.

14. Grabber, J. H.; Ralph, J.; Hatfield, R. D. Ferulate Cross-Links Limit the Enzymatic Degradation of Synthetically Lignified Primary Walls of Maize. *J. Agric. Food Chem.* **1988**, *46*, 2609–2614.

15. Hespell, R. B. Extraction and Characterization of Hemicellulose from the Corn Fiber Produced by Corn Wet-Milling Processes. *J. Agric. Food Chem.* **1998**, *46*, 2615–2619.

16. Iiyama, K.; Lam, T. B. T.; Stone, B. A. Covalent Cross-Links in the Cell Wall. *Plant Physiol.* **1994**, *104*, 315–320.

17. Ishii, T. Structure and Functions of Feruloylated Polysaccharides. *Plant Sci.* **1997**, *127*, 111–127.

18. Izydorczyk, M. S.; Biliaderis C. G. Cereal Arabinoxylans: Advances in Structure and Physicochemical Properties. *Carbohydr. Polym.* **1995**, *28*, 33–48.

19. Kraemer, E. O. Molecular Weight of Celluloses and Cellulose Derivates. *Ind. Eng. Chem.* **1938**, *30*, 1200–1203.

20. Lapierre, C.; Pollet, B.; Ralet, M. C.; Saulnier, L. The Phenolic Fraction of Maize Bran: Evidence for Lignin-Heteroxylan Association. *Phytochemistry* **2001**, *57*, 765–772.

21. McCleary, B. V.; Glenie-Holmes, M. Measurement of (1-3),(1-4)-β-D-Glucan in Barley Malt. *J. Inst. Brew.* **1985**, *91*, 285–295.

22. Mead, D. J.; Fouss, R. M. Viscosities of Solutions of Polyvinyl Chloride. *J. Am. Chem. Soc.* **1942**, *64*, 277–282.

23. Micard, V.; Grabber, J. H.; Ralph, J.; Renard, C. M. G. C.; Thibault, J. F. Deydrodiferulic Acids from Sugar-Beet Pulp. *Phytochemistry* **1997**, *44*, 1365–1368.

24. Montgomery, R.; Smith, F. Structure of Corn Hull Hemicellulose. III. Identification of the Methylated Aldobiuronic Acid Obtained from Methyl Corn Hull Hemicellulose. *J. Am. Chem. Soc.* **1957**, *79*, 695–697.

25. Ng, A.; Greenshields, R. N.; Waldron, K. W. Oxidative Cross-Linking of Corn Bran Hemicellulose: Formation of Ferulic Acid Dehydrodimers. *Carbohydr. Res.* **1997**, *303*, 459–462.

26. Ralph, J.; Quideau, S.; Grabber, J. H.; Hatfield, R. D. Identification and Synthesis of New Ferulic Acid Dehydrodimers Present in Grass Cell Walls. *J. Chem. Soc. Perkin Trans.* **1994**, *1*, 3485–3498.

27. Rao, M. V. S. Viscosity of Dilute to Moderately Concentrated Polymer Solutions. *Polymer* **1993**, *34*, 592–596.

28. Ross-Murphy, S. B. *Biophysical Methods in Food Research;* Blackwell: Oxford, 1984.

29. Rouau, X.; Cheynier, V.; Surget, A.; Gloux, D.; Barron, C.; Meudec, E.; Montero, J. L.; Criton, M. A Dehydrotrimer of Ferulic Acid from Maize Bran. *Phytochemistry* **2003**, *63*, 899–903.

30. Saulnier, L.; Vigouroux, J.; Thibault, J.-F. Isolation and Partial Characterization of Feruloylated Oligosaccharides from Maize Bran. *Carbohydr. Res.* **1995a,** *272,* 241–253.

31. Saulnier, L.; Marot, C.; Chanliaud, E.; Thibault, J.-F. Cell Wall Polysaccharide Interactions in Maize Bran. *Carbohydr. Polym.* **1995b,** *26,* 279–287.

32. Saulnier, L.; Crépeau, M. J.; Lahaye, M.; Thibault, J. F.; García-Conesa, M. T.; Kroon, P. A.; Williamson, G. Isolation Ans Structural Determination of Two 5-5'-Diferuloyl Oligosaccharides Indicate that Maize Heteroxylans are Covalently Cross-Linked by Oxidatively Coupled Ferulates. *Carbohydr. Res.* **1999,** *320,* 82–92.

33. Singh, V.; Doner, L. W.; Johnston, D. B.; Hicks, K. B.; Eckhoff, S. R. Comparison of Coarse and Fine Corn Fiber for Corn Fiber Gum Yields and Sugar Profiles. *Cereal Chem.* **2000,** *77,* 560–561.

34. Suguwara, M.; Suzuki, T.; Totsuka, A.; Takeuchi, M.; Ueki, K. Composition of Corn Hull Dietary Fiber. *Starch/Staerke* **1994,** *46,* 335–337.

35. Turgut, N. D.; Silva, R. V. Effect of Water Stress on Plant Growth and Thymol and Carvacrol Concentrations in Mexican Oregano Grown under Controlled Conditions. *J. Appl. Hortic.* **2005,** *7,* 20–22.

36. Vansteenkiste, E.; Babot, C.; Rouau, X.; Micard, V. Oxidative Gelation of Feruloylated Arabinoxylan as Affected by Protein. Influence on Protein Enzymatic Hydrolysis. *Food Hydrocolloids* **2004,** *18,* 557–564.

37. Waldron, K. W.; Parr, A. J.; Ng, A.; Ralph, J. Cell Wall Esterified Phenolic Dimers: Identification and Quantification by Reverse Phase High Performance Liquid Chromatography and Diode Array Detection. *Phytochem. Anal.* **1996,** *7,* 305–312.

38. Whistler, R. L. Hemicelluloses. In *Industrial Gums, Polysaccharides and their Derivatives;* Whistler, R. L., BeMiller, J. N., Eds.; Academic Press: Orlando, 1993; pp 295–308.

39. Whistler, R. L.; Corbett, W. M. Oligosaccharides from Partial Hydrolysis of Corn Fiber Hemicellulose. *J. Am. Chem. Soc.* **1955,** *77,* 628–6330.

40. Whistler R. L.; BeMiller, J. N. Hydrolisis of Components from Methylated Corn Fiber Gum. *J. Am. Chem. Soc.* **1956,** *78,* 1163–1165.

41. Wolf, M. J.; MacMasters, M. M.; Cannon, J. A.; Rosewell, E. C.; Rist, C. E. Preparation and Some Properties of Hemicellulose from Corn Hulls. *Cereal Chem.* **1953,** *30,* 451–470.

42. Woo, D. H. Stabilization of the Emulsion Prepared with Dietary Fiber from Corn Hull. *Food Sci. Technol.* **2001,** *10,* 348–353.

PART V
Polymer Technology and Science

CHAPTER 13

ETHYLENE VINYL ACETATE COPOLYMER/GRAPHENE OXIDE NANOCOMPOSITE FILMS PREPARED VIA A SOLUTION CASTING METHOD

SAJJAD SEDAGHAT[*]

Department of Chemistry, College of Science Shahr-e-Qods Branch, Islamic Azad University, Tehran, Iran

[*]*Corresponding author. E-mail: sajjadsedaghat@yahoo.com*

CONTENTS

ABSTRACT

In this text, ethylene vinyl acetate/graphene oxide (EVA/GO) nanocomposite films were prepared via solution method for packaging industries. Mechanical measurements showed that Young modulus and tensile strength of EVA have improved with introducing a small amount of GO.

The morphological studies of prepared nanocomposites were investigated using scanning electron microscope (SEM) and X-ray diffraction (XRD) methods and the results of the permeability measurements showed that prepared films have good resistance against oxygen.

13.1   INTRODUCTION

Graphene, a single layer of carbon atoms in a hexagonal lattice, has recently gained much attention due to its novel electronic and mechanical properties.[1] Graphene is usually prepared by the reduction of its precursor graphene oxide (GO),[2] which is a typical pseudo two-dimensional oxygen-containing solid in bulk form, possesses functional groups, including hydroxyls, epoxides, and carboxyls.[3–6] Both graphene and GO papers show very high mechanical properties with good biocompatibility, and they have potential application as biomaterials.[7–9] The chemical groups of GO have been found to be a feasible and effective means of improving the dispersion of graphene. Additionally, functional side groups bound to the surface of GO or graphene sheets may improve the interfacial inter-action between GO/graphene and the matrix.[10] On the other hand, as a hydrophilic biopolymer with $-NH_2$ and $-OH$ in each unit, chitosan can be protonated to polycationic material in acid media, which favors the interaction between polymer chains and GO sheets. Thus, a good disper-sion of GO in chitosan solution is expected. In recent years, polymer/graphene nanocomposites have attracted much attention in science and also in industries. These nanocomposites have improved properties in comparison of pure polymers and even microcomposites. The improve-ment usually occurs in tensile modulus, mechanical strength, permeability, and flame retardancy.[11–15] Depending on the level of diffusion of polymeric chains between GO layers it is possible to have intercalated or exfoliated nanocomposites. In intercalated state, the polymeric chains enter free space between graphene layers and increase the layer distance.

While in exfoliated state, the graphene layers are completely sepa-rated and disperse into polymeric matrix. Graphene with high electrical, mechanical, and thermal properties is used as proper nanoparticles to improve the properties of polymeric matrix. One of the major applications of a carbon nanoparticle, such as graphene, is as polymeric matrix. The nanocomposite is a multiphase material in which at least one phase should be in nanometric scale.[14–17]

The polymeric nanocomposites could be used in high tech applications. Some of improved properties of polymer/graphene nanocomposites are: improved mechanical and thermal properties in comparison of net polymer, having electrical conductivity, increased barrier property against water steam and gases, enhanced chemical stability and improved biocompatibility.

13.2 EXPERIMENTAL

13.2.1 MATERIALS

Ethylene vinyl acetate (EVA) with 18 and 28% vinyl acetate was purchased from Hyundai Petrochemical of Korea. Toluene was obtained from industrial sector and used as solvent. Sulfuric acid, sodium permanganate, potassium nitrate, hydrogen peroxide, hydrochloric acid, and ethanol were prepared from Merck and used as obtained without any purification. Natural false of graphite was purchased from Dae-Jung South Korea. The scanning electron microscope (SEM) images were taken from fracture surface of samples by using Japan Hitachi S-4160. The X-ray diffraction (XRD) test was done by XRD STOE, Germany with 40 kV. The melt strength was measured by MFI Gotche Testinig Machinery, GT-7100-MI. The permeability apparatus consists of two parts: (i) bottom part where there is a circular plate with surface area of 8.2 cm² on which the film will be put. Behind this part, there is a metal mesh that is responsible for holding the film and on the bottom of this mesh, there is a path to bubble. (ii) In top part, there is a gas path for film which is connected to the gas cylinder. Furthermore, there is a pressure gage that shows the pressure. All the parts are completely sealed so that only gas can pass through the film.

13.2.2 SYNTHESIS OF GRAPHENE OXIDE (GO)

GO was synthesized according to modified Hummers method[18] and that isbriefly described in subsequent text.

13.2.2.1 PREPARATION OF NANOCOMPOSITE FILMS

In order to prepare nanocomposite films, various amounts of GO were introduced into EVA 5 wt.% solution. Then the mixture was stirred for 3 h at 80°C. The homogeneous solution was cast on the glass plate and dried at ambient temperature for 24 h. Then the plate was immersed into 1 wt.% sodium hydroxide solution for neutralization. After that, the film was removed from the plate and dried at room temperature.

13.3 RESULTS

13.3.1 PERMEABILITY MEASUREMENTS

The diffusion coefficient as a function of GO loading for EVA18 and 28 are illustrated in Figures 13.1 and 13.2. As it can be seen, in both EVA introducing a small amount of GO into polymeric matrix dramatically reduced permeability of films probably due to good barrier properties of GO.

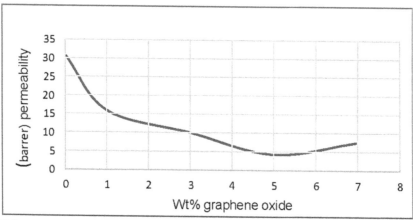

FIGURE 13.1   Diffusion coefficient as a function of wt.% graphene oxide (GO) ethylene vinyl acetate (EVA)18/GO nanocomposites.

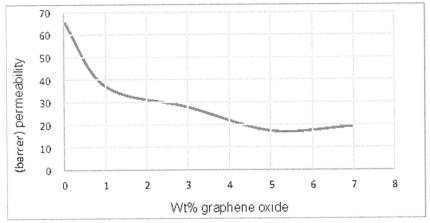

FIGURE 13.2 Diffusion coefficient as a function of wt.% EVA28/GO nanocomposites.

Comparison between Figures 13.1 and 13.2 shows that the amount of vinyl acetate monomer in EVA can influence the permeability of nano-composites. In Figure 13.3, the permeability of pure EVA was determined. Regarding the permeability test, increasing vinyl acetate causes the increase in permeability due to reduction in crystallinity of polymeric matrix.

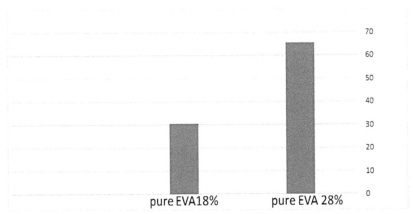

FIGURE 13.3 Effect of vinyl acetate monomer content on oxygen permeability of pure polymer films.

13.3.2 MORPHOLOGICAL CHARACTERIZATION

13.3.2.1 X-RAY DIFFRACTION (XRD)

As can be seen (Figs. 13.4–13.7) from XRD patterns of graphite and GO the interlayer d-spacing is equal to 0.33 and 1.06 nm, respectively, which corresponds to $2\theta = 8.38$. The XRD results confirmed that the GO layers separated and the polymer chains diffused between them and finally caused elimination of GO peak.

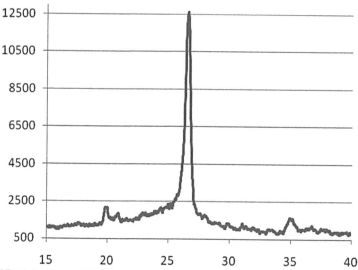

FIGURE 13.4   X-ray diffraction (XRD) of graphite powder.

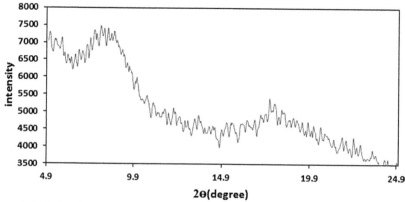

FIGURE 13.5   The XRD of pure GO.

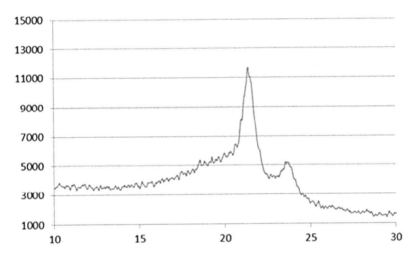

FIGURE 13.6   The XRD of pure EVA18 film.

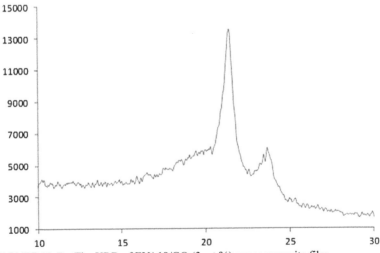

FIGURE 13.7   The XRD of EVA18/GO (3 wt.%) nanocomposite film.

13.3.2.2   SCANNING ELECTRON MICROSCOPY (SEM)

Figures 13.8 and 13.9 illustrated SEM images from fracture surface of EVA18/GO and EVA28/GO with 3 wt.%·GO. As can be seen, there is no phase diffraction between polymer matrix and nanofiller. On the other hand, there is a good interaction between the two phases.

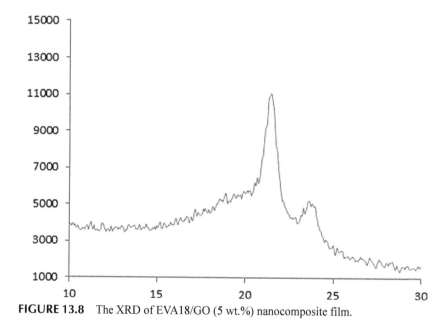

FIGURE 13.8   The XRD of EVA18/GO (5 wt.%) nanocomposite film.

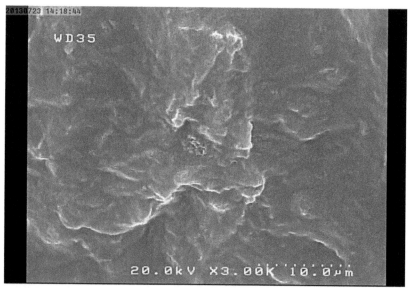

FIGURE 13.9   Scanning electron microscopy (SEM) image of EVA28/GO (3 wt.%) nanocomposite film with 3000 magnification.

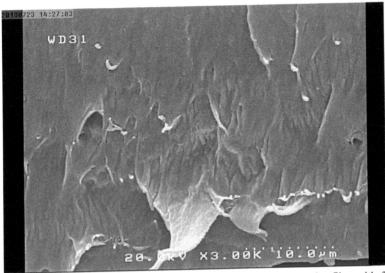

FIGURE 13.10 SEM image of EVA18/GO (3 wt.%) nanocomposite film with 3000 magnification.

13.3.3 MECHANICAL PROPERTIES

The effect of filler amount on the mechanical properties of prepared nano-composites such as tensile strength, tensile modulus, and elongation at break was investigated (Figs. 13.10–13.16). Tensile modulus is one of the most important mechanical properties in polymeric nanocomposites.

As illustrated in Figure 13.10 tensile modulus increased as GO loading increased. This behavior is due to the good interaction between matrix and filler and also proper dispersion of filler into matrix. As illustrated tensile strength of nanocomposites was increased as GO loading increased up to 5 wt.% and after that reduced probably due to formation agglomeration of nanofillers into matrix which acts as stress concentration area.

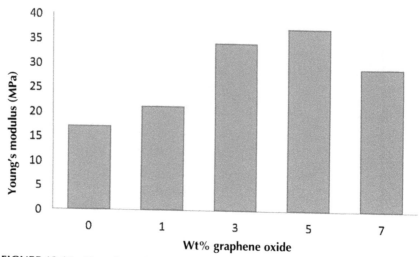

FIGURE 13.11   Young's modulus of the EVA28/GO nanocomposite films as a function of the GO content.

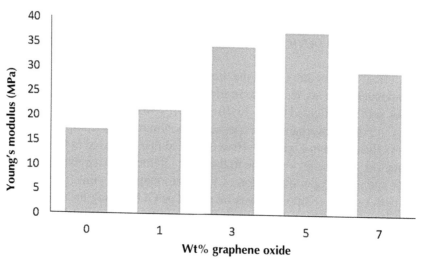

FIGURE 13.12   Young's modulus of the EVA18/GO nanocomposite films as a function of the GO content.

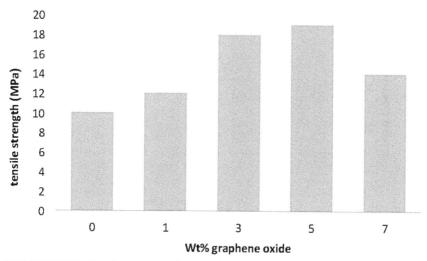

FIGURE 13.13 Tensile strength of prepared EVA28/GO nanocomposites as a function of GO content.

The elongation at break as a function of GO loading is shown in Figures 13.14 and 13.15. As we expected, incorporation of filler into polymer matrix has reduced elongation of polymer. This is because of reduction of deformation ability of polymer due to presence of filler which makes the matrix more brittle.

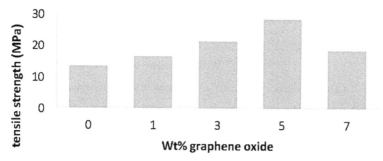

FIGURE 13.14 Tensile strength of prepared EVA18/GO nanocomposites as a function of GO content.

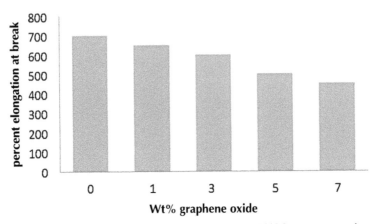

FIGURE 13.15 Elongation at break of prepared EVA28/GO nanocomposites as a function of GO content.

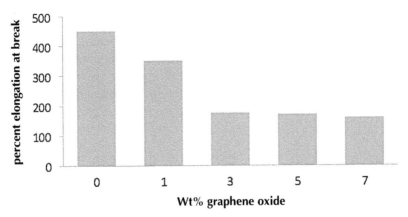

FIGURE 13.16 Elongation at break of prepared EVA18/GO nanocomposites as a function of GO content.

Investigation of mechanical properties of the prepared nanocomposites show that increasing GO content up to 5 wt.%, the modulus and tensile increase while increasing GO loading to 7 wt.%, reduced them due to filler agglomeration formation. Furthermore, introducing GO into matrix reduces elongation at break.

13.4   CONCLUSION

GO was synthesized via modified Hummer's method and the results confirmed this synthesis. The prepared GO was used to prepare impermeable EVA/GO nanocomposite for packaging. The properties of prepared film were investigated. The permeability test revealed that the permeability of EVA against gases such as CO_2 and O_2 are relatively high, especially when vinyl acetate is high. Introducing small amounts of GO into polymeric matrix dramatically reduced their permeability against O_2 and CO_2. Furthermore, the mechanical strength of prepared films which was prepared from pure EVA is not enough for packaging. Therefore, it is necessary to improve EVA mechanical and barrier properties. The results confirmed that the mechanical strength as well as Young's modulus was improved by incorporating GO into matrix; however, elongation at break reduced probably due to restriction effect of GO.

KEYWORDS

- ethylene vinyl acetate (EVA) copolymer
- graphene oxide (GO)
- mechanical properties
- permeability

REFERENCES

1. Ramanathan, T.; Abdala, A. A.; Stankovich, S.; Dikin, D. A.; Herrera-Alonso, M.; Piner, R. D. Functionalized Graphene Sheets for Polymer Nanocomposites. *Nat. Nanotechnol.* **2008**, *3*(6), 327–331.
2. Chen, W. F.; Yan, L. F.; Bangal, P. R. Preparation of Graphene by the Rapid and Mild Thermal Reduction of Graphene Oxide Induced by Microwaves. *Carbon* **2010**, *48*(4), 1146–1152.
3. McAllister, M. J.; Li, J. L.; Adamson, D. H.; Schniepp, H. C.; Abdala, A. A.; Liu, J. Single Sheet Functionalized Graphene by Oxidation and Thermal Expansion of Graphite. *Chem. Mater.* **2007**, *19*(18), 4396–4404.

4. Niyogi, S.; Bekyarova, E., Itkis, M. E.; McWilliams, J. L.; Hamon, M. A.; Haddon, R. C. Solution Properties of Graphite and Graphene. *J. Am. Chem. Soc.* **2006,** *128*(24), 7720–7721.

5. Stankovich, S.; Dikin, D. A.; Piner, R. D.; Kohlhaas, K. A.; Kleinhammes, A.; Jia, Y. Synthesis of Graphene-Based Nanosheets via Chemical Reduction of Exfoliated Graphite Oxide. *Carbon* **2007,** *45*(7), 1558–1565.

6. Vickery, J. L.; Patil, A. J.; Mann, S. Fabrication of Graphene-Polymer Nanocomposites with Higher-Order Three-Dimensional Architectures. *Adv. Mater.* **2009,** *21*(21), 2180–2184.

7. Stankovich, S.; Dikin, D. A.; Dommett, G. H. B.; Kohlhaas, K. M.; Zimney, E. J.; Stach, E. A. Graphene-Based Composite Materials. *Nature* **2006,** *442*(7100), 282–286.

8. Wakabayashi, K.; Pierre, C.; Dikin, D. A.; Ruoff, R. S.; Ramanathan, T.; Brinson, L. C. Polymer–Graphite Nanocomposites: Effective Dispersion and Major Property Enhancement via Solid-State Shear Pulverization. *Macromolecules* **2008,** *41*(6), 1905–1908.

9. Wang, S. R.; Tambraparni, M.; Qiu, J. J.; Tipton, J., Dean, D. Thermal Expansion of Graphene Composites. *Macromolecules* **2009,** *42*(14), 5251–5255.

10. Coleman, J. N.; Cadek, M.; Blake, R.; Nicolosi, V.; Ryan, K. P.; Belton, C. High-Performance Nanotube-Reinforced Plastics: Understanding the Mechanism of Strength Increase. *Adv. Funct. Mater.* **2004,** *14*(8), 791–798.

11. Novoselov, K. S.; Geim, A. K.; Morozov, S. V.; Jiang, D.; Zhang, Y.; Dubonos, S. V.; Grigorieva, I. V.; Firsov, A. A. Electric Field Effect in Atomically Thin Carbon Films. *Science* **2004,** *306,* 666–669.

12. Kuilla, T.; Bhadra, S.; Yao, D.; Kim, N. H.; Bose, S.; Lee, J. H. Recent Advances in Graphene Based Polymer Composites. *Prog. Polym. Sci.* **2010,** *35,* 1350–1375.

13. Kim, H.; Abdala, A. A.; Macosko, C. W. Graphene/Polymer Nanocomposites. *Macromolecules* **2010,** *43,* 6515–6530.

14. Potts, J. R.; Dreyer, D. R.; Bielawski, C. W.; Ruoff, R. S. Graphene-Based Polymer Nanocomposites. *Polymer* **2011,** *52,* 5–25.

15. Park, S.; An, J.; Jung, I.; Piner, R. D.; An, S. J.; Li, X.; Velamakanni, A.; Ruoff, R. S. Colloidal Suspensions of Highly Reduced Graphene Oxide in a Wide Variety of Organic Solvents. *Nano Lett.* **2009,** *9,* 1593–1597.

16. Shen, J.; Hu, Y.; Shi, M.; Lu, X.; Qin, C.; Li, C.; Ye, M. Fast and Facile Preparation of Graphene Oxide and Reduced Graphene Oxide Nanoplatelets. *Chem. Mater.* **2009,** *21,* 3514–3520.

17. Kuila, T.; Bose, S.; Mishra, A. K.; Khanra, P.; Kim, N. H.; Lee, J. H. Chemical Functionalization of Graphene and its Applications. *Prog. Mater. Sci.* **2012,** *57,* 1061–1105.

18. Hummers, W. S.; Offeman, R. E. Preparation of Graphitic Oxide. *J. Am. Chem. Soc.* **1958,** *80,* 1339–1339.

CHAPTER 14

GUM ARABIC: A REMARKABLE BIOPOLYMER FOR FOOD AND BIOMEDICAL APPLICATIONS

ADRIANA P. GEROLA[1], ADLEY F. RUBIRA[2], EDVANI C. MUNIZ[2,3], AND ARTUR J. M. VALENTE[4,*]

[1]Chemistry Department, Federal University of Santa Catarina, Florianópolis, Santa Catarina 88040-900 Brazil

[2]Chemistry Department, Maringá State University, Maringá, Paraná 87020-900 Brazil

[3]Postgraduate Program on Materials Science and Engineering, Federal University of Technology, Paraná (UTFPR-LD), Londrina Paraná 86036-370 Brazil

[4]CQC, Department of Chemistry, University of Coimbra, 3004-535 Coimbra, Portugal[*]

[*]Corresponding author. E-mail: avalente@ci.uc.pt

CONTENTS

ABSTRACT

Gum Arabic (GA or Acacia Gum) is a polysaccharide obtained from acacia trees. Although some of its properties and applications have been known for centuries, due to its remarkable multifunctional properties, GA still has a high and increasing demand, in particular for food additives, pharmaceutical, cosmetic, printing, and textile industries. In this chapter, we will focus on the work carried out involving GA in the last 4 years, with a particular emphasis for applications related to biomedical/pharmaceutics and food industries.

14.1   INTRODUCTION

Gum Arabic (GA or Acacia Gum) is the exudate from the *Acacia senegal* and *Acacia seyal* trees (Fig. 14.1). GA is a branched-chain polysaccharide with the main chain of $(1\rightarrow3)$-β-d-galactopyranosyl units and side chains containing *L*-arabinofuranosyl, *L*-rhamnopyranosyl, *D*-galactopyranosyl, and *D*-glucopyranosyl uronic acid units (Fig. 14.2).[1,2] These compounds constitute around 90% of the total gum, which corresponds to the so-called arabinogalactan fraction. A smaller fraction is a higher molecular weight arabinogalactan protein, while the smallest fraction, denominated as a glycoprotein, possesses the highest protein content.[3,4] The composition of GA is, however, dependent on its origin, that is, it is an exudate of *Acacia senegal* or *Acacia seyal*.[5,6] A comprehensive study on the composition of GA was recently published.[7]

  GA is biocompatible, nontoxic, water soluble, shows pH stability and gelling characteristics, and is a relatively inexpensive polysaccharide, being the most widely used natural gum in a wide range of industries, including food, cosmetic, and pharmaceuticals.[8] GA is, since 1973, in the list of compounds "Generally Recognized as Safe" (GRAS) of the United States Food and Drug Administration and is classified as an "emulsifier, stabilizer, thickener, and gelling agent" additive, with the number E414, by the Food Standard Agency of the European Union. However, the use of GA as a food additive has some restrictions.[9]

FIGURE 14.1 *Acacia* tree.

Besides the use of GA in biomedical, pharmaceutical, and food indus-tries, which will be discussed in detail in the following sections, GA has been used with success in many other applications. An interesting applica-tion of GA is as a chemical admixture in concrete. The application of small amounts (ca. 0.1–0.6%) of GA to cement may lead to two different effects: has a significant influence on the cement dispersion in the concrete and acts as water-reducing agent.[10,11]

Sun et al.[12] studied the use of GA to control the size of Ru–Zn nanopar-ticle (NP) catalyst for the selective hydrogenation of benzene to cyclo-hexene. They found that the hydrogenation yield increases from *ca.* 40[13] to 59.6%[12] by using Ru–Zn NPs obtained in the presence of GA, without compromising the reusability of the catalyst.

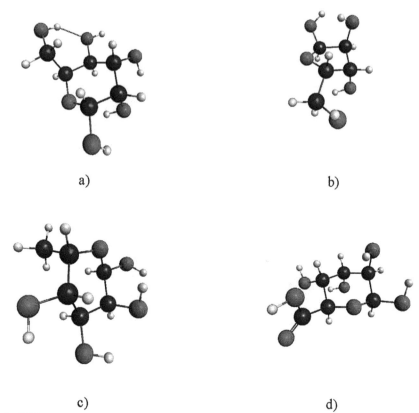

a)   b)

c)   d)

FIGURE 14.2 Molecular structures of (a) D-galactopyranose, (b) L-arabinofuranose, (c) L-rhamnopyranose, and (d) D-glucuronic acid.

The performance of batteries and electroactive compounds has also been improved by using GA. Despite the attractiveness of silicon (Si) to be used as a anode material in lithium ion batteries, due to its high specific capacity (1 order of magnitude higher than graphite),[14] its application is quite limited by the mechanical and physical drawbacks arisen during the lithiation/delithiation processes.[15] The use of GA has shown to be highly efficient to overcome those problems by acting as a binder to immobilize Si electrode materials and thus allowing, for example, volume change without cracking and, simultaneous, electrical properties and electrode performance are not affected.[16]

GA has been successfully used to improve the mechanical properties of polyaniline (PANi), a conjugated polymer with interesting properties which allows it to be used in a wide spectrum of applications, including biomedical uses (such as artificial muscles),[17] electronics,[18] electro-chromic devices,[19] and the development of chemical sensors and biosen-sors.[20,21] However, this polymer has very poor mechanical properties, limiting its use in the pure form. A methodology to solve this problem is to produce blends with neutral or ionic polymers, such as GA, which have good mechanical and processing properties.[22] GA has been used also for preventing the occurrence of cross-linking between PANi chains, leading to a higher doped PANi structure; this allows to obtain highly stable elec-troactive PANi-based nanoblends.[23,24]

The applicability of GA as corrosion inhibitor has also been tested. Umoren et al.[25] have studied the corrosion behavior of mild steel and aluminum, when in contact with sulfuric acid solutions (0.1 M) at different temperatures. They have shown that by adding GA (0.5 g/l) to the sulfuric acid solution, the corrosion rates of mild steel and Al decrease, leading to maximum inhibition efficiencies of 38% ((at 60°C) and 77% (30°C), respectively. More recently, the application of GA as a corrosion inhibitor of pipeline steel in HCl (1 M) has been tested and the maximum inhibition efficiency was found to be 92%.[26]

The topic of wastewater treatment is also an important area where GA has been tested. Environmental concerns lead to the replacement of many well-established adsorbents by gum-based biopolymers.[27,28] Aero-gels and hydrogels are common adsorbents; the former are characterized by high surface area while the latter are cross-linked polymeric struc-tures capable of absorbing and retaining large amounts of water.[29,30] Wang et al.[31] reported that aerogels composed by GA and sodium montmoril-lonite can be successfully synthesized by using a freeze-drying process (in contrast with the supercritical drying process commonly used to obtain, for example, silica-based aerogels) for further application as adsorbent. On the other hand, Ibrahim et al.[32] showed that GA and carboxymethyl cellulose blend hydrogels can be used for dye sorption. The use of GA for the efficient removal (in general, higher than 70%) of metal ions Ag(I) and Zn(II) has also been described by Das et al.[33] They showed that composites constituted by sulfonated GA and mushroom (*Pleurotus platypus*) powder. The obtained composite is able to sorb up to *ca.* 288 and 138 mg/g of Zn(II) and Ag(I), respectively, and allows a regeneration up to 4 cycles.

14.2 BIOMEDICAL AND PHARMACEUTICAL APPLICATIONS

GA has been extensively used for oral and topical pharmaceutical applications due to its suspending and emulsifying properties.[34] GA is a polysaccharide indigestible to humans and animals, and it has been contributed for the reduction of obesity[35] and has shown potential for the attenuation of the development of nephropathy in type 1 diabetes. In fact, studies carried out in rats showed that the addition of GA to the rats' diet leads to a significant decrease in the serum triglyceride, total cholesterol, and high-density lipoprotein (HDL) cholesterol, when compared with the control group.[36] This is in agreement with previous studies, where the effect of GA on hyperlipidemia in Sudanese patients was tested.[37] In this study, no significant difference in the HDL, between the study (patients received atorvastatin plus GA) and control (patients only received atorvastatin) groups, was found; however, in the study group, the level of total cholesterol and low-density lipoprotein (LDL) has been reduced; once the LDL is a major risk factor for a developing ischemic heart disease, it is suggested that GA consumption may have a beneficial effect on the prevention of that disease.[37]

GA shows an amazing wide range of applications, which includes its use in type IV gypsum die materials to improve the abrasion resistance,[38] the use of GA as a medium for the production of chitosan–gelatin nanofibers with improved strength and suitable for the tissue engineering applications,[39] and be used as a probe for control and monitoring of the kidney stone formation.[40]

In the last decade, GA, an anionic polysaccharide, has been widely used in coacervation.[41–43] Complex coacervation is the spontaneous liquid/liquid phase separation in colloidal systems driven by the electrostatic interaction between two oppositely charged polyelectrolytes.[44,45] The coacervates, which correspond to the concentrated phase, can be used for different applications, including micro- and nanoencapsulation, with applications in pharmaceutical industry.[46,47]

Chitosan has been one of the most popular biopolymers for the development of pH-sensitive delivery systems due to its biocompatibility, biodegradability, low toxicity, abundant availability, and acceptable production cost. For those reasons, coacervates between chitosan and GA have been studied in detail.[48–50,51] By blending chitosan with GA, the resulting coacervate, at optimum pH and weight ratio, exhibited relatively low water uptake, erosion, and water permeability than chitosan membranes; furthermore, chitosan–GA films demonstrated good mucoadhesive properties, which makes these films good candidates for drug delivery.[52] Also in this topic, Tsai et al.[53] have studied the effect of GA addition to chitosan/pectin complex solutions; they found that the solution viscosity significantly decreases in the presence of GA, as a consequence of intermolecular entanglements. The obtained complex was tested as a matrix for the release of insulin to the physiological pH.

Huang et al.[45,54,55] have investigated the coacervation between GA and o-carboxymethyl chitosan (o-CMC. In order to stabilize the coacervate, the cross-linking process was promoted by using genipin as cross-linker.[56] More recently, they have reported the synthesis of a genipin-cross-linked o-CMC–GA coacervate. In that study, authors showed that coacervates obtained at different acidic pH values were prepared and, consequently, cross-linked by using genipin. The cross-linking intends to improve the stability and retard the release kinetics of drugs and thus to increase the applicability of these gels. The obtained gels were then loaded with bovine serum albumin (BSA) and the release for the three different simulated media (gastric, intestine, and colon), corresponding to the three different pH values (1.2, 6.8, and 7.4, respectively), was analyzed. The cross-linked o-CMC–GA gel showed a high resistance to the simulated gastric fluid (SGF; with a cumulative BSA release equal to 17%) followed by a significant release (ca. 80%) at pH mimetizing the colon fluid. These results are quite promising for an intestine-/colon-targeted delivery system. It is worth noticing that, in general, chitosan–GA coacervates are unstable to certain simulated gastrointestinal fluids, which limits its application as an oral delivery matrix.[57]

GA and GA-based hydrogels are versatile for a wide range of applications such as for removal of the metal ions and dyes from aqueous solutions,[58–60] graphite exfoliation,[1] wound dressing,[61] drug delivery,[34,62–64] and surface modification of NPs.[65–68] In the following paragraphs, these two future applications will be further discussed.

GA has also been successfully used to develop hydrogels showing a responsive behavior toward different types of exterior stimuli.[65,69,70]

Gerola et al.[71,72] have synthesized pH-responsive hydrogels based on modified Gum Arabic (M-GA) with different degrees of methacrylation for a colon-specific drug delivery of curcumin (CUR).

CUR {(1E,6E)-1,7-bis(4-hydroxy-3-methoxyphenyl)-1,6-heptadiene-3,5-dione} is a yellow polyphenolic compound, with a very low solubility in an aqueous media, occurring in dried rhizomes of turmeric (*Curcuma longa*).[73] CUR has been used as a natural food coloring agent[74] and it is associated with the numerous pharmacological applications,[75] including healing, antimicrobial, antiviral, anti-inflammatory, antioxidant, antiproliferative,[76] hypocholesterolemic, and anticarcinogenic properties.[77] In addition, CUR has shown significant activity in the treatment of chronic diseases such as type II diabetes, rheumatoid arthritis, multiple sclerosis, and Alzheimer's disease.[78]

The low solubility of CUR in simulated gastric and intestinal fluids (SGF—pH 1.2 and SIF—pH 6.8, respectively) is a limiting factor for the bioavailability of CUR for in vitro and in vivo release tests. Therefore, the authors used the inclusion complexation with cyclodextrins (CD) as an alternative strategy to overcome this problem. CD is a cyclic oligosaccharide in the shape of truncated cones made up of units of α-glucopyranose linked by α-1,4 glucoside bonds. The main feature of CD is the cavity that enables them to form inclusion complexes with guest molecules of appropriate size, shape, and polarity. Host–guest-like complexes may improve the properties of the guest molecule, such as solubility enhancement, bioavailability, and stability improvement.[79]

The formation of supramolecular host–guest complexes CUR/α-CD and CUR/β-CD, in different molar ratios, r, provides a considerable increase in the solubility of CUR in biological fluids. The higher solubility efficiency (SE = solubility of CUR complexed with CD/solubility of CUR in pure water) was observed for the CUR/α-CD complex with $r = 0.25$, corresponding to an increase in the solubility of the CUR of 172 times.[71] The formation of these complexes was confirmed by ^{1}H NMR. The stoichiometry of the α-CD and β-CD were 1:1 and 1:2, respectively, while the binding constants were 344 mol^{-1} l for α-CD and 7.2×10^7 mol^{-2} l^2 for β-CD. The β-CD complexes showed the highest binding constants; however, the greater α-CD solubility guarantees the highest amount of CUR solubilized in aqueous environments. Since solubilization of CUR in the aqueous media is a prerequisite for release studies in biological fluids, the CUR/α-CD complex at $r = 0.25$ was chosen for the preparation of the hydrogels.

For the preparation of composite hydrogels, GA was initially modified with glycidyl methacrylate (GMA) in the aqueous acid medium and different percentages of GMA (x), with $x = 1.0, 1.5, 2.0, 3.0, 4.0$, and 6.0. The relative substitution degree (DS_R) of the carboxyl groups showed a linear increase with the GMA percentage up to 2% and for %GMA ≥ 3%; the DS_R values remains approximately constant.[72]

Modified GA with different methacrylation degree (M-GA$_x$) was used for the synthesis of a GA-based pH-responsive hydrogel (HM-GA$_x$). The effect of the DS_R on the swelling index (SI) of the HM-GA$_x$ was evaluated in SIF and SGF at 37°C. The results showed that both pH and DS affect the SI of the hydrogels. The highest SI is found for a lower percentage of GMA in SIF.

The composite hydrogels of M-GA$_x$ containing CUR/α-CD complex (r=0.25) were obtained with the well-defined structure for x>2. CUR release kinetics, such as host–guest complexes, from HM-GA$_x$ were responsive to pH and DS_R, and the amount of released CUR increases by decreasing the DS_R and by increasing the pH. In general, the release kinetics showed similar profiles, characterized by one-step release. However, for the release of CUR from HM-GA$_{2.0}$ in SIF, we verified the well-defined two-step mechanism (Fig. 14.3). The pH effect on the release profiles is associated with deprotonation of the carboxyl groups of the polysaccharide chain in SIF, which leads to swelling of the hydrogel and erosion. On the other hand, the higher DS_R (HM-GA$_{6.0}$) seems to be more effective in avoiding hydrogel erosion/degradation in SIF (Fig. 14.3).

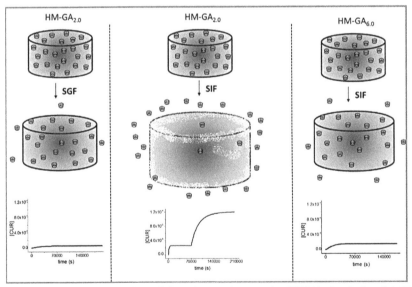

FIGURE 14.3   Representation of the release of curcumin (CUR)/α-(cyclodextrin) CD complex from HM-GA$_x$.

The effects of DS_R and pH on the surface morphology of HM-GA in the absence and presence of CUR complexes were evaluated. A porous structure was observed for the swollen hydrogels in SGF, and a decrease in a pore size was observed with increasing methacrylation degree. The hydrogels swollen in SIF showed surface erosion/degradation, except for HM-GA$_{6.0}$. Furthermore, the incorporation of CUR complexes within HM-GA$_{2.0}$ led to a significant increase of pore size, justifying the different mechanism and the release rate compared to the gels with a higher DS_R.

Additionally, hydrogels prepared from M-GA$_x$ showed non-cytotoxicity against Caco-2 cells. Together, with the results discussed above, this fact shows that M-GA$_x$ can be used for a colon-specific drug delivery, with the rate constant and the cumulative drug release controlled by the cross-linking degree. It is interesting to note that, more recently, Sarika et al.[80] demonstrated that the conjugation of GA and CUR leads to an increase in the CRU solubility and stability, in the physiological media, by forming spherical nanomicelles; the anticancer activity of the conjugate micelles against the human hepatocellular carcinoma (HepG2) and human breast carcinoma (MCF-7) cells have been tested. The best results were obtained for HepG2 which was explained by the synergetic effect of GA. Following the same objectives, Sarika and Nirmala[81] developed nanogels formed by GA aldehyde cross-linked with gelatin. The obtained nanogels were loaded with CUR and in vitro cytotoxicity and intracellular uptake studies toward MCF-7 cells were carried out.

For the same objective, that is, the delivery of drugs into simulated intestinal fluid, Reis et al. developed a GA hydrogel cross-linked with N,N'-dimethylacrylamide and methacrylic acid. These compounds were used due to their excellent ability to undergo hydrogelation at conditions that do not affect the structure of the polysaccharide. The matrix shows a pH-responsive structural change, being more expressive to basic environmental as measured through the release kinetic profiles of the potassium diclofenac.[82] It should be noticed that the use of N,N'-dimethylacrylamide as cross-linker was previously tested by Bossoni et al.[63] for the synthesis of GA hydrogels for the encapsulation and release of Vitamin B$_{12}$ and potassium diclofenac.

However, GA has also been used as a cross-linker. Sarika et al. used GA aldehyde, obtained through the oxidation of GA by using sodium metaperiodate, to cross-link gelatin—a product obtained by a partial denaturation of collagen. The gelatin-based hydrogel has been tested as scaffold;

the results against L-929 and HepG2 cells showed that scaffold seems to be suitable for spheroid cell culture.[8]

Au nanoparticles (AuNPs) have been extensively used in the biomedical applications, including drug delivery, cancer treatment, and biomedical imaging and sensing, due to their excellent compatibility with the human organism, low toxicity, and tunable stability[83] However, bare AuNPs are unstable in physiological media, forming large aggregates and consequently the clearance rate from the body increases.[84] GA has been used as a biocompatible stabilizing agent for the synthesis of the hybrid NPs.[85] Owing to its well-steric stabilization effect and also the presence of the functional groups (e.g., carboxylate and amine), it can be sorbed on the surface of colloids and be used as emulsifiers or capping agents.[85] Hybrid GA–AuNPs are much more stable at different physiological pHs, than native AuNPs, contributing to increase their potential for the biomedical applications.[86] Barros et al.[86] showed that GA–AuNPs promote selective changes in the viability, proliferation, and morphology of melanoma cells. Moreover, the higher stability of GA–AuNPs allows the use of these particles as an X-ray contrast agent, showing a more than threefold X-ray attenuation coefficient, when compared with the common contrast agent clinically available for the X-ray computed tomography.[84] The coating of AuNPs with GA is also opening new horizons for tumor-target drug delivery, once it will allow electrostatic interactions between hybrid AuNPs and cationic drugs. An example is the use of epirubicin, a cationic anticancer drug, for the treatment of lung cancer. The adsorption of epirubicin to AuNPs does not allow an efficient delivery of the drug to the tumor site; however, after coating NPs with GA, the uptake of drug by cancer cell increases.[87] A similar approach has been applied to radioactive gold NPs (198-AuNPs). Several authors found that GA-coated radioactive AuNPs show higher bioavailability in tumor (e.g., liver and prostate) and, consequently, turning the radiotherapy more effective.[88–90]

AgNPs have been employed in as antimicrobial agents, wound dressings, and coatings of medical devices.[91] Similarly to AuNPs, GA capped-silver NPs have attracted the interest of researchers interested in improving the efficiency of AgNPs for biomedical applications, including their role as antiglycating agent[92] or as a surgical implants surface coating material, with antibacterial properties against multidrug-resistant bacteria.[93]

GA has also been used as a coating agent of magnetic nanoparticles (MNPs) and, in particular, iron oxide NPs. Fe_2O_3 NPs are also useful for biomedical applications such as, magnetic resonance imaging (MRI) and magnetic particle imaging.[94] For MRI, GA can be adsorbed to MNPs surface physically or covalently[95]; the latter, has the advantage of producing GA layers with higher stability. Among other MNPs applications, the use of GA-coated Fe_2O_3 NPs for purification of human antibodies deserves attention. Alves et al. have demonstrated the hybrid GA–MNPs work as cation exchange adsorbents and its feasibility for antibody purification of different cell lines.[96]

14.3 FOOD APPLICATIONS

GA is widely used in the food industry due to its emulsifying, thickener, binding, and self-enhancing properties. In this section, we will describe some of the recent advances and applications of GA as a food additive.

In the previous section, the coacervation principles were summarized and here we only highlight the applications of coacervates for food applications. Proteins and polysaccharides are the most frequently used hydrocolloids in the food industry, and their interaction can provide products such as complexes coacervates, which can be used as ingredients (additives) and biomaterials or in microencapsulation systems.[97,98]

The encapsulation of food ingredients has been widely used due to its efficiency against deterioration improvement, for example, the storage, volatile losses, or interaction with other ingredients; besides, it also enables the production of functional ingredients in the form of powdered products with new properties.[99,100] Wang et al. have used coacervates of GA and BSA for encapsulation and protection of CUR. Besides potential application of anticancer drug, CUR is considered a functional ingredient, which provides a wide range of benefits, including antioxidant activity; however, as described before, CUR is poorly soluble in water, it is susceptible to light and is unstable and undergoes rapid hydrolytic degradation in neutral or basic conditions.[101] GA coacervate was obtained at the electrical equivalence pH of 3.9, and a CUR encapsulation efficiency of 92% was assured.[102]

This strategy has been followed by different researchers for encapsulation of different food additives.[103] For example, sulforaphane, an isothiocyanate derivative obtained from, for example, broccoli seed extracts, with interest for nutraceutical and pharmaceutical industries due to its anticancer and chemopreventive effects[104,105] has been encapsulated in gelatin–GA coacervates to protect the extract from modification due to the changes in pH and temperature.[106] Protein isolate–GA complex coacervates have been used for the encapsulation of probiotic bacteria[107] and fatty acids[107,108] and thus their viability in food during manufacture, storage, and delivery to the targeted site of gastrointestinal tract increases.[109]

Another approach for encapsulation of food ingredients (e.g., flavors, essential oils (EOs), and colorant) by using GA is through the spray-drying technique. Spray-drying consists of the conversion of fluid foods into powders, resulting in easy storage and transportation of active materials. The process is mainly based on the addition of active compound to a polysaccharide-containing aqueous solution, in which the active material is dissolved, dispersed, or emulsified.[110]

Honey is a well-known natural food with health benefits[111,112]; however, the formation of crystals of glucose is the major limitation for the food processing.[113] Such drawback can be overcome by converting honey from a liquid into a powder, resulting in its stability and ease of handling. That can be achieved by the addition of carrier agents such as GA and malto-dextrin, which prevents problems related with honey hygroscopicity, thus obtaining a final stable and processable honey product.[114]

Cano-Higuita et al.[110,115] have studied the effect of ternary blends, involving GA, maltodextrin, and modified starch (MS) as wall materials, in the microencapsulation of CUR by spray-drying. The obtained CUR-containing microcapsules are more soluble in water (an increase of 85% was found) and more resistant to light upon storage than the native CUR.

Carotenoids are an important family of compounds due to its provitamin A activity and antioxidant properties. β-carotene is, probably, the better known member of this family and it can be obtained in many fruits and vegetables. As an antioxidant, β-carotene protects against free-radical damage[116]; however, this action is frequently limited, once it easily suffers degradation due to the different kind of factors, including humidity, temperature, light, oxygen, and interactions with other food ingredients. Therefore, microencapsulation by freeze-drying, an alternative method to the spray-drying,[117] and using GA as a wall material has been tested. The effect of relative humidity in the encapsulated β-carotene showed that, in a 70 days period, the degradation follows a first-order kinetics with higher ability for food applications, for example, in pastry, than the native β-carotene.[118]

In order to improve the water solubility and bioavailability, β-carotene can be dissolved in the oil phase of oil in water (O/W) emulsions, and later incorporated into food products. However, β-carotene degradation is also dependent on emulsion properties, including, for example, emulsifiers and food products. GA is an extremely important and commonly employed emulsifying agent which maintains its function independently of external conditions (low pH, high ionic strength, etc.).[119–121] Among all different components of GA, it is known that the arabinogalactan protein (AGP) complex is the main component responsible for GA ability to stabilize emulsions.

The use of emulsions in beverage differs from other food emulsions because they are consumed in highly diluted form rather than in their original concentrated ones. GA-containing emulsions showed to be effective for increasing the stability of β-carotene and thus to contribute for increasing the shelf life of food products.[122]

GA has shown a huge versatility as encapsulating agent of EOs. EOs are volatile aromatic liquids derived from plants or plant parts, including leaves, flowers, fruits, seeds, and so forth.[123,124] EOs are important for their antioxidant, antibacterial, antifungal, and insecticidal properties and have been widely used as functional ingredients in food, pharmaceutical, and cosmetic products.[123] However, the industrial processing of EOs is limited by their high volatility, instability, and poor water solubility. Thus, most of these compounds are encapsulated in matrices to increase their protection, control the release of active compounds, reduce vaporization, and promote the handling.[125] For example, jasmine EO has been entrapped in GA–gelatin coacervates[98]; the obtained nanocapsules show, in terms of structural features and volatile flavor compounds, a good heat resistance when submitted at 80°C, which can be used as a deliver vehicle flavors and fragrances.

The mixture of maltodextrin with GA has been reported to be effective in the microencapsulation of cardamom oil using a spray drier.[126,127] Following the same procedure, Alves et al.[128] studied the encapsulation of EO obtained from fruits of *Pterodon emarginatus* (a plant widely distributed in the state of Minas Gerais—Brazil). Different emulsion formulations were tested. For the best formulation, encapsulation efficiencies of ca. 98% with high storage stability (retention of more than 70% of some EO's components for 45 days) were obtained. GA has also been used in combination with MS and maltodextrin for the encapsulation of fennel oleoresin by freeze-drying[129] The best results in terms of encapsulation, redispersibility, and storage stability of the final encapsulated product were found for GA–MS mixture.

The development of low-cost nanoemulsions containing GA for incorporation of EOs was studied by Luo et al.[130] The formation of nanoemulsions, with droplets smaller than 200 nm in diameter, can prevent the creaming and precipitation of EOs droplets and potentially enhance their antimicrobial activity. By using clove bud oil, they developed a novel self-emulsification procedure that showed stability during storage and with potential for applications as antimicrobial food preservatives or sanitizers in postharvest washing solutions for fresh produce production.

Ribeiro et al.[131] reported the incorporation of lemongrass (Cymbopogon citratus) EOs in GA, previously modified with sodium trimetaphosphate (a cross-linking agent). The modification of GA allowed an increase of encapsulation efficiency from 85% (obtained for the unmodified GA) to 97%, justified by the better viscosity and swelling properties of the modified GA.

Anthocyanins are one of the most commonly used food additive due to their brilliant red color and health benefits, which includes antioxidant, anti-inflammatory, and antidiabetic effects.[132–134] However, in the same way as carotenoids, this type of flavonoids suffer degradation upon storage, leading to color fading and bioactivity loss, which decreases its application for food and beverage products.[135] It is known that the thermal stability of anthocyanins is significantly improved by GA.[136] The effect of the presence of GA on the stability of anthocyanin in a beverage model (aqueous citric acid solution) in the presence of ascorbic acid was tested by Chung et al.[137] They found that the anthocyanin stability increases by increasing the amount of GA (from 0.05 to 1.5%). This has been justified by interactions occurring between anthocyanin and glycoprotein fractions of the GA through hydrogen bonding.

The microencapsulation of anthocyanins (obtained from barberry and saffron petal's extracts) by using GA as wall material, through the spray-drying and freeze-dried processes, has been proved to be efficient in the stabilization of anthocyanins, and allowing its use, in the powder form, as food additive.[138–140]

GA can also be used as stabilizing of preservatives used in food and beverage products. ε-polylysine (PL) has a high antimicrobial activity, nontoxicity, and is water soluble; these properties makes the ε-PL a potential food additive for preventing spoilage caused by microbial growth.[141] However, the cationic nature of this polymer may cause clouding or sedimentation in food products, as a consequence of electrostatic interactions with anionic components existing in the food. GA can be used to interact with PL to form electrostatic complexes. Studies on apple juice beverage demonstrated that the obtained PL–GA complexes improve the stability of PL without compromising its antimicrobial effect.[142]

Fish oil is the richest dietary source of polyunsaturated fatty acids, such as eicosapentaenoic acid (EPA) and docosahexaenoic acid (DHA), also known as Omega-3 fatty acids, providing potential health benefits when consumed regularly at appropriate concentrations.[143] The poor water solubility and the high susceptibility to oxidation of these functional lipids are significant drawbacks for their use in food formulations.[144,145] The encapsulation of fish oil by using polysaccharides is a viable approach for increasing the bioavailability of Omega-3 in food and beverages. Ilyasoglu and El[146] have demonstrated that by using sodium caseinate-GA complexes provide an appropriate matrix for encapsulation (with efficiencies up to ca. 80%) and deliver fatty acids in fruit juices. Another issue, normally associated with fatty acids, is the strong odor that may prevent the consumption. Li et al.[147] developed an emulsion composed by GA, casein, and β-CD for encapsulation of fish oil by using spray-drying. Other dietary supplements, such as vitamin C,[148] vitamin E,[149] and iodine,[150] have been processed with GA for the production of nutritional bioactive foods. Iron is an essential trace element necessary for the human well-being and its deficiency is a major cause of anemia.[151] The addition of iron as food additive is therefore a logical strategy to prevent diseases.[152] Addition of iron to milk and milk-related products is an effective nutritional strategy to correct dietary iron deficiency. The microencapsulation of iron for further use as additive will prevent the iron from oxidation and the perception of bitter flavor caused by iron salt, and increase the bioavailability.[153] Gupta et al.[154] have used a modified solvent evaporation for encapsulation iron in a GA-based blend. They showed that the encapsulation efficiency is higher than 90% and the resulting microcapsules are stable. The addition of those iron-containing microcapsules to milk resulted in an improvement of bioavailability and flavor when compared with milk fortified with just iron salt.

The aroma in food and beverages is a key point for the market success; with the time, many molecules responsible for flavoring suffer degradation leading to a decrease in the odor level or even to a bad smell. The emulsifying properties of GA are important for aroma compounds retention.[155] For example, volatile sulfur compounds are used for contributing the aromas in wine and coffee. For example, 2-furfurylthiol (2FFT), dimethyl disulfide (DMDS), and thiophene (THIO) are important sulfur contributors to roasted coffee aromas. However, they are volatile and suffer oxidation in contact with air and light which gives negativity to the consumer acceptance of coffee products. The microencapsulation of those compounds in GA, by using the spray-drying method, prevents the oxidation/degradation of sulfur compounds and thus preventing the loss of product.[156]

The consumption of fruits and vegetables is essential for the human diet once they provide vitamins, minerals, and fibers bringing benefits for the human health and well-being. However, these fresh commodities are perishable which limits their storage and shelf-life lifetimes. The use of edible films is one of the strategies to control postharvest decay and improve visual and tactile features of fruits and vegetables.[157,158] GA, the least viscous and most soluble among the hydrocolloids,[159] has been used as main component of edible coatings. Al-Juhaimi has used GA solution to produce edible coating for tomato. They showed that GA (at 5–20% solutions) contributes for the preservation of organoleptic qualities of tomato during the storage period (16 days).[160] Papaya is a fruit highly susceptible to various postharvest diseases, in particular anthracnose (caused by fungi in the genus Colletotrichum) which can cause ca. 50% losses of the production.[161] The use of GA as a matrix for the encapsulation of ginger oil (2%) and ginger extract (1.5%), and the use of composite film as an edible coating showed to be an effective biofungicide for papaya (for 10 days), by keeping the quality of papaya in terms of firmness and peel color.[162] Mango is a high demanding fruit due to its delicious taste and nutritional properties. However, similar to papaya, sensitivity to chilling injury, short shelf life, and postharvest diseases limit the market and, consequently, decrease the potential trade profit. In order to increase the shelf time, by keeping all the nutritional and physical properties of the fruit, edible films are constituted by mixing GA to calcium chloride.[163] The application of the obtained edible film to mango will allow to preserve mango fruit quality during low temperature (6°C) storage during ca. 1 month.[164,165] The beneficial effects of GA on postharvest quality and shelf life have been used for other perishable products such as cherries[166] and green bell peppers.[167] In both cases, the application of GA coating allows to extend the shelf life up to 15 days, at 2°C, and 21 days, respectively, with a significant degradation delay as measured using weight, color, and firmness. The application of AgNPs with GA, in the latter case, also permits to improve the microbial aging.

The application of GA in blends for food packaging can also be found in the literature. Poly(lactic acid) (PLA) is a very attractive biopolymer for food packaging application because it shows good mechanical, thermal, and barrier properties; however, PLA is a very brittle material with less than 10% elongation at break.[168] This is overcome by using different blends or composites. Tripathi and Katiyar[169] have reported synthesis of lactic acid-grafted GA (LA-g-GA) by polycondensation and consequent preparation of LA-g-GA-containing PLA composite films. These films show a significant decrease in the oxygen and water vapor transmission rates, when compared with PLA, suggesting the potential of GA for the production of packaging materials.

14.4 ACKNOWLEDGMENT

This work was supported by the *Fundação para a Ciência e a Tecnologia* (FCT), Portuguese agency for scientific research, through the programs UID/QUI/UI0313/2013, UID/EQU/00102/2013, and COMPETE, and by the Conselho Nacional de Desenvolvimento Científico e Tecnológico (CNPq—Brazil) through the project 00336/2014-6.

KEYWORDS

- **gum arabic**
- **biomedical**
- **pharmaceutics**
- **food**

REFERENCES

1. Gils, P. S.; Ray, D.; Sahoo, P. K. Designing of Silver Nanoparticles in Gum Arabic Based Semi-IPN Hydrogel. *Int. J. Biol. Macromol.* **2010,** *46,* 237–244. DOI: 10.1016/j.ijbiomac.2009.12.014.
2. Patel, S.; Goyal, A. Applications of Natural Polymer Gum Arabic: A Review. *Int. J. Food Prop.* **2015,** *18,* 986–998. DOI: 10.1080/10942912.2013.809541.
3. Dhenadhayalan, N.; Mythily, R.; Kumaran, R. Fluorescence Spectral Studies of Gum Arabic: Multi-emission of Gum Arabic in Aqueous Solution. *J. Lumin.* **2014,** *155,* 322–329. DOI: 10.1016/j.jlumin.2014.06.022.
4. Connolly, S.; Fenyo, J.-C.; Vandevelde, M.-C. Effect of a Proteinase on the Macromolecular Distribution of Acacia Senegal Gum. *Carbohydr. Polym.* **1988,** *8,* 23–32. DOI: 10.1016/0144-8617(88)90033-1.
5. Randall, R. C.; Phillips, G. O.; Williams, P. A. The Role of the Proteinaceous Component on the Emulsifying Properties of Gum Arabic. *Food Hydrocoll.* **1988,** *2,* 131–140. DOI: 10.1016/S0268-005X(88)80011-0.
6. Osman, M. E.; Menzies, A. R.; Williams, P. A.; Phillips, G. O.; Baldwin, T. C. The Molecular Characterisation of the Polysaccharide Gum from Acacia Senegal. *Carbohydr. Res.* **1993,** *246,* 303–318. DOI: 10.1016/0008-6215(93)84042-5.
7. Gashua, I. B.; Williams, P. A.; Baldwin, T. C. Molecular Characteristics, Association and Interfacial Properties of Gum Arabic Harvested from Both Acacia Senegal and Acacia Seyal. *Food Hydrocoll.* **2016,** *61,* 514–522. DOI: 10.1016/j.foodhyd.2016.06.005.
8. Sarika, P. R.; Cinthya, K.; Jayakrishnan, A.; Anilkumar, P. R.; James, N. R. Modified Gum Arabic Cross-Linked Gelatin Scaffold for Biomedical Applications. *Mater. Sci. Eng. C.* **2014,** *43,* 272–279. DOI: 10.1016/j.msec.2014.06.042.
9. Food Additives, (n.d.). https://webgate.ec.europa.eu/foods_system/main/index.cfm?event=substance.view&identifier=00159 (accessed March 2, 2017).
10. 10. Zhao, C.; Zhao, Q.; Zhang, Y.; Zhou, M. The Effect of Gum Arabic on the Dispersion of Cement Pastes. **2015,** 483–494. DOI: 10.1007/978-3-319-13948-7_47.
11. 11. Mbugua, R.; Salim, R.; Ndambuki, J. Effect of Gum Arabic karroo as a Water-Reducing Admixture in Concrete. *Materials (Basel)* **2016,** *9,* E80. DOI: 10.3390/ma9020080.
12. 12. Sun, H.-J.; Chen, J.-J.; Huang, Z.-X.; Liu, Z.-Y.; Liu, S.-C. Selective Hydrogenation of Benzene to Cyclohexene over the Nano-sized Ru-Zn Catalyst Modified by Arabic Gum. *Chinese J. Inorg. Chem.* **2016,** *32,* 202–210.
13. 13. Liu, Z.; Sun, H.; Wang, D.; Guo, W.; Zhou, X.; Liu, S.; et al. Selective Hydrogenation of Benzene to Cyclohexene over Ru-Zn Catalyst with Nano-sized Zirconia as Dispersant. *Chinese J. Catal.* **2010,** *31,* 150–152. DOI: 10.1016/S1872-2067(09)60039-5.
14. Etacheri, V.; Marom, R.; Elazari, R.; Salitra, G.; Aurbach, D. Challenges in the Development of Advanced Li-Ion Batteries: A Review. *Energy Environ. Sci.* **2011,** *4,* 3243–3262. DOI: 10.1039/c1ee01598b.

15. Fu, K.; Xue, L.; Yildiz, O.; Li, S.; Lee, H.; Li, Y.; et al. Effect of CVD Carbon Coatings on Si@CNF Composite as Anode for Lithium-Ion Batteries. *Nano Energy.* **2013,** *2,* 976–986. DOI: 10.1016/j.nanoen.2013.03.019.

16. Ling, M.; Xu, Y.; Zhao, H.; Gu, X.; Qiu, J.; Li, S.; et al. Dual-Functional Gum Arabic Binder for Silicon Anodes in Lithium Ion Batteries. *Nano Energy* **2015,** *12,* 178–185. DOI: 10.1016/j.nanoen.2014.12.011.

17. Guimard, N. K.; Gomez, N.; Schmidt, C. E. Conducting Polymers in Biomedical Engineering. *Prog. Polym. Sci.* **2007,** *32,* 876–921. DOI: 10.1016/j. progpolymsci.2007.05.012.

18. Morrin, A.; Ngamna, O.; O'Malley, E.; Kent, N.; Moulton, S. E.; Wallace, G. G.; et al. The Fabrication and Characterization of Inkjet-Printed Polyaniline Nanoparticle Films. *Electrochim. Acta.* **2008,** *53,* 5092–5099. DOI: 10.1016/j.electacta.2008.02.010.

19. Argun, A. A.; Aubert, P.-H.; Thompson, B. C.; Schwendeman, I.; Gaupp, C. L.; Hwang, J.; et al. Multicolored Electrochromism in Polymers: Structures and Devices. *Chem. Mater.* **2004,** *16,* 4401–4412. DOI: 10.1021/cm049669l.

20. Sen, T.; Mishra, S.; Shimpi, N. G. Synthesis and Sensing Applications of Polyaniline Nanocomposites: A Review. *RSC Adv.* **2016,** *6,* 42196–42222. DOI: 10.1039/C6RA03049A.

21. Valente, A. J. M.; Burrows, H. D.; Polishchuk, A. Y.; Domingues, C. P.; Borges, O. M. F.; Eusébio, M. E. S.; et al. Permeation of Sodium Dodecyl Sulfate Through Polyaniline-Modified Cellulose Acetate Membranes. *Polymer (Guildf)* **2005,** *46,* 5918–5928. DOI: 10.1016/j.polymer.2005.05.103.

22. Cerqueira, D. A.; Valente, A. J. M.; Filho, G. R.; Burrows, H. D. Synthesis and Properties of Polyaniline–Cellulose Acetate Blends: The Use of Sugarcane Bagasse Waste and the Effect of the Substitution Degree. *Carbohydr. Polym.* **2009,** *78,* 402–408. DOI: 10.1016/j.carbpol.2009.04.016.

23. Cornelsen, P. A.; Quintanilha, R. C.; Vidotti, M.; Gorin, P. A. J.; Simas-Tosin, F. F.; Riegel-Vidotti, I. C. Native and Structurally Modified Gum Arabic: Exploring the Effect of the Gum's Microstructure in Obtaining Electroactive Nanoparticles. *Carbohydr. Polym.* **2015,** *119,* 35–43. DOI: 10.1016/j.carbpol.2014.11.020.

24. Quintanilha, R. C.; Orth, E. S.; Grein-Iankovski, A.; Riegel-Vidotti, I. C.; Vidotti, M. The Use of Gum Arabic as "Green" Stabilizer of Poly(aniline) Nanocomposites: A Comprehensive Study of Spectroscopic, Morphological and Electrochemical Properties. *J. Colloid Interface Sci.* **2014,** *434,* 18–27. DOI: 10.1016/j.jcis.2014.08.006.

25. Umoren, S. A. Inhibition of Aluminium and Mild Steel Corrosion in Acidic Medium Using Gum Arabic. *Cellulose.* **2008,** *15,* 751–761. DOI: 10.1007/s10570-008-9226-4.

26. Bentrah, H.; Rahali, Y.; Chala, A. Gum Arabic as an Eco-Friendly Inhibitor for API 5L X42 Pipeline Steel in HCl Medium. *Corros. Sci.* **2014,** *82,* 426–431. DOI: 10.1016/j. corsci.2013.12.018.

27. Sand, A.; Yadav, M.; Behari, K. Graft Copolymerization of 2-Acrylamidoglycolic Acid on to Xanthan Gum and Study of Its Physicochemical Properties. *Carbohydr. Polym.* **2010,** *81,* 626–632. DOI: 10.1016/j.carbpol.2010.03.022.

28. Singh, V.; Kumari, P.; Pandey, S.; Narayan, T. Removal of Chromium (VI) Using Poly(methylacrylate) Functionalized Guar Gum. *Bioresour. Technol.* **2009,** *100,* 1977–1982. DOI: 10.1016/j.biortech.2008.10.034.

29. Vareda, J. P.; Valente, A. J. M.; Durães, L. Heavy Metals in Iberian Soils: Removal by Current Adsorbents/Amendments and Prospective for Aerogels. *Adv. Colloid Interface Sci.* **2016,** *237,* 28–42. DOI: 10.1016/j.cis.2016.08.009.

30. Fajardo, A.; Pereira, A.; Rubira, A.; Valente, A.; Muniz, E. Stimuli-Responsive Polysaccharide-Based Hydrogels. In *Polysaccharide Hydrogels*; Pan Stanford: Singapore, 2015: pp 325–366. DOI: 10.1201/b19751-10.

31. Wang, L.; Sánchez-Soto, M.; Abt, T. Properties of Bio-based Gum Arabic/Clay Aerogels. *Ind. Crops Prod.* **2016,** *91,* 15–21. DOI: 10.1016/j.indcrop.2016.05.001.

32. Ibrahim, S. M.; Mousaa, I. M.; Ibrahim, M. S. Characterization of Gamma Irradiated Plasticized Carboxymethyl Cellulose (CMC)/Gum Arabic (GA) Polymer Blends as Adsorbents for Dyestuffs. *Bull. Mater. Sci.* **2014,** *37,* 603–608.

33. Das, D.; Vimala, R.; Das, N. Removal of Ag(I) and Zn(II) Ions from Single and Binary Solution Using Sulfonated form of Gum Arabic-Powdered Mushroom Composite Hollow Semispheres: Equilibrium, Kinetic, Thermodynamic and Ex Situ Studies. *Ecol. Eng.* **2015,** *75,* 116–122. DOI: 10.1016/j.ecoleng.2014.11.037.

34. Kipping, T.; Rein, H. Continuous Production of Controlled Release Dosage Forms Based on Hot-Melt Extruded Gum Arabic: Formulation Development, In Vitro Characterization and Evaluation of Potential Application Fields. *Int. J. Pharm.* **2016,** *497,* 36–53. DOI: 10.1016/j.ijpharm.2015.11.021.

35. Musa, H.; Fedail, J.; Ahmed, A.; Musa, T.; Sifaldin, A. Effect of Gum Arabic on Oxidative Stress Markers in the Liver of High Fat Diet Induced Obesity in Mice. In *Gums and Stabilisers for the Food Industry 18: Hydrocolloid Functionality for Affordable and Sustainable Global Food Solutions*; Williams, P., Phillips, G., Eds.; Royal Society of Chemistry, Thomas Graham House, Science Park, Cambridge CB4 4WF: Cambs, England, 2016: pp 256–263.

36. Musa, H.; Ahmed, A.; Fedail, J.; Musa, T.; Sifaldin, A. Gum Arabic Attenuates the Development of Nephropathy in Type 1 Diabetes. In Gums and Stabilisers for the Food Industry 18: Hydrocolloid Functionality for Affordable and Sustainable Global Food Solutions; Williams, P., Phillips, G., Eds.; Royal Society of Chemistry, Thomas Graham House, Science Park, Cambridge CB4 4WF: Cambs, England, 2016; pp. 245–255.

37. Mohamed, R. E.; Gadour, M. O.; Adam, I. The Lowering Effect of Gum Arabic on Hyperlipidemia in Sudanese Patients. *Front. Physiol.* **2015,** *6.* DOI: 10.3389/fphys.2015.00160.

38. Tripathi, A.; Gupta, A.; Bagchi, S.; Mishra, L.; Gautam, A.; Madhok, R. Comparison of the Effect of Addition of Cyanoacrylate, Epoxy Resin, and Gum Arabic on Surface Hardness of Die Stone. *J. Prosthodont.* **2016,** *25,* 235–240. DOI: 10.1111/jopr.12314.

39. Tsai, R.-Y.; Kuo, T.-Y.; Hung, S.-C.; Lin, C.-M.; Hsien, T.-Y.; Wang, D.-M.; et al. Use of Gum Arabic to Improve the Fabrication of Chitosan–Gelatin-Based Nanofibers for Tissue Engineering. *Carbohydr. Polym.* **2015,** *115,* 525–532. DOI: 10.1016/j.carbpol.2014.08.108.

40. Gangu, K. K.; Tammineni, G. R.; Dadhich, A. S.; Mukkamala, S. B. Control of Phase and Morphology of Calcium Oxalate Crystals by Natural Polysaccharide. *Gum Arabic Mol. Cryst. Liq. Cryst.* **2014,** *591,* 114–122. DOI: 10.1080/15421406.2013.836880.

41. Espinosa-Andrews, H.; Sandoval-Castilla, O.; Vázquez-Torres, H.; Vernon-Carter, E. J.; Lobato-Calleros, C. Determination of the Gum Arabic–Chitosan Interactions by Fourier Transform Infrared Spectroscopy and Characterization of the Microstructure and Rheological Features of Their Coacervates. *Carbohydr. Polym.* **2010,** *79,* 541–546. DOI: 10.1016/j.carbpol.2009.08.040.

42. Anvari, M.; Chung, D. Effect of Cooling–Heating Rate on Sol–Gel Transformation of Fish Gelatin–Gum Arabic Complex Coacervate Phase. *Int. J. Biol. Macromol.* **2016,** *91,* 450–456. DOI: 10.1016/j.ijbiomac.2016.05.096.

43. Anvari, M.; Chung, D. Dynamic Rheological and Structural Characterization of Fish Gelatin—Gum Arabic Coacervate Gels Cross-Linked by Tannic Acid. *Food Hydrocoll.* **2016,** *60,* 516–524. DOI: 10.1016/j.foodhyd.2016.04.028.

44. Faraj, J. A.; Dorati, R.; Schoubben, A.; Worthen, D.; Selmin, F.; Capan, Y.; et al. Development of a Peptide-Containing Chewing Gum as a Sustained Release Antiplaque Antimicrobial Delivery System. *AAPS PharmSciTech* **2007,** *8,* E177–E185. DOI: 10.1208/pt0801026.

45. Huang, G.-Q.; Xiao, J.-X.; Jia, L.; Yang, J. Characterization of O-Carboxymethyl Chitosan—Gum Arabic Coacervates as a Function of Degree of Substitution. *J. Dispers. Sci. Technol.* **2016,** *37,* 1368–1374. DOI: 10.1080/01932691.2015.1101609.

46. George, M.; Abraham, T. E. Polyionic Hydrocolloids for the Intestinal Delivery of Protein Drugs: Alginate and Chitosan—a Review. *J. Control. Release.* **2006,** *114,* 1–14. DOI: 10.1016/j.jconrel.2006.04.017.

47. Tan, C.; Xie, J.; Zhang, X.; Cai, J.; Xia, S. Polysaccharide-Based Nanoparticles by Chitosan and Gum Arabic Polyelectrolyte Complexation as Carriers for Curcumin. *Food Hydrocoll.* **2016,** *57,* 236–245. DOI: 10.1016/j.foodhyd.2016.01.021.

48. Espinosa-Andrews, H.; Báez-González, J. G.; Cruz-Sosa, F.; Vernon-Carter, E. J. Gum Arabic−Chitosan Complex Coacervation. *Biomacromolecules* **2007,** *8,* 1313–1318. DOI: 10.1021/bm0611634.

49. Moschakis, T.; Murray, B. S.; Biliaderis, C. G. Modifications in Stability and Structure of Whey Protein-Coated o/w Emulsions by Interacting Chitosan and Gum Arabic Mixed Dispersions. *Food Hydrocoll.* **2010,** *24,* 8–17. DOI: 10.1016/j.foodhyd.2009.07.001.

50. Espinosa-Andrews, H.; Enríquez-Ramírez, K. E.; García-Márquez, E.; Ramírez-Santiago, C.; Lobato-Calleros, C.; Vernon-Carter, J. Interrelationship Between the Zeta Potential and Viscoelastic Properties in Coacervates Complexes. *Carbohydr. Polym.* **2013,** *95,* 161–166. DOI: 10.1016/j.carbpol.2013.02.053.

51. Roldan-Cruz, C.; Carmona-Ascencio, J.; Vernon-Carter, E. J.; Alvarez-Ramirez, J. Electrical Impedance Spectroscopy for Monitoring the Gum Arabic–Chitosan Complexation Process in Bulk Solution. *Colloids Surf. A Physicochem. Eng. Asp.* **2016,** *495,* 125–135. DOI: 10.1016/j.colsurfa.2016.02.004.

52. Sakloetsakun, D.; Preechagoon, D.; Bernkop-Schnürch, A.; Pongjanyakul, T. Chitosan–Gum Arabic Polyelectrolyte Complex Films: Physicochemical, Mechanical and Mucoadhesive Properties. *Pharm. Dev. Technol.* **2015,** 1–10. DOI: 10.3109/10837450.2015.1035727.

53. Tsai, R.-Y.; Chen, P.-W.; Kuo, T.-Y.; Lin, C.-M.; Wang, D.-M.; Hsien, T.-Y.; et al. Chitosan/Pectin/Gum Arabic Polyelectrolyte Complex: Process-Dependent

Appearance, Microstructure Analysis and Its Application. *Carbohydr. Polym.* **2014**, *101*, 752–759. DOI: 10.1016/j.carbpol.2013.10.008.

54. Huang, G. Q.; Xiao, J. X.; Wang, S. Q.; Qiu, H. W. Rheological Properties of O-Carboxymethyl Chitosan—gum Arabic Coacervates as a Function of Coacervation pH. *Food Hydrocoll.* **2015**, *43*, 436–441. DOI: 10.1016/j.foodhyd.2014.06.015.

55. Huang, G.-Q.; Xiao, J.-X.; Jia, L.; Yang, J. Complex Coacervation of O-Carboxymethylated Chitosan and Gum Arabic. *Int. J. Polym. Mater. Polym. Biomater.* **2015**, *64*, 198–204. DOI: 10.1080/00914037.2014.936591.

56. Huang, G.-Q.; Cheng, L.-Y.; Xiao, J.-X.; Wang, S.-Q.; Han, X.-N. Genipin-Cross-linked O-Carboxymethyl Chitosan-Gum Arabic Coacervate as a pH-Sensitive Delivery System and Microstructure Characterization. *J. Biomater. Appl.* **2016**, *31*, 193–204. DOI: 10.1177/0885328216651393.

57. Avadi, M. R.; Sadeghi, A. M. M.; Mohammadpour, N.; Abedin, S.; Atyabi, F.; Dinarvand, R.; et al. Preparation and Characterization of Insulin Nanoparticles Using Chitosan and Arabic Gum with Ionic Gelation Method. *Nanomed. Nanotechnol. Biol. Med.* **2010**, *6*, 58–63. DOI: 10.1016/j.nano.2009.04.007.

58. Paulino, A. T.; Belfiore, L. A.; Kubota, L. T.; Muniz, E. C.; Tambourgi, E. B. Efficiency of Hydrogels Based on Natural Polysaccharides in the Removal of Cd^{2+} Ions from Aqueous Solutions. *Chem. Eng. J.* **2011**, *168*, 68–76. DOI: 10.1016/j.cej.2010.12.037.

59. Paulino, A. T.; Guilherme, M. R.; Reis, A. V.; Campese, G. M.; Muniz, E. C.; Nozaki, J. Removal of Methylene Blue Dye from an Aqueous Media Using Superabsorbent Hydrogel Supported on Modified Polysaccharide. *J. Colloid Interface Sci.* **2006**, *301*, 55–62. DOI: 10.1016/j.jcis.2006.04.036.

60. Favaro, S. L.; de Oliveira, F.; Reis, A. V.; Guilherme, M. R.; Muniz, E. C.; Tambourgi, E. B. Superabsorbent Hydrogel Composed of Covalently Crosslinked Gum Arabic with Fast Swelling Dynamics. *J. Appl. Polym. Sci.* **2008**, *107*, 1500–1506. DOI: 10.1002/app.27140.

61. Singh, B.; Dhiman, A. Design of Acacia Gum–Carbopol–Cross-Linked-Polyvinylimidazole Hydrogel Wound Dressings for Antibiotic/Anesthetic Drug Delivery. *Ind. Eng. Chem. Res.* **2016**, *55*, 9176–9188. DOI: 10.1021/acs.iecr.6b01963.

62. Rana, V.; Rai, P.; Tiwary, A. K.; Singh, R. S.; Kennedy, J. F.; Knill, C. J. Modified Gums: Approaches and Applications in Drug Delivery. *Carbohydr. Polym.* **2011**, *83*, 1031–1047. DOI: 10.1016/j.carbpol.2010.09.010.

63. Bossoni, R.; Riul, A.; Valente, A. J. M.; Rubira, A. F.; Muniz, E. C. Release of Vitamin B 12 and Diclofenac Potassium from N,N-Dimethylacrylamide-Modified Arabic Gum Hydrogels—the Partition-Diffusion Model. *J. Braz. Chem. Soc.* **2014**. DOI: 10.5935/0103-5053.20140090.

64. Nishi, K. K.; Jayakrishnan, A. Self-Gelling Primaquine–Gum Arabic Conjugate: An Injectable Controlled Delivery System for Primaquine. *Biomacromolecules* **2007**, *8*, 84–90. DOI: 10.1021/bm060612x.

65. Paulino, A. T.; Guilherme, M. R.; Mattoso, L. H. C.; Tambourgi, E. B. Smart Hydrogels Based on Modified Gum Arabic as a Potential Device for Magnetic Biomaterial. *Macromol. Chem. Phys.* **2010**, *211*, 1196–1205. DOI: 10.1002/macp.200900657.

66. Chockalingam, A.; Babu, H.; Chittor, R.; Tiwari, J. Gum Arabic Modified Fe3O4 Nanoparticles Cross Linked with Collagen for Isolation of Bacteria. *J. Nanobiotechnol.* **2010,** *8,* 30. DOI: 10.1186/1477-3155-8-30.

67. Williams, D. N.; Gold, K. A.; Holoman, T. R. P.; Ehrman, S. H.; Wilson, O. C. Surface Modification of Magnetic Nanoparticles Using Gum Arabic. *J. Nanopart. Res.* **2006,** *8,* 749–753. DOI: 10.1007/s11051-006-9084-7.

68. Kong, H.; Yang, J.; Zhang, Y.; Fang, Y.; Nishinari, K.; Phillips, G. O. Synthesis and Antioxidant Properties of Gum Arabic-Stabilized Selenium Nanoparticles. *Int. J. Biol. Macromol.* **2014,** *65,* 155–162. DOI: 10.1016/j.ijbiomac.2014.01.011.

69. Reis, A. V.; Guilherme, M. R.; Cavalcanti, O. A.; Rubira, A. F.; Muniz, E. C. Synthesis and Characterization of pH-Responsive Hydrogels Based on Chemically Modified Arabic Gum Polysaccharide. *Polymer (Guildf)* **2006,** *47,* 2023–2029. DOI: 10.1016/j.polymer.2006.01.058.

70. Kaith, B. S.; Ranjta, S. Synthesis of pH—Thermosensitive Gum Arabic Based Hydrogel and Study of Its Salt-Resistant Swelling Behavior for Saline Water Treatment. *Desalin. Water Treat.* **2010,** *24,* 28–37. DOI: 10.5004/dwt.2010.1145.

71. Gerola, A. P.; Silva, D. C.; Jesus, S.; Carvalho, R. A.; Rubira, A. F.; Muniz, E. C.; et al. Synthesis and Controlled Curcumin Supramolecular Complex Release from pH-Sensitive Modified Gum-Arabic-Based Hydrogels. *RSC Adv.* **2015,** *5,* 94519–94533. DOI: 10.1039/C5RA14331D.

72. Gerola, A. P.; Silva, D. C.; Matsushita, A. F. Y.; Borges, O.; Rubira, A. F.; Muniz, E. C.; et al. The Effect of Methacrylation on the Behavior of Gum Arabic as pH-Responsive Matrix for Colon-Specific Drug Delivery. *Eur. Polym. J.* **2016,** *78,* 326–339. DOI: 10.1016/j.eurpolymj.2016.03.041.

73. Patra, D.; Barakat, C. Synchronous Fluorescence Spectroscopic Study of Solvatochromic Curcumin Dye. *Spectrochim. Acta Part A Mol. Biomol. Spectrosc.* **2011,** *79,* 1034–1041. DOI: 10.1016/j.saa.2011.04.016.

74. Micucci, M.; Aldini, R.; Cevenini, M.; Colliva, C.; Spinozzi, S.; Roda, G.; et al. *Curcuma Longa* L. as a Therapeutic Agent in Intestinal Motility Disorders. 2: Safety Profile in Mouse. *PLoS One* **2013,** *8,* e80925. DOI: 10.1371/journal.pone.0080925.

75. Anand, P.; Kunnumakkara, A. B.; Newman, R. A.; Aggarwal, B. B. Bioavailability of Curcumin: Problems and Promises. *Mol. Pharm.* **2007,** *4,* 807–818. DOI: 10.1021/mp700113r.

76. Aggarwal, B. B.; Sung, B. Pharmacological Basis for the Role of Curcumin in Chronic Diseases: An Age-Old Spice with Modern Targets. *Trends Pharmacol. Sci.* **2009,** *30,* 85–94. DOI: 10.1016/j.tips.2008.11.002.

77. Luanpitpong, S.; Talbott, S. J.; Rojanasakul, Y.; Nimmannit, U.; Pongrakhananon, V.; Wang, L.; et al. Regulation of Lung Cancer Cell Migration and Invasion by Reactive Oxygen Species and Caveolin-1. *J. Biol. Chem.* **2010,** *285,* 38832–38840. DOI: 10.1074/jbc.M110.124958.

78. Monroy, A.; Lithgow, G. J.; Alavez, S. Curcumin and Neurodegenerative Diseases. *BioFactors* **2013,** *39,* 122–132. DOI: 10.1002/biof.1063.

79. Valente, A. J. M.; Söderman, O. The Formation of Host-Guest Complexes Between Surfactants and Cyclodextrins. *Adv. Colloid Interface Sci.* **2014,** *205,* 156–176. DOI: 10.1016/j.cis.2013.08.001.

80. Sarika, P. R.; James, N. R.; Kumar, P. R. A.; Raj, D. K.; Kumary, T. V. Gum Arabic-Curcumin Conjugate Micelles with Enhanced Loading for Curcumin Delivery to Hepatocarcinoma Cells. *Carbohydr. Polym.* **2015,** *134,* 167–174. DOI: 10.1016/j. carbpol.2015.07.068.

81. Sarika, P. R.; Nirmala, R. J. Curcumin Loaded Gum Arabic Aldehyde-Gelatin Nanogels for Breast Cancer Therapy. *Mater. Sci. Eng. C.* **2016,** *65,* 331–337. DOI: 10.1016/j.msec.2016.04.044.

82. Reis, A. V.; Moia, T. A.; Sitta, D. L. A.; Mauricio, M. R.; Tenório-Neto, E. T.; Guilherme, M. R.; et al. Sustained Release of Potassium Diclofenac from a pH-Responsive Hydrogel Based on Gum Arabic Conjugates into Simulated Intestinal Fluid. *J. Appl. Polym. Sci.* **2016,** *133,* n/a. DOI:10.1002/app.43319.

83. Cabuzu, D.; Cirja, A.; Puiu, R.; Grumezescu, A. Biomedical Applications of Gold Nanoparticles. *Curr. Top. Med. Chem.* **2015,** *15,* 1605–1613. DOI: 10.2174/1568026 615666150414144750.

84. Shahidi, S.; Iranpour, S.; Iranpour, P.; Alavi, A. A.; Mahyari, F. A.; Tohidi, M.; et al. A New X-Ray Contrast Agent Based on Highly Stable Gum Arabic-Gold Nanoparticles Synthesised in Deep Eutectic Solvent. *J. Exp. Nanosci.* **2015,** *10,* 911–924. DOI: 10.1080/17458080.2014.933493.

85. Kattumuri, V.; Katti, K.; Bhaskaran, S.; Boote, E. J.; Casteel, S. W.; Fent, G. M.; et al. Gum Arabic as a Phytochemical Construct for the Stabilization of Gold Nanoparticles: In Vivo Pharmacokinetics and X-ray-Contrast-Imaging Studies. *Small* **2007,** *3,* 333–341. DOI: 10.1002/smll.200600427.

86. Ribeiro de Barros, H.; Cardoso, M. B.; Camargo de Oliveira, C.; Cavichiolo Franco, C. R.; de Lima Belan, D.; Vidotti, M.; et al. Stability of Gum Arabic-Gold Nanoparticles in Physiological Simulated pHs and Their Selective Effect on Cell Lines. *RSC Adv.* **2016,** *6,* 9411–9420. DOI: 10.1039/C5RA24858B.

87. Renuga Devi, P.; Senthil Kumar, C.; Selvamani, P.; Subramanian, N.; Ruckmani, K. Synthesis and Characterization of Arabic Gum Capped Gold Nanoparticles for Tumor-Targeted Drug Delivery. *Mater. Lett.* **2015,** *139,* 241–244. DOI: 10.1016/j. matlet.2014.10.010.

88. Gamal-Eldeen, A. M.; Moustafa, D.; El-Daly, S. M.; El-Hussieny, E. A.; Saleh, S.; Khoobchandani, M.; et al. Photothermal Therapy Mediated by Gum Arabic-Conjugated Gold Nanoparticles Suppresses Liver Preneoplastic Lesions in Mice. *J. Photochem. Photobiol. B Biol.* **2016,** *163,* 47–56. DOI: 10.1016/j.jphotobiol.2016.08.009.

89. Alizadeh, M.; Qaradaghi, V. In *Simulation of Radioactive Gold Nanoparticles Functionalized with Gum Arabic Glycoprotein in Liver Cancer Therapy, 2015 41st Annual Northeast Biomedical Engineering Conference, IEEE,* 2015; pp 1–2. DOI: 10.1109/ NEBEC.2015.7117068.

90. Axiak-Bechtel, S.; Upendran, A.; Lattimer, J.; Kelsey, J.; Cutler, C.; Selting, K.; et al. Gum Arabic-Coated Radioactive Gold Nanoparticles Cause No Short-Term Local or Systemic Toxicity in the Clinically Relevant Canine Model of Prostate Cancer. *Int. J. Nanomed.* **2014,** *9,* 5001. DOI: 10.2147/IJN.S67333.

91. Aziz, S.; Aziz, S.; Akbarzadeh, A. Advances in Silver Nanotechnology: An Update on Biomedical Applications and Future Perspectives. *Drug Res. (Stuttg)* **2017,** *67,* 198–203. DOI: 10.1055/s-0042-112810.

92. Ashraf, J. M.; Ansari, M. A.; Choi, I.; Khan, H. M.; Alzohairy, M. A. Antiglycating Potential of Gum Arabic Capped-Silver Nanoparticles. *Appl. Biochem. Biotechnol.* **2014**, *174*, 398–410. DOI: 10.1007/s12010-014-1065-1.

93. Ansari, M. A.; Khan, H. M.; Khan, A. A.; Cameotra, S. S.; Saquib, Q.; Musarrat, J. Gum Arabic Capped-Silver Nanoparticles Inhibit Biofilm Formation by Multi-Drug Resistant Strains of Pseudomonas Aeruginosa. *J. Basic Microbiol.* **2014**, *54*, 688–699. DOI: 10.1002/jobm.201300748.

94. Zhang, L.; Yu, F.; Cole, A. J.; Chertok, B.; David, A. E.; Wang, J.; et al. Gum Arabic-Coated Magnetic Nanoparticles for Potential Application in Simultaneous Magnetic Targeting and Tumor Imaging. *AAPS J.* **2009**, *11*, 693–699. DOI: 10.1208/s12248-009-9151-y.

95. Palma, S. I. C. J.; Carvalho, A.; Silva, J.; Martins, P.; Marciello, M.; Fernandes, A. R.; et al. Covalent Coupling of Gum Arabic onto Superparamagnetic Iron Oxide Nanoparticles for MRI Cell Labeling: Physicochemical and In Vitro Characterization. *Contrast Media Mol. Imaging* **2015**, *10*, 320–328. DOI: 10.1002/cmmi.1635.

96. Alves, B. M.; Borlido, L.; Rosa, S. A. S. L.; Silva, M. F. F.; Aires-Barros, M. R.; Roque, A. C. A.; et al. Purification of Human Antibodies from Animal Cell Cultures Using Gum Arabic Coated Magnetic Particles. *J. Chem. Technol. Biotechnol.* **2015**, *90*, 838–846. DOI: 10.1002/jctb.4378.

97. Dong, D.; Qi, Z.; Hua, Y.; Chen, Y.; Kong, X.; Zhang, C. Microencapsulation of Flaxseed Oil by Soya Proteins-Gum Arabic Complex Coacervation. *Int. J. Food Sci. Technol.* **2015**, *50*, 1785–1791. DOI: 10.1111/ijfs.12812.

98. Lv, Y.; Yang, F.; Li, X.; Zhang, X.; Abbas, S. Formation of Heat-Resistant Nanocapsules of Jasmine Essential Oil Via Gelatin/Gum Arabic Based Complex Coacervation. *Food Hydrocoll.* **2014**, *35*, 305–314. DOI: 10.1016/j.foodhyd.2013.06.003.

99. Ré, M. I. Microencapsulation by Spray. *Dry. Technol.* **1998**, *16*, 1195–1236. DOI: 10.1080/07373939808917460.

100. Desai, K. G. H.; Jin Park, H. Recent Developments in Microencapsulation of Food Ingredients. *Dry. Technol.* **2005**, *23*, 1361–1394. DOI: 10.1081/DRT-200063478.

101. Wang, Y.-J.; Pan, M.-H.; Cheng, A.-L.; Lin, L.-I.; Ho, Y.-S.; Hsieh, C.-Y.; et al. Stability of Curcumin in Buffer Solutions and Characterization of Its Degradation Products. *J. Pharm. Biomed. Anal.* **1997**, *15*, 1867–1876. DOI: 10.1016/S0731-7085(96)02024-9.

102. Shahgholian, N.; Rajabzadeh, G. Fabrication and Characterization of Curcumin-Loaded Albumin/Gum Arabic Coacervate. *Food Hydrocoll.* **2016**, *59*, 17–25. DOI: 10.1016/j.foodhyd.2015.11.031.

103. Rao, P. S.; Bajaj, R. K.; Mann, B.; Arora, S.; Tomar, S. K. Encapsulation of Antioxidant Peptide Enriched Casein Hydrolysate Using Maltodextrin–Gum Arabic Blend. *J. Food Sci. Technol.* **2016**, *53*, 3834–3843. DOI: 10.1007/s13197-016-2376-8.

104. Do, D. P.; Pai, S. B.; Rizvi, S. A. A.; D'Souza, M. J. Development of Sulforaphane-Encapsulated Microspheres for Cancer Epigenetic Therapy. *Int. J. Pharm.* **2010**, *386*, 114–121. DOI: 10.1016/j.ijpharm.2009.11.009.

105. Fahey, J.; Talalay, P. Antioxidant Functions of Sulforaphane: A Potent Inducer of Phase II Detoxication Enzymes. *Food Chem. Toxicol.* **1999**, *37*, 973–979. DOI: 10.1016/S0278-6915(99)00082-4.

106. Wu, Y.; Zou, L.; Mao, J.; Huang, J.; Liu, S. Stability and Encapsulation Efficiency of Sulforaphane Microencapsulated by Spray Drying. *Carbohydr. Polym.* **2014,** *102,* 497–503. DOI: 10.1016/j.carbpol.2013.11.057.

107. Eratte, D.; McKnight, S.; Gengenbach, T. R.; Dowling, K.; Barrow, C. J.; Adhikari, B. P. Co-encapsulation and Characterisation of Omega-3 Fatty Acids and Probiotic Bacteria in Whey Protein Isolate–Gum Arabic Complex Coacervates. *J. Funct. Foods.* **2015,** *19,* 882–892. DOI: 10.1016/j.jff.2015.01.037.

108. Eratte, D.; Wang, B.; Dowling, K.; Barrow, C. J.; Adhikari, B. P. Complex Coacervation with Whey Protein Isolate and Gum Arabic for the Microencapsulation of Omega-3 Rich Tuna Oil. *Food Funct.* **2014,** *5,* 2743–2750. DOI: 10.1039/C4FO00296B.

109. Ross, R. P.; Desmond, C.; Fitzgerald, G. F.; Stanton, C. Overcoming the Technological Hurdles in the Development of Probiotic Foods. *J. Appl. Microbiol.* **2005,** *98,* 1410–1417. DOI: 10.1111/j.1365-2672.2005.02654.x.

110. Cano-Higuita, D. M.; Velez, H. A. V.; Telis, V. R. N. Microencapsulation of Turmeric Oleoresin in Binary and Ternary Blends of Gum Arabic, Maltodextrin and Modified Starch. *Ciência E Agrotecnologia.* **2015,** *39,* 173–182.

111. Erejuwa, O. O.; Sulaiman, S. A.; Ab Wahab, M. S. Honey: A Novel Antioxidant. *Molecules.* **2012,** *17,* 4400–4423. DOI: 10.3390/molecules17044400.

112. Koç, A. N.; Silici, S.; Kasap, F.; Hörmet-Öz, H. T.; Mavus-Buldu, H.; Ercal, B. D. Antifungal Activity of the Honeybee Products Against Candida spp. and Trichosporon spp. *J. Med. Food.* **2011,** *14,* 128–134. DOI: 10.1089/jmf.2009.0296.

113. Suhag, Y.; Nanda, V. Optimisation of Process Parameters to Develop Nutritionally Rich Spray-Dried Honey Powder with Vitamin C Content and Antioxidant Properties. *Int. J. Food Sci. Technol.* **2015,** *50,* 1771–1777. DOI: 10.1111/ijfs.12841.

114. Suhag, Y.; Nayik, G. A.; Nanda, V. Effect of Gum Arabic Concentration and Inlet Temperature During Spray Drying on Physical and Antioxidant Properties of Honey Powder. *J. Food Meas. Charact.* **2016,** *10,* 350–356. DOI: 10.1007/s11694-016-9313-4.

115. Cano-Higuita, D. M.; Malacrida, C. R.; Telis, V. R. N. Stability of Curcumin Microencapsulated by Spray and Freeze Drying in Binary and Ternary Matrices of Maltodextrin, Gum Arabic and Modified Starch. *J. Food Process. Preserv.* **2015,** *39,* 2049–2060. DOI: 10.1111/jfpp.12448.

116. Johnson, E. The Role of Carotenoids in Human Health. *Nutr. Clin. Care* **2002,** *5,* 56–65.

117. Boiero, M. L.; Mandrioli, M.; Vanden Braber, N.; Rodriguez-Estrada, M. T.; García, N. A.; Borsarelli, C. D.; et al. Gum Arabic Microcapsules as Protectors of the Photoinduced Degradation of Riboflavin in Whole Milk. *J. Dairy Sci.* **2014,** *97,* 5328–5336. DOI: 10.3168/jds.2013-7886.

118. Mahfoudhi, N.; Hamdi, S. Kinetic Degradation and Storage Stability of β-Carotene Encapsulated by Freeze-Drying Using Almond Gum and Gum Arabic as Wall Materials. *J. Food Process. Preserv.* **2015,** *39,* 896–906. DOI: 10.1111/jfpp.12302.

119. Hu, Q.; Gerhard, H.; Upadhyaya, I.; Venkitanarayanan, K.; Luo, Y. Antimicrobial Eugenol Nanoemulsion Prepared by Gum Arabic and Lecithin and Evaluation of Drying Technologies. *Int. J. Biol. Macromol.* **2016,** *87,* 130–140. DOI: 10.1016/j.ijbiomac.2016.02.051.

120. Chivero, P.; Gohtani, S.; Yoshii, H.; Nakamura, A. Assessment of Soy Soluble Poly-saccharide, Gum Arabic and OSA-Starch as Emulsifiers for Mayonnaise-Like Emulsions. *LWT Food Sci. Technol.* **2016,** *69,* 59–66. DOI: 10.1016/j.lwt.2015.12.064.

121. Soleimanpour, M.; Koocheki, A.; Kadkhodaee, R. Influence of Main Emulsion Components on the Physical Properties of Corn Oil in Water Emulsion: Effect of Oil Volume Fraction, Whey Protein Concentrate and Lepidium Perfoliatum Seed Gum. *Food Res. Int.* **2013,** *50,* 457–466. DOI: 10.1016/j.foodres.2012.04.001.

122. Liu, Y.; Hou, Z.; Yang, J.; Gao, Y. Effects of Antioxidants on the Stability of β-Carotene in O/W Emulsions Stabilized by Gum Arabic. *J. Food Sci. Technol.* **2014.** DOI: 10.1007/s13197-014-1380-0.

123. Dorman, H. J. D.; Deans, S. G. Antimicrobial Agents from Plants: Antibacterial Activity of Plant Volatile Oils. *J. Appl. Microbiol.* **2000,** *88,* 308–316. DOI: 10.1046/j.1365-2672.2000.00969.x.

124. Sacchetti, G.; Maietti, S.; Muzzoli, M.; Scaglianti, M.; Manfredini, S.; Radice, M.; et al. Comparative Evaluation of 11 Essential Oils of Different Origin as Functional Antioxidants, Antiradicals and Antimicrobials in Foods. *Food Chem.* **2005,** *91,* 621–632. DOI: 10.1016/j.foodchem.2004.06.031.

125. Ribeiro, A.; Arnaud, P.; Frazao, S.; Venancio, F.; Chaumeil, J. Development of Vegetable Extracts by Microencapsulation. *J. Microencapsul.* **1997,** *14,* 735–742.

126. Zhang, L.; Mou, D.; Du, Y. Procyanidins: Extraction and Micro-Encapsulation. *J. Sci. Food Agric.* **2007,** *87,* 2192–2197. DOI: 10.1002/jsfa.2899.

127. Bhandari, B. R.; Dumoulin, E. D.; Richard, H. M. J.; Noleau, I.; Lebert, A. M. Flavor Encapsulation by Spray Drying: Application to Citral and Linalyl Acetate. *J. Food Sci.* **1992,** *57,* 217–221. DOI: 10.1111/j.1365-2621.1992.tb05459.x.

128. Alves, S. F.; Borges, L. L.; dos Santos, T. O.; de Paula, J. R.; Conceição, E. C.; Bara, M. T. F. Microencapsulation of Essential Oil from Fruits of Pterodon Emarginatus Using Gum Arabic and Maltodextrin as Wall Materials: Composition and Stability. *Dry. Technol.* **2014,** *32,* 96–105. DOI: 10.1080/07373937.2013.816315.

129. Chranioti, C.; Tzia, C. Arabic Gum Mixtures as Encapsulating Agents of Freeze-Dried Fennel Oleoresin Products. *Food Bioprocess Technol.* **2014,** *7,* 1057–1065. DOI: 10.1007/s11947-013-1074-z.

130. Luo, Y.; Zhang, Y.; Pan, K.; Critzer, F.; Davidson, P. M.; Zhong, Q. Self-Emulsification of Alkaline-Dissolved Clove Bud Oil by Whey Protein, Gum Arabic, Lecithin, and Their Combinations. *J. Agric. Food Chem.* **2014,** *62,* 4417–4424. DOI: 10.1021/jf500698k.

131. Ribeiro, F. W. M.; Laurentino, L. da S.; Alves, C. R.; Bastos, M. do S. R.; da Costa, J. M. C.; Canuto, K. M.; et al. Chemical Modification of Gum Arabic and Its Application in the Encapsulation of Cymbopogon Citratus Essential Oil. *J. Appl. Polym. Sci.* **2015,** *132,* n/a. DOI: 10.1002/app.41519.

132. Lila, M. Anthocyanins and Human Health: An In Vitro Investigative Approach. *J. Biomed. Biotechnol.* **2004,** *5,* 306–313.

133. Pérez-Ramírez, I. F.; Castaño-Tostado, E.; Ramírez-de León, J. A.; Rocha-Guzmán, N. E.; Reynoso-Camacho, R. Effect of Stevia and Citric Acid on the Stability of Phenolic Compounds and In Vitro Antioxidant and Antidiabetic Capacity of a Roselle

(*Hibiscus sabdariffa* L.) Beverage. *Food Chem.* **2015**, *172*, 885–892. DOI: 10.1016/j.foodchem.2014.09.126.

134. Kuntz, S.; Kunz, C.; Herrmann, J.; Borsch, C. H.; Abel, G.; Fröhling, B.; et al. Anthocyanins from Fruit Juices Improve the Antioxidant Status of Healthy Young Female Volunteers Without Affecting Anti-inflammatory Parameters: Results from the Randomised, Double-Blind, Placebo-Controlled, Cross-Over Anthonia (Anthocyanins in Nutrition Investigation Alliance) Study. *Br. J. Nutr.* **2014**, *112*, 925–936. DOI: 10.1017/S0007114514001482.

135. Hubbermann, E. M.; Heins, A.; Stöckmann, H.; Schwarz, K. Influence of Acids, Salt, Sugars and Hydrocolloids on the Colour Stability of Anthocyanin Rich Black Currant and Elderberry Concentrates. *Eur. Food Res. Technol.* **2006**, *223*, 83–90. DOI: 10.1007/s00217-005-0139-2.

136. Guan, Y.; Zhong, Q. The Improved Thermal Stability of Anthocyanins at pH 5.0 by Gum Arabic. *LWT Food Sci. Technol.* **2015**, *64*, 706–712. DOI: 10.1016/j.lwt.2015.06.018.

137. Chung, C.; Rojanasasithara, T.; Mutilangi, W.; McClements, D. J. Enhancement of Colour Stability of Anthocyanins in Model Beverages by Gum Arabic Addition. *Food Chem.* **2016**, *201*, 14–22. DOI: 10.1016/j.foodchem.2016.01.051.

138. Akhavan Mahdavi, S.; Jafari, S. M.; Assadpoor, E.; Dehnad, D. Microencapsulation Optimization of Natural Anthocyanins with Maltodextrin, Gum Arabic and Gelatin. *Int. J. Biol. Macromol.* **2016**, *85*, 379–385. DOI: 10.1016/j.ijbiomac.2016.01.011.

139. Mahdavee Khazaei, K.; Jafari, S. M.; Ghorbani, M.; Hemmati Kakhki, A. Application of Maltodextrin and Gum Arabic in Microencapsulation of Saffron Petal's Anthocyanins and Evaluating Their Storage Stability and Color. *Carbohydr. Polym.* **2014**, *105*, 57–62. DOI: 10.1016/j.carbpol.2014.01.042.

140. Rajabi, H.; Ghorbani, M.; Jafari, S. M.; Sadeghi Mahoonak, A.; Rajabzadeh, G. Retention of Saffron Bioactive Components by Spray Drying Encapsulation Using Maltodextrin, Gum Arabic and Gelatin as Wall Materials. *Food Hydrocoll.* **2015**, *51*, 327–337. DOI: 10.1016/j.foodhyd.2015.05.033.

141. Yoshida, T.; Nagasawa, T. ε-Poly-l-lysine: Microbial Production, Biodegradation and Application Potential. *Appl. Microbiol. Biotechnol.* **2003**, *62*, 21–26. DOI: 10.1007/s00253-003-1312-9.

142. Chang, Y.; McLandsborough, L.; McClements, D. J. Antimicrobial Delivery Systems Based on Electrostatic Complexes of Cationic ε-Polylysine and Anionic Gum Arabic. *Food Hydrocoll.* **2014**, *35*, 137–143. DOI: 10.1016/j.foodhyd.2013.05.004.

143. McClements, D. J. Design of Nano-laminated Coatings to Control Bioavailability of Lipophilic Food Components. *J. Food Sci.* **2010**, *75*, R30–R42. DOI: 10.1111/j.1750-3841.2009.01452.x.

144. Tan, C.; Nakajima, M. β-Carotene Nanodispersions: Preparation, Characterization and Stability Evaluation. *Food Chem.* **2005**, *92*, 661–671. DOI: 10.1016/j.foodchem.2004.08.044.

145. Tatar, F.; Tunç, M. T.; Dervisoglu, M.; Cekmecelioglu, D.; Kahyaoglu, T. Evaluation of Hemicellulose as a Coating Material with Gum Arabic for Food Microencapsulation. *Food Res. Int.* **2014**, *57*, 168–175. DOI: 10.1016/j.foodres.2014.01.022.

146. Ilyasoglu, H.; El, S. N. Nanoencapsulation of EPA/DHA with Sodium Caseinate–Gum Arabic Complex and Its Usage in the Enrichment of Fruit Juice. *LWT Food Sci. Technol.* **2014,** *56,* 461–468. DOI: 10.1016/j.lwt.2013.12.002.

147. Li, J.; Xiong, S.; Wang, F.; Regenstein, J. M.; Liu, R. Optimization of Microencapsulation of Fish Oil with Gum Arabic/Casein/Beta-Cyclodextrin Mixtures by Spray Drying. *J. Food Sci.* **2015,** *80,* C1445–C1452. DOI: 10.1111/1750-3841.12928.

148. Al-Ismail, K.; El-Dijani, L.; Al-Khatib, H.; Saleh, M. Effect of Microencapsulation of Vitamin C with Gum Arabic, Whey Protein Isolate and Some Blends on Its Stability. *J. Sci. Ind. Res.* **2016,** *75,* 176–180.

149. Ozturk, B.; Argin, S.; Ozilgen, M.; McClements, D. J. Formation and Stabilization of Nanoemulsion-Based Vitamin E Delivery Systems Using Natural Biopolymers: Whey Protein Isolate and Gum Arabic. *Food Chem.* **2015,** *188,* 256–263. DOI: 10.1016/j.foodchem.2015.05.005.

150. Ganie, S. A.; Ali, A.; Mazumdar, N. Iodine Derivatives of Chemically Modified Gum Arabic Microspheres. *Carbohydr. Polym.* **2015,** *129,* 224–231. DOI: 10.1016/j.carbpol.2015.04.044.

151. Tripathi, B.; Platel, K. Iron Fortification of Finger Millet (*Eleucine coracana*) Flour with EDTA and Folic Acid as Co-fortificants. *Food Chem.* **2011,** *126,* 537–542. DOI: 10.1016/j.foodchem.2010.11.039.

152. Boccio, J.; Monteiro, J. B. Fortificación de Alimentos Con Hierro y Zinc: Pros y Contras Desde un Punto de Vista Alimenticio y Nutricional. *Rev. Nutr.* **2004,** *17,* 71–78. DOI: 10.1590/S1415-52732004000100008.

153. de Souza, J. R. R.; Feitosa, J. P. A.; Ricardo, N. M. P. S.; Trevisan, M. T. S.; de Paula, H. C. B.; Ulrich, C. M.; et al. Spray-Drying Encapsulation of Mangiferin Using Natural Polymers. *Food Hydrocoll.* **2013,** *33,* 10–18. DOI: 10.1016/j.foodhyd.2013.02.017.

154. Gupta, C.; Chawla, P.; Arora, S.; Tomar, S. K.; Singh, A. K. Iron Microencapsulation with Blend of Gum Arabic, Maltodextrin and Modified Starch Using Modified Solvent Evaporation Method—Milk Fortification. *Food Hydrocoll.* **2015,** *43,* 622–628. DOI: 10.1016/j.foodhyd.2014.07.021.

155. Savary, G.; Hucher, N.; Petibon, O.; Grisel, M. Study of Interactions Between Aroma Compounds and Acacia Gum Using Headspace Measurements. *Food Hydrocoll.* **2014,** *37,* 1–6. DOI: 10.1016/j.foodhyd.2013.10.026.

156. Uekane, T. M.; Costa, A. C. P.; Pierucci, A. P. T. R.; da Rocha-Leão, M. H. M.; Rezende, C. M. Sulfur Aroma Compounds in Gum Arabic/Maltodextrin Microparticles. *LWT Food Sci. Technol.* **2016,** *70,* 342–348. DOI: 10.1016/j.lwt.2016.03.003.

157. Cisneros-Zevallos, L. The Use of Controlled Postharvest Abiotic Stresses as a Tool for Enhancing the Nutraceutical Content and Adding Value of Fresh Fruits and Vegetables. *J. Food Sci.* **2003,** *68,* 1560–1565. DOI: 10.1111/j.1365-2621.2003.tb12291.x.

158. Aloui, H.; Khwaldia, K.; Licciardello, F.; Mazzaglia, A.; Muratore, G.; Hamdi, M.; et al. Efficacy of the Combined Application of Chitosan and Locust Bean Gum with Different Citrus Essential Oils to Control Postharvest Spoilage Caused by Aspergillus Flavus in Dates. *Int. J. Food Microbiol.* **2014,** *170,* 21–28. DOI: 10.1016/j.ijfoodmicro.2013.10.017.

159. Nisperos-Carriedo, M. *Edible Coatings and Films to Improve the Food Quality;* Technomic Publish Co. Inc.: Lancaster, PA, 1994.

160. Al-Juhaimi, F. Y. Physicochemical and Sensory Characteristics of Arabic Gum-Coated Tomato (*Solanum Lycopersicum* L.) Fruits During Storage. *J. Food Process. Preserv.* **2014,** *38,* 971–979. DOI: 10.1111/jfpp.12053.

161. Zahid, N.; Ali, A.; Manickam, S.; Siddiqui, Y.; Maqbool, M. Potential of Chitosan-Loaded Nanoemulsions to Control Different Colletotrichum spp. and Maintain Quality of Tropical Fruits During Cold Storage. *J. Appl. Microbiol.* **2012,** *113,* 925–939. DOI: 10.1111/j.1365-2672.2012.05398.x.

162. Ali, A.; Hei, G. K.; Keat, Y. W. Efficacy of Ginger Oil and Extract Combined with Gum Arabic on Anthracnose and Quality of Papaya Fruit During Cold Storage. *J. Food Sci. Technol.* **2016,** *53,* 1435–1444. DOI: 10.1007/s13197-015-2124-5.

163. Singh, Z.; Singh, R. K.; Sane, V. A.; Nath, P. Mango—Postharvest Biology and Biotechnology. *CRC Crit. Rev. Plant Sci.* **2013,** *32,* 217–236. DOI: 10.1080/07352689.2012.743399.

164. Khaliq, G.; Muda Mohamed, M. T.; Ghazali, H. M.; Ding, P.; Ali, A. Influence of Gum Arabic Coating Enriched with Calcium Chloride on Physiological, Biochemical and Quality Responses of Mango (*Mangifera indica* L.) Fruit Stored Under Low Temperature Stress. *Postharvest Biol. Technol.* **2016,** *111,* 362–369. DOI: 10.1016/j.postharvbio.2015.09.029.

165. Khaliq, G.; Muda Mohamed, M. T.; Ali, A.; Ding, P.; Ghazali, H. M. Effect of Gum Arabic Coating Combined with Calcium Chloride on Physico-chemical and Qualitative Properties of Mango (*Mangifera indica* L.) Fruit During Low Temperature Storage. *Sci. Hortic. (Amsterdam)* **2015,** *190,* 187–194. DOI: 10.1016/j.scienta.2015.04.020.

166. Mahfoudhi, N.; Hamdi, S. Use of Almond Gum and Gum Arabic as Novel Edible Coating to Delay Postharvest Ripening and to Maintain Sweet Cherry (*Prunus avium*) Quality During Storage. *J. Food Process. Preserv.* **2015,** *39,* 1499–1508. DOI: 10.1111/jfpp.12369.

167. Hedayati, S.; Niakousari, M. Effect of Coatings of Silver Nanoparticles and Gum Arabic on Physicochemical and Microbial Properties of Green Bell Pepper (*Capsicum annuum*). *J. Food Process. Preserv.* **2015,** *39,* 2001–2007. DOI: 10.1111/jfpp.12440.

168. Martelo, L.; Jiménez, A.; Valente, A. J.; Burrows, H. D.; Marques, A. T.; Forster, M.; et al. Incorporation of Polyfluorenes into Poly(lactic Acid) Films for Sensor and Optoelectronics Applications. *Polym. Int.* **2012,** *61,* 1023–1030. DOI: 10.1002/pi.4176.

169. Tripathi, N.; Katiyar, V. PLA/Functionalized-Gum Arabic Based Bionanocomposite Films for High Gas Barrier Applications. *J. Appl. Polym. Sci.* **2016,** *133,* n/a. DOI: 10.1002/app.43458.

PART VI
Case Studies

NUCLEAR FUSION AND THE AMERICAN NUCLEAR COVER-UP IN SPAIN: PALOMARES DISASTER (1966)

FRANCISCO TORRENS[1,*] AND GLORIA CASTELLANO[2]

[1]Institut Universitari de Ciència Molecular, Universitat de València, Edificid'Instituts de Paterna, 22085, E-46071 València, Spain

[2]Departamento de Ciencias Experimentales y Matemáticas, Facultad de Veterinaria y Ciencias Experimentales, Universidad Católica de Valencia San Vicente Mártir, Guillem de Castro-94, E-46001 València, Spain

*Corresponding author. E-mail: torrens@uv.es

CONTENTS

ABSTRACT

All elements that encircle people are the product of a nuclear fusion. Fusion drives stars, for example, the Sun, inside which all elements heavier than H are created. If people could use the energy of the stars on the Earth, fusion could become the key of an unlimited clean energy. On January 17, 1966, a US Air Force B-52 collided with its refueling plane, killing seven airmen, and dropping four H-bombs; conventional explosives in two detonated on impact with earth, blowing them to bits and scattering radioactive plutonium (Pu) (mutagen, carcinogen) over the farming town of Palomares, population 2000.

15.1 INTRODUCTION

The energy which people get from the Sun is generated at its core in fusion reactions, in which H is turned into He.[1] If the reactions could be used on Earth, humanity would be able to provide itself with an abundance of energy. The problem of maintaining practical nuclear fusion (NF) was one of the central problems of physics for 60 years, and only now it becomes clear that a solution is in fact possible.

15.2 FUSION POWER

Fusion power is expected to be a major source of energy in the long-range future and commercialization is expected in 21st century.[2] The thermo-nuclear (fusion) power plants will be far more purer and safer than the fission plants. The fusion process is based on the principle that when the nuclei of two light atoms are forced together to form one or two nuclei with smaller mass, energy is released. The fuels to be used in fusion reactors are the gases D_2, T_2, or ^3He. The D, a heavy isotope of H, is found in ordinary sea water. It can be extracted relatively cheaply as heavy water D_2O. When used in a fusion reactor, D in seawater will provide an inexhaustible supply of energy for the world. The problem lies in getting the process started. Fission is easy by comparison (which is why people already have fission plants, whereas scientists are not even certain, after six decades of research that the fusion process could be made to work in a controlled fashion on Earth). The fusion that can be accomplished on Earth was already demonstrated in the uncontrolled fury of H-bomb. In addition, it can work in a controlled sense is known from the fact that this is the process which provides heat and light of the Sun. By controlled, one means that it continues at a steady rate, and does not run away or build up into an explosion. Fusion fuels are, in general, common, accessible, and cheap; they burn cleanly, which is not true of fission fuels. Fusion has high efficiencies (85–90% in certain configurations) and greater potential safety than fission.

For fusion to occur, two light nuclei must smash together at high enough speeds ($3.5 \cdot 10^7$ km·h^{-1}) to cause them to fuse. In gases, the velocity of movement, energy and temperature of particles are all mixed up together. As a result, one way to get gas particles to fuse is to heat them up to extremely high temperatures (HTs) (tens of millions of degrees). Thus, another term for fusion is controlled thermonuclear reaction (CTR). In the Sun, the process takes place at $1.4 \cdot 10^7$°C and high pressure. The raw materials are H-nuclei, which are H-atoms that were ionized or stripped of their electrons. At such temperatures, all gases are ionized, leaving a group of charged particles [mainly protons (p$^+$) and electrons (e$^-$)] that, if allowed to come together, would makeup an electrically neutral gas. Such a gas is called plasma. The responsiveness of plasma to electromagnetic forces was applied in the magnetohydrodynamic (MHD) power generator. The characteristic is put to work in CTR, that is, to confine the plasma to a particular working volume. A physical container cannot be used (at those temperatures, any substance would vaporize and, more important, the reaction would be quenched immediately). Therefore, the approach was to use magnetic containment. The objective is to hold the gas together long enough to bring it up to the temperature and pressure required to initiate the fusion process. However, the main difficulty is that artificially confined high temperature (HT) plasmas are extremely unstable. Because of the great complexity of the behavior of hot plasmas in magnetic field (MF), a large number of approaches are investigated. Basically, the magnetic confinement schemes fall into open and closed systems. A closed system might be shaped like a doughnut, with the plasma circulating via the hollow interior. The advantage is that no end exists out of which the plasma could escape. The fact that the ring is curved, however, means that MF is not an even one, being more concentrated at the inner portion, which leads to greater difficulty in maintaining an even confinement. The international favorites are two devices called the stellarator and the tokamak. It is in the later devices that the hopeful developments mentioned earlier took place, which was accomplished by Soviet scientists. The news electrified the scientific world. Quickly thereafter, a number of American and other machines were adapted to the configuration, whose major advantage, oddly enough, is its simplicity. It is this characteristic that enables it, in both an economic and technical sense, to be made larger. Its disadvantage is that it is less flexible and harder to use in experiments that require measurements. In essence, what is done in such machines is to pass a large current through the plasma contained in the doughnut (torus), thus heating it, as in the resistive coils of a toaster. At the same time, the plasma must be held together and prevented from squirming out of shape and, thus, touching the container walls, which is done maintaining or increasing the strength of shaped MFs.

It is expected that practically, economical fusion plants will have to be approached in stepwise fashion. First, demonstration of scientific feasibility must come. After feasibility is demonstrated, the successful approach or approaches will then have to be duplicated in demonstration or prototype plants that will be far larger and more expensive than the present class of research devices. The consensus is that people will see fusion reactor plants in 21st century. The final result, hopefully, will be plants that could supply people with an inexhaustible source of power with minimum impact on the environment. It is claimed that they will be so safe that they could be located right inside urban areas, which would eliminate the need for transmission of power from outlying areas and make possible to use the excess heat for space heating, distillation of sewage, and so forth, which normally must be dissipated into great quantities of cooling water. The CTR is not a panacea, for there will be some concentration of T_2, a radioactive material. The advantage that fusion reactors present over fission devices in this respect is that T_2 could be recycled right at the plant, thus, eliminating one major source of worry today, shipment of radioactive materials. Another major advantage is that there will be no radioactive waste materials that must be stored.

15.3 NUCLEAR FUSION (NF)

NF is light atomic nuclei combination to form heavier others.[3] When they get compressed enough, H nuclei can be fused to produce He, releasing energy. Gradually, on forming heavier nuclei via a series of NF reactions (NFRes), all elements are created from scratch.

A Strong Squeeze   To fuse even the lightest nuclei, for example, H, is tremendously difficult. Huge temperature and pressure are needed, as NF occurs naturally in only extreme places (e.g., Sun and stars). In order that two nuclei fuse, one should overcome the forces that enable them to join. Nuclei consist of protons (p^+) and neutrons (n^0) blocked together via the strong nuclear force (SNF), which is dominant at tiny nuclear scale and much weaker outside. As p^+ ions are cationic, their electrical charges repel each other, slightly separating. However, SNF cohesion is more powerful, so that the nucleus stays joined. As SNF acts in the short range, its combined force is greater for smaller than bigger nuclei. For a heavy nucleus, for example, ^{238}U, mutual attraction will not be so strong between nucleons on the opposite sides of the nuclei. Repulsive electric force already is perceived at the largest distances and becomes stronger than the biggest nuclei, because it includes the entire nucleus and the greatest number of cations. Equilibrium net effect is that the energy needed for maintaining nucleus joint, averaged by nucleon, rises with atomic weight till Ni/Fe, which is stable and, later, decays for the biggest nuclei. Big nuclei NF takes place with relative easiness as it can be interrupted by a small hit. In NF, the energy barrier one should overcome is lower for H isotope with a p^+. The H arises in types: normal H atoms with a p^+ encircled by an electron (e^-), D (heavy H: 1 p^+/1 e^-/1 n^0), and T with two added n^0. The simplest NFRe is: $H,D \rightarrow T,n^0$.

Nuclear Fusion (NF) Reactors (NFRs)   Physicists replicate extreme NFR conditions in order to generate energy. However, they are at decades. The most advanced NFRs are uncommercial in energy. The NF is an energy-producing Holy Grail. Compared with fission, NFRes are clean and efficient. Few atoms are needed for producing enormous amounts of energy (Einstein's equation $E = mc^2$), few residues are generated but not as harmful as super-heavy elements coming from fission. The NF does not produce greenhouse effect gases, promising an independent and reliable energy source assuming its fuel, H/D, be manufactured. However, it will produce certain radioactive by-products, for example, n^0, which is released in the main NFRes and should be eliminated. At HTs, the main difficulty is to control burning gases, so that although NF is achieved, monstrous NFRs function for only a few seconds every time. An international team built a greater NFR (International Thermonuclear Experimental Reactor

(ITER)), which analyzes commercializing NF viability.

Stardust Stars are natural NFRs. Bethe described how they shined converting H (p^+) to He nuclei (2 p^+/2 n^0). In the transfer, additional particles (positrons and neutrinos) take part, so that two original p^+ become n^0. Inside stars, heavier elements gradually form by steps through NF cooking similar to a recipe. Bigger nuclei form via a series of steps from burning first H, later He, then elements lighter than Fe and, finally, heavier than Fe. Stars, for example, Sun, shine because they are in their greatest part H that fuses with He, which takes place slowly enough in order that heavy elements take place in only small amounts. In the biggest stars, NFRe speeds up because of C/N/O intervention in later NFRes. Heavier elements take place more quickly. Once He is present, C is obtained from it (NF of 3 ^{4}He via unstable ^{8}Be). Once a little of C is obtained, it is combined with He into O/Ne/Mg, which is slow; transformations take the greatest part of a star life. Elements heavier than Fe take place in slightly different NFRes, gradually constructing nuclear sequences in rising periodic table order.

The First Stars Some of the first light elements were not created in stars but proper big bang ball of fire. At first, the universe was so hot that not even atoms were stable. As it was cooling down, H were the first to condense together with a small amount of He/Li and a tiny amount of Be, which were the first ingredients of all stars and things. All heavier elements were created inside and around the stars, and thrown to the space via stars that exploded (supernovae). However, people have not understood how the first stars lighted. The first star did not contain heavy elements, only H, and it could not cool down quickly enough to collapse and light its NF mechanism. The NF is a fundamental energy source in universe. If people achieve to exploit it, energy worries will finish. However, it means to use stars enormous potential here on the Earth, is not easy.

15.4 NUCLEAR FUSION REACTION

NF is a process via which two light atomic nuclei merge to form a heavier nucleus, so that its mass is lower than the sum of the masses of both initial nuclei, that is, the final nucleus is more stable than the initial nuclei, which is fulfilled for only small nuclei.[4] After Einstein equation, energy and mass are equivalent: $E = mc^2$. If the final nucleus presents less mass than both initial nuclei, the mass defect was transformed into released energy, which could be used in the same way as fossil fuels combustion. The reaction that takes place in a fusion reactor happens between two H-isotopes, D, and T, in the following way (cf. Fig. 15.1):

$$D\left(^{2}H\right) + T\left(^{3}H\right) \rightarrow {}^{4}He\,(3.5\text{MeV}) + n\,(14.1\text{MeV}) \qquad (15.1)$$

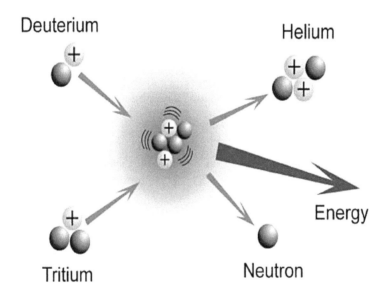

FIGURE 15.1   Scheme of a fusion reaction.

15.5   BASIC NUCLEAR FUSION

Verdú Martín and S. Morató Rafet organized a Basic Course on NF and proposed questions (Qs) and answers (As) on introducing NF nuclear physics.[5,6]

Q1.   What is NF?
A1.   It is a subject that studies the atomic nucleus.
Q2.   What does it permit?

A2. It permits to obtain the energy and some applications, for example, in medicine, and so forth.

They presented magnetic confinement NF (MCF) problems (Ps).
P1.   The instabilities of plasma.
P2.   The discontinuities or interruptions of MF.

They presented Qs on inertial confinement NF (ICF).
Q3.   Why NF?
Q4.   When NF?
Q5.   Where NF?

They proposed the following challenges (Cs) of NF.
C1.   In MCF: superconductors.
C2.   In ICF: lasers and their lack of homogeneity.

Fuel is a D_2/T_2 mixture, reacting as:

$$D_2 + T_2 \rightarrow D^+ + T^+ \rightarrow He^{2+} \ (3.5 MeV),$$
$$n \ (14 MeV) \rightarrow He, D_2, T_2, \quad \text{impurities}$$

$$(15.2)$$

The reaction above needs $T = H_1^3$:

$$H_1^2 + H_1^3 \rightarrow He_2^4 + n_0^1 \tag{15.3}$$

The $T = H_1^3$ is not available on Earth and must be prepared in the plant by:

$$n_0^1 + Li_3^6 \rightarrow He_2^4 + H_1^3 \tag{15.4}$$

but Li_3^6 must also be prepared.

Materials are degraded by irradiations that generate vacancies and interstitial particles in crystals.

They proposed Qs and A on NF in perspective.
Q6.   What is important in NF?
Q7.   How has this way been produced?
Q8.   What have the steps been?
Q9.   Why have these been the steps?
Q10. (Newton, 1717). Are the big bodies and light convertible, and can the bodies receive part of their activity from the light particles that enter the system?

Q11. (Einstein, 1905). Does the inertia of a body depend on its energy content?

A11. (Einstein, 1905). $E^2 = (pc)^2 + (m_0c^2)^2$.

They proposed Q and A on the present of MCF.

Q12. Can a new Fukushima occur in ITER?

A12. Not at such a scale; there is 4 g of D_2 versus15,000 kg of U in the nuclear reactor.

They proposed Q and A on the present of ICF.

Q12. Is the objective gotten?

A12. No; in 2013, the breakeven ($E_{released} > E_{absorbed}$) is gotten but ignition (energy gain) is not reached.

15.6 THE AMERICAN NUCLEAR COVER-UP IN SPAIN: PALOMARES DISASTER (1966)

Howard proposed the following Qs and As on Palomares disaster (1966, *cf.* Fig. 15.2).[7]

FIGURE 15.2   Palomares disaster (1966).

Q1. Palomares was the worst nuclear weapons, why do few outside Spain know about it?

Q2. Therefore, what did exactly happen?

A2. On January 17, 1966, a US Air Force B-52 collided with its refueling plane, killing seven airmen and dropping four H-bombs;

conventional explosives in two detonated on impact with earth, blowing them to bits and scattering radioactive plutonium (Pu) (mutagen and carcinogen) over the farming town of Palomares, population 2000.

Q3. Therefore, what was of greatest significance in early 1966?

A3. In addition, seven airmen, and eight more were killed in a Palomares supply plane crash, people in Palomares suffered potentially fatal radioactive exposures.

Q4. Why then were *ca.* 5000 barrels of hot soil/crops shipped away for burial in South Carolina?

Q5. Why today is Pu found throughout the food chain in Palomares?

Q6. Why is radioactivity evident downwind, in neighboring Villaricos?

Q7. Around the 50th anniversary, will Americans continue to cover their ears and avert their eyes?

Q8. What do Americans (concerned citizens everywhere) need to know?

A8. Of crucial concern, many Spanish injuries, fatalities, and miscarriages were attributed to disaster.

Q9. What had to be done?

Q10. Therefore, who did what?

Q11. What now?

Q12. What do I see?

A12. Foremost: resilience.

Meghani raised the following Qs on seeing the obscured.

Q13. What does return me to acute shame?

Q14. How could US military officials do this?

Q15. How could the highest authorities in Washington countenance it?

Q16. Who works in fields laden with Pu?

Q17. How, despite this, or to remedy this, do gender/sexual outcast often adopt liminal spaces, turning dangerous criminalized inner cities, for example, into lively nighttime gathering places and cultural centers?

Q18. How far does that radiation extend?

Q19. How did it arrive if, as Ambassador Duke claimed in 1966, no trace whatsoever of radioactivity has ever been found?

A19. Howard proposed additional Qs/A on American nuclear cover-up and Palomares disaster.

Q20. Why those Spanish and US Governments' attitudes?

Q21. Why the international attitude of benevolence and magnanimity on the incident?

Q22. Why the attitude of Spanish Monarchy in Exile?

Q23. What did it happen?

A23. People went hungry.

Q24. Do you repent?

A24. He presented the following conclusion (C).

C1. Spanish and US Governments' attitudes were to minimize and trivialize the disaster.

He raised additional Qs.

Q25. Were you surprised of the reply of Spanish and US Governments?

Q26. Was Spanish Minister and US Ambassador's bath most known information of the situation?

15.7 ACKNOWLEDGMENT

Francisco Torrens belongs to the Institut Universitari de Ciència Molecular, Universitat de València. Gloria Castellano belongs to the Departamento de Ciencias Experimentales y Matemáticas, Facultad de Veterinaria y Ciencias Experimentales, Universidad Católica de Valencia San Vicente Mártir. The authors thank support from Generalitat Valenciana (Project No. PROMETEO/2016/094) and Universidad Católica de Valencia San Vicente Mártir (Project No. PRUCV/2015/617).

KEYWORDS

- Science
- Franco dictatorship
- technology
- nuclear energy
- Operation Palomares
- Operation Head Start

REFERENCES

1. Voronov, G. S. *Storming the Fortress of Fusion;* Mir: Moscow, Russia, 1988.
2. Novikov, Y. *Environmental Protection;* Mir: Moscow, Russia, 1990.
3. Baker, J. *50 Physics Ideas You Really Need to Know;* Quercus: London, UK, 2007.
4. Pérez Martín, S. Ed. *Curso Básico de Ciencia y Tecnología Nuclear;* Jóvenes Nucleares–Sociedad Nuclear Española: Madrid, Spain, 2013.
5. Verdú Martín, G. J.; Morató Rafet, S. *Curso Básico de Fusión Nuclear;* València, Spain, Dec. 2, 2016, Universidad Politécnica de Valencia: València, Spain, 2016.
6. Fernández-Cosials, K.; Barbas Espa, A. Eds., Curso Básico de Fusión Nuclear; Jóvenes Nucleares–Sociedad Nuclear Española: Madrid, Spain, 2017.
7. Howard, J. *White Sepulchres: Palomares Disaster Semicentennial Publication;* Universitat de València: València, 2016.

CHAPTER 16

HYALURONIC ACID TRANSPORT PROPERTIES AND ITS MEDICAL APPLICATIONS IN VOICE DISORDERS

EDUARDA F. G. AZEVEDO[1,*], MARIA L. G. AZEVEDO[2],
ANA C. F. RIBEIRO[1], ALEŠ MRÁČEK[3,4], LENKA GŘUNDĚLOVÁ[4],
AND ANTONÍN MINAŘÍK[3,4]

[1]*Coimbra Chemistry Centre, Department of Chemistry, University of Coimbra, 3004-535, Coimbra, Portugal*

[2]*Otorhinolaryngology Service, Centro Hospitalar Baixo Vouga, Aveiro, Portugal, E-mail: luisaazevedo.md@gmail.com*

[3]*Department of Physics and Materials Engineering, Faculty of Technology, Tomas Bata University in Zlin, Czech Republic, E-mail: mracek@ft.utb.cz*

[4]*Centre of Polymer Systems, Tomas Bata University in Zlin, Czech Republic*

*Corresponding author. E-mail: edy.gil.azevedo@gmail.com

CONTENTS

ABSTRACT

Hyaluronic acid (HA) is a natural linear polysaccharide occurring in a wide range of molecular weights. Due to hydrogen bonding, it behaves in solution as an extended, randomly entangled coil, forming a continuous polymer network. The chains entangle each other on this structure if in very low concentration solutions, leading to a mild viscosity (molecular weight dependent), rather than higher than expected viscosity due to greater HA chain entanglement that is shear-dependent if HA solutions are higher concentrations. Ionic effects and electrostatic on HA are considered as a function of counter ion type and valency, as solution properties affect the hydrogen bonding and electrostatic interaction between the solution and HA, resulting in a change in HA chain stiffness. Some of the functions of the polysaccharide are connected to its rheological properties (concentration and molecular weight dependent) and make HA ideal for lubrication in biomedical applications and is known as a "pseudoplastic" material.

Hyaluronic acid is naturally present, and abundant, in all biological fluids of some bacteria and all vertebrates, therefore in the human body. It is synthesized and catabolized by most cells, playing an important role in the regulation of the transport of fluids and solute in the intercellular. It is also recognized by its prominent viscoelastic properties, acting as a lubricant and shock absorber, as well as by its biocompatibility, biodegradability, and non-immunogenicity properties, which allow it to be relevant in pharmaceutical and medical applications, as well described in this chapter.

16.1 HYALURONIC ACID (HA)

16.1.1 STRUCTURE AND CHEMICAL PROPERTIES

Hyaluronic acid (HA) is a natural linear polysaccharide also called hyaluronan (sodium salt of HA), occurring in a wide range of molecular weights (in organism usually from 0.5 kDa to 2.5 MDa; prepared by biotechnological ways can be in the range from 5 kDa to 2.8 MDa). Hyaluronic acid was first obtained from the vitreous humor of cattle eyes, and largely studied by Meyer et al.,[1–7] after the precipitation of diluted native vitreous humor by dilute acetic acid, using the Mörner method of preparation.[8]

This linear polysaccharide consists of a disaccharide repeating sequence of D-glucuronic acid and N-acetyl-D-glucosamine, linked via alternating β-(1→4) and β-(1→3) glycosidic bonds (Fig. 16.1).[9-16]

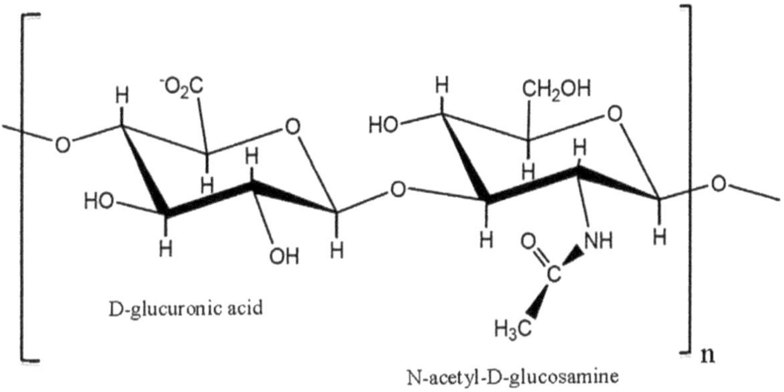

FIGURE 16.1 Structure of the disaccharide repeating unit in hyaluronic acid (HA) salt.

The sequence can be repeated more than 10,000 times including several thousand sugar molecules in the backbone, to form a macromolecule with an extended length of 2–25 μm in size.

Hyaluronic acid is a member of the glycosaminoglycan (GAG) family. The GAGs are unbranched single-chain polymers of disaccharide units, containing N-acetyl-D-glucosamine and hexose, with its second sugar being an HA (except in keratin sulfate, which contains galactose instead). Hyaluronic acid differs from other GAGs (chondroitin sulfate, heparin/ heparansulfate, and keratan sulfate) because it contains no sulfate groups or epimerized uronic acid residues and its polysaccharide chain is larger.[9]

Hyaluronic acid is a polyanion that can self-associate and bind to water molecules (when not bound to other molecules) giving it a stiff, viscous quality similar to gelatin (depends on concentration and molecular weight). Due to hydrogen bonding, the molecule behaves in solution as an extended, randomly entangled coil, forming a continuous polymer network.

Hyaluronic acid axial hydrogen atoms form a nonpolar face (relatively hydrophobic) and the equatorial side chains form a more polar face (hydrophilic) leading to a twisted ribbon structure for HA called as coiled structure.[10,17,18] Each disaccharide unit is rotated by 180° to previous and following unit of the chain. The third following HA unit is on the same plane because two rotations represent a twist of 360°. This arrangement causes the formation of helix chain stabilized by hydrogen bond (secondary structure of HA, Fig. 16.2).[17-19]

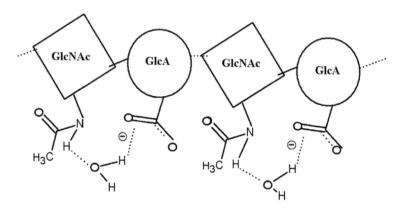

GlcNAc - N-acetyl-D-glucosamine,   GlcA - D-glucuronic acid

Figure 16.2   The secondary structure of HA in water solution (trans conformation).

Some studies show the existence of hydrophobic domains created by eight C–H groups alternating on the sides of the simple helix (Fig. 16.3A). This effect causes twisting of chain and creates hydrophobic domains enabling association with HA chains.[17-19] Two HA molecules can generate antiparallel structure (double helix—tertiary structure, Fig. 16.3B).

Other studies describe behaviors of HA in water solution with sodium and ammonium salts where HA chain form four times twisted left-handed helix (Fig 16.4). This chain twisting is based on six dimers of HA.[20]

The chains entangle each other on this structure if in very low concentration solutions, leading to a mild viscosity (molecular weight dependent), rather than higher than expected viscosity due to greater HA chain entanglement that is shear-dependent if HA solutions are higher concentrations. This rheological property (concentration and molecular weight dependent) has made HA ideal for lubrication in biomedical applications and is known as a "pseudoplastic" material.[10]

The molecular self-association and its entanglement of the random coils in a solution of HA occur by forming antiparallel double helices, bundles, and ropes.[11] The hydrogen bonding between adjacent saccharides occurred alongside mutual electrostatic repulsion among carboxyl groups, thus stiffening HA networks.

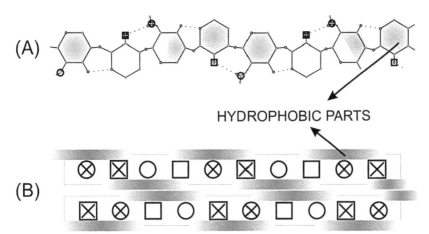

Figure 16.3   Secondary and tertiary structures of HA.

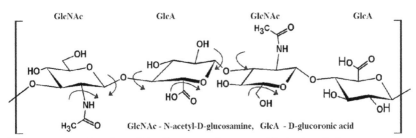

Figure 16.4   Explanation of HA dimers rotations caused by ionic solutions.

    The conformation of a wide variety of HA structures was obtained by tapping mode atomic force microscopy in the air, atomic force micros-copy (AFM).[21] Intermolecular association of extended chains resulting from extensional forces, relaxed chains with a possible helical bias and condensed forms of HA exhibiting varying degrees of condensation into globular, pearl necklace forms to thick rod-like structures, result in the formation of networks and twisted fibers, in which the chain direction is not necessarily parallel to the fiber direction (Fig. 16.5).[21]

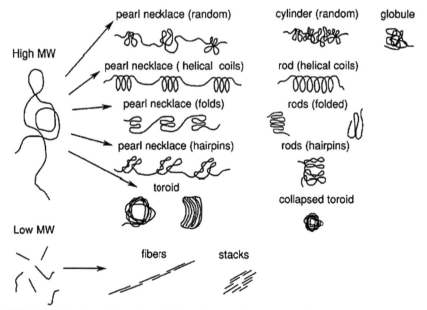

FIGURE 16.5   Possible modes of HA polyelectrolyte condensation.

Viscoelasticity of HA can be tied to these molecular interactions, also dependent on concentration and molecular weight.[10,12]

Ionic effects and electrostatic on HA are considered as a function of counterion type and valency. Studies revealed the effect of electrostatic shielding and also the profound effect of alkaline pH on HA chain stiffness, showing that solution properties affect the hydrogen bonding and electrostatic interaction between the solution and HA, resulting in a change in HA chain stiffness.[11,22] The hydrodynamic radius of HA depends on pH, as the volume occupied by HA can change dramatically due to increased electrostatic interactions and hydrogen bond formation resulting in the reduction of the hydrodynamic radius of HA.[11]

Some of the functions of the polysaccharide are connected with the unique physical and chemical characteristics of the network such as its rheological properties, flow resistance, osmotic pressure, exclusion properties, and filter effect.[13]

16.1.2 DIFFUSION COEFFICIENTS

Studies of diffusion coefficients of Na-HA in aqueous solutions and on the molecular mechanics calculations,[23,24] showed that the diffusion of this polysaccharide in aqueous solutions is strongly affected by the presence of new different species resulting from various equilibria (for example, aggregation due to interactions between the flexible and stiff segments or intermolecular hydrodynamic interactions of each of the part with the solvent) and, consequently, to the decreasing of the diffusion coefficients with the increase in concentration. The effect of those interactions on the diffusion of Na-HA, confirmed by analysis of the dependence of diffusion on concentration as well as by molecular mechanic calculations, is mainly due to the variation of thermodynamic factor values, F_T (due to the nonideality in thermodynamic behavior) and secondarily, to the electrophoretic effect in the mobility factor, F_M. Considering that the mutual diffusion coefficient, D, is a product of two factors (kinetic and thermodynamic), the mobility of these ions in diffusion varies much less with concentration than it does with their gradient of the chemical potential. This difference is due to the fact that in diffusion the ions move in the same direction. Consequently, the presence of interionic effects leads to a small electrophoretic effect. Considering the effect of viscosity on diffusion of this polyelectrolyte in the Gordon equation, results are closer to the experimental data, therefore, the behavior of the polyelectrolyte seems to depend on the viscosity change of the solution, thus the variation of the viscosity is much greater than the variation of the diffusion coefficient for the same interval of concentrations.[24] Diffusion coefficients, together with viscosity data measured for aqueous solutions of Na-HA provide transport data necessary to model the diffusion in pharmaceutical, engineering, and biological applications.[23,24]

Mráček et al.[25] studied the influence of Hofmeister series ions on HA behavior and HA film swelling and characterized the mutual diffusion coefficient of swelling and its mean value of the amorphous polymer film swelling in a thermodynamically compatible solvent (below the glass transition temperature), by the interference of monochromatic light in the wedge arrangement, observing a sharp increase of diffusion coefficient with polymer concentration[26] and diffusion process of Na-HA and Na-HA-n-alkyl derivatives films swelling.[27]

16.1.3 PHYSIOLOGICAL FUNCTION AND MECHANISM OF BIOSYNTHESIS

Hyaluronic acid is naturally present and abundant in all biological fluids of some bacteria and all vertebrates, therefore in the human body.[28] It is found in the highest concentrations in the extracellular matrix (ECM), especially, of soft connective tissues, of eye vitreous humor and of joints (synovial fluids and cartilage). It exists as a pool associated with the cell surface, another bound to other matrix components, and a largely mobile pool.[28,29]

Hyaluronic acid forms the structural basis of the pericellular matrix (plays a complex role in cell adhesion/de-adhesion, and cell shape changes associated with proliferation and locomotion), extruding directly through the plasma membrane by one of the three HA synthases and anchored to the cell surface by the synthase or cell surface receptors such as CD44 or RHAMM. Aggregating proteoglycans and other HA-binding proteins contribute to the material and biological properties of the matrix and regulate cell and tissue function (Fig. 16.6).[30] Native high molecular weight HA (nHA) and oligosaccharides of HA (oHA) provoke distinct biological effects upon binding to CD44, influencing cell adhesion and signaling for tumor progression and inflammation processes.[31]

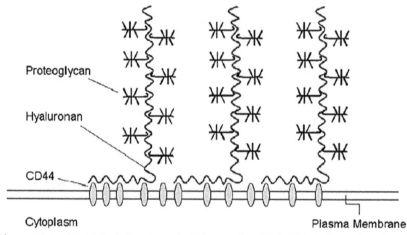

Figure 16.6   Model depicting the pericellular matrix with the HA chains anchored to the cell surface via CD44 and the associated aggregating proteoglycans.

Hyaluronic acid synthesis can create and support plasma membrane protrusions. The HA-dependent pericellular matrix is a multifunctional regulator of cell adhesion, cell shape, and behavior. Hyaluronic acid-binding proteoglycans and other associated proteins in the cell coat contribute to structural morphologies and material properties of the pericellular matrix and add complexity to the biological functions of the matrix.[30]

The various conformation of HA structures depend in the physiological environments in which HA is found, thus, in liquid connective tissues and extracellular matrix of cartilage the relaxed, weakly helical, coiled conformations of HA (where dilute), or partially condensed conformations (where excluded volume effects are significant) are found. In the absence of extensional forces or specific protein-induced conformational changes, intracellular HA, being in a crowded environment is likely to adopt a condensed form. The crowded environment of HA bound to the cell surface may also explain the observed difference between free HA and cell surface HA in binding to crystals of calcium-(R,R)-tartrate tetrahydrate. In tight intercellular spaces or in protein-HA complexes or bound to cell surface receptors may undergo forced extension and subsequent fibrillar association under the influence of fluid flow.[21]

Hyaluronic acid is synthesized and catabolized by most cells. Its mechanism of biosynthesis, via an enzymatic process is unique because it is not covalently linked to a protein backbone during synthesis[32] and is produced at the inner side of the plasma membrane (rather than in the endoplasmic reticulum and Golgi apparatus), by adding monosaccharides to the reducing end of the chain.[13] It is normally synthesized at the plasma membrane by HA synthases, one or more of the three HA synthases (HAS1, HAS2, and HAS3), a class of integral membrane proteins, in the mammalian genomes,[29] using cytosolic UDP-GlcUA and UDP-GlcNAc substrates. Dynamic HA metabolism can regulate cytosolic UDP-GlcNAc levels to sustain functions of O-GlcNAc transferase.

Hyaluronic acid produced and secreted by cells like fibroblasts, keratinocytes, or chondrocytes, grows out from the cell surface and, for example, fibroblasts, surround themselves with a coat of HA.[13] The rate of biosynthesis is regulated by various factors such as growth factors, hormones, inflammatory mediators, etc. The responsible enzyme, HA synthase, is a phosphoprotein and the regulation of the synthetic rate is apparently via phosphorylation. Hyaluronic acid production increases in proliferating cells and the polymer may play a role in mitosis. Some proteins, the hyaladherins, specifically recognize the HA structure and this kind of interactions bind HA with proteoglycans to stabilize the structure of the matrix, and with cell surfaces to modify cell behavior. Because of the striking physicochemical properties of HA solutions, various physiological functions have been assigned to it, including lubrication, water homeostasis, filtering effects, and regulation of plasma protein distribution.[28]

Playing an important role in the regulation of the transport of fluids and solute in the intercellular processes,[27] it is one of the most hygroscopic molecules in nature and when hydrated, it can contain up to 1000-fold more water than its own weight, which is particularly important for its moisturizing ability. It is also recognized by its prominent viscoelastic properties, acting as a lubricant and shock absorber,[27,33] as well as by its biocompatibility, biodegradability, and non-immunogenicity properties, which allows it to be relevant in pharmaceutical and medical applications.[34–54]

16.1.4   MECHANISM OF CATABOLISM AND ELIMINATION

Degradation of HA can occur via enzymatic or nonenzymatic reactions. Most cells catabolize HA from its glycocalyces by an endosomal uptake pathway (a receptor-mediated endocytosis and lysosomal degradation). Three types of enzymes (hyaluronidase, β-D-glucuronidase, and β-N-acetylhexosaminidase), found in the intercellular space and in serum, are involved in enzymatic degradation of HA. Hyaluronidase cleaves high molecular weight HA into smaller fragments while the other two enzymes degrade the fragments by removing nonreducing terminal sugars.[10,35,55] The cleavage can occur in a single glycosidic bond on the HA backbone causing fragmentation, or the enzyme can remove a single monosaccharide unit from the HA backbone.[55] The degradation occurs either locally or after transport by lymph flow from the tissues to lymph nodes which degrade much of it.[13] The remainder enters the general circulation and is removed from the blood, mainly, by the endothelial cells of the liver sinusoids.[28]

Nonenzymatic mechanisms of HA degradation include thermal or shear stress. Increasing temperature of HA solutions result in degradation and decreased viscosity exponentially as a function of temperature.[55] Ultrasonication degrades HA in a nonrandom fashion, the high molecular weight HA chains degrade slower than low molecular weight HA chains subjected to ultrasonication.[11,55]

Hydrolysis and degradation by oxidants are nonenzymatic degradation pathways for HA,[35,55,56] occurring in a random fashion often resulting in disaccharide fragment production. Acidic hydrolysis degrades the HA moiety and alkaline hydrolysis occurs on N-acetylglucosamine units.[55] Degradation via oxidation, by reactive oxygen species (ROS) can occur in cells as a consequence of aerobic respiration, and ROS such as superoxide anions, hydroxyl radicals, and hypochlorite can cause HA chain cleavage. Reactive oxygen species are proposed to be involved in several inflammatory and degenerative processes such as arthritis.[11,55]

Besides the enzymatic degradation and nonenzymatic degradation pathways, two more pathways are engaged in HA catabolism: turnover (internalization and degradation within the tissue) and release from the tissue matrix, drainage into the vasculature, and clearance via lymph nodes, liver, and kidney.[57]

The concentration of HA in the human body varies from a high concentration of 4 g/kg in umbilical cord, 2–4 g/dm^3 in synovial fluid, 0.2 g/kg in dermis, and about 10 mg/dm^3 in thoracic lymph, to a low concentration of 10–100 $\mu g/dm^3$ in normal serum.[10,11] The largest amount of HA is found in the ECM of skin and musculoskeletal tissue.

Elevated HA levels in blood in certain diseases are indicating that both an impaired catabolism in the liver and an increased synthesis in the peripheral tissues modify the HA level.[36,58] The serum level is regulated by the influx of the polysaccharide from the peripheral tissues via lymph and its receptor-mediated clearance by liver endothelial cells. Elevated serum levels of HA can indicate certain diseases, such as cirrhosis, rheumatoid arthritis, mesothelioma, scleroderma, and more rare diseases, such as Wilms tumor, cutaneous hyaluronanosis, and Werner's syndrome. The metabolism of HA has made this polysaccharide an interesting clinical marker for a number of pathological conditions.[36,58] Studies proved that HA is also essential for heart development.[37,59]

Greater levels of structure and cross-linking of HA contribute to processes such as ovulation and inflammation, thus knowledge of pericellular matrix structure and function under different biological conditions and disease states aids in formulating therapeutic intervention strategies and bioengineering approaches.[30]

Depending on the location in the body, most of the HA is catabolized within days. The normal half-life of HA is 1–3 weeks in inert tissues such as cartilage, 1–2 days in the epidermis of skin, and 2–5 min. in blood.

16.2 BIOMEDICAL APPLICATION OF HYALURONIC ACID

The HA has several applications in medicine. The first biomedical application of HA occurred in the late 1950s when it was used for a vitreous replacement during eye surgery. Initially, HA was isolated from umbilical cords and shortly afterward, from rooster combs. Currently, the HA that is used in medicine is extracted from rooster combs or made by bacteria in the laboratory.[38]

The viscoelastic matrix of HA can act as a strong biocompatible support material and is, therefore, commonly used as a growth scaffold in surgery, wound healing, and embryology (embryo implantation).[39]

The FDA has approved the use of HA as a viscoelastic tool during certain procedures to help replace natural fluids, due to its distinct physico-chemical properties in ophthalmological surgery, laryngeal surgery, osteo-arthritis, sores in the mouth, as a filler in plastic surgery, applied to the skin for healing wounds, burns, skin ulcers, and to prevent the effects of aging. Hyaluronic acid can be taken by mouth or injected into the affected area and works by acting as a cushion and lubricant affecting the way the body responds to injury. The medical applications and the commonly used commercial preparations containing HA used in clinical procedures are summarized in the literature.[39]

A concentrated solution of HA (10 mg/cm^3) has become a device in ophthalmic surgery, through its tissue protective and rheological proper-ties.[40] In cataract surgery, the role of HA in facilitating procedures and protecting the corneal endothelium is well established. Some benefit has also been gained with the use of HA in penetrating keratoplasty, trabecu-lectomy, retinal reattachment, and trauma surgery, although its efficacy in these indications is less well-defined in the literature.[60]

Hyaluronic acid helps to retain synovial fluid (lubricates joints and tendon sheaths and carries metabolites to and from the avascular articular cartilage) in the joint cavity when pressure is raised and acts, at least in part, by exerting osmotic pressure at the interface between synovial matrix and a concentration polarization layer.[41]

In pathological conditions, the concentration and molecular weight of HA are reduced, resulting in synovial fluid of lower elasticity and viscosity. The factors which contribute to the low concentrations of HA are dilu-tional effects, reduced HA synthesis, and free radical degradation. When viscoelasticity of synovial fluid is reduced, the transmission of mechanical force to cartilage may increase its susceptibility to damage. Therefore, the restoration of the normal articular homeostasis is the rationale for HA administration into osteoarthritic joints. Moreover, being HA, a physiolog-ical component, it is very likely that it may be deprived of adverse reac-tions, also after repeated administrations.[42] Administration of purified high molecular weight HA into orthopedic joints can restore the desirable rheological properties and alleviate some of the symptoms of osteoarthritis.

The restoration of viscoelastic properties, such as cushioning, lubrication, and elasticity,[61] while the restoration of joint rheology, anti-inflammatory and antinociceptive effects, normalization of endogenous HA synthesis, and chondroprotection, explain why the clinical efficacy is maintained for several months.[43–45,62,63]

In addition to its lubricating and cushioning properties, the anti-inflammatory activity and a possible disease-modifying effect for HA in animals has prompted its investigation as a treatment in osteoarthritis and, to a much lesser extent, in rheumatoid arthritis. Hyaluronic acid as weekly intra-articular injections for 3–7 weeks, improved knee pain and joint motion in patients with osteoarthritis, distinguishing it from other therapies by providing a sustained effect after treatment discontinuation, together with its very good tolerability profile.[60] Injection of HA into osteoarthritic joints restores the viscoelasticity, augments the flow of joint fluid, normalizes endogenous HA synthesis and improves joint function.[39,46] Hyaluronic acid easily forms polyelectrolyte complexes with chitosan and chondroitin sulfate and the synergy of chitosan with HA develops enhanced performances in regenerating hyaline cartilage, typical results being structural integrity of the hyaline-like neocartilage and reconstitution of the subchondral bone, with positive cartilage staining for collagen-II and GAG in the treated sites.[47]

The anti-inflammatory effect of HA intra-articular injections can modify the knee or hip joint metabolism in patients with osteoarthritis resulting in a decrease in proinflammatory T cells concentrations,[48] reducing synovial inflammation and restoring the rheological properties of synovial fluid, so it regulates the articular milieu. The presence of activated T cells in patients with osteoarthritis confirms the immunological/inflammatory involvement in osteoarthritis.[48]

In the ear, studies performed in rats about the potential ototoxicity of HA 1.9%, showed that three months after HA exposure, all auditory brainstem response (ABR) thresholds were normal, although viscous material was present in the middle ear, concluding that HA causes reversible changes in inner ear function but lacks ototoxic adverse effects.[49]

The inner ear is one of the most challenging organs for drug delivery, mainly, because of the blood-perilymph barrier. Therefore, local rather than systemic drug delivery methods are being developed for inner ear therapy. A suitable drug delivery system should be easily injectable and provide sustained release of the drug to minimize the number of transtympanic injection. Recent studies revealed that a combination of PEGylated liposomes and HA gel sustained the delivery of drugs to the inner ear without any negative effect on the hearing thresholds. The shear-thinning behavior of HA ensured a suitable injectability, thus its non-thixotropic, viscoelastic, and mucoadhesive properties provided a fast recovery of high initial viscosity after injection, as well as a prolonged residence time in the middle ear in contact with the round window. The drug delivery to the perilymph lasts for up to 30 days, while the round window acted as a reservoir for liposomes.[50]

A few cases of large traumatically perforations of the tympanic membrane (occupying more than one quadrant) were treated with HA 1%, without delay after the accident, and the margins of the perforations were noticeably restored. The size of the perforation was immediately reduced to roughly half of its original size and immediate restoration and covering of a traumatically ruptured tympanic membrane improve the healing potential of the drum and counteract middle ear infection.[51]

Another medical application of the HA is in the larynx, the aim of our study. Vocal folds are soft laryngeal connective tissues with distinct layered structures and complex multicomponent matrix compositions that endow phonatory and respiratory functions. Its unique vibrational mechanical properties render the delicate tissue susceptible to injuries that can potentially lead to scar formation and voice impairment.[52] Regenerative medicine strategies, established numerous efforts to attempted repair and restore the functional outcomes of injured vocal folds. Engineering principles and the advance of biotechnologies have achieved in creating state-of-the-art scaffolds with various techniques ranging from injectable hydrogels, electrospun fibers, and synthetic polymers to stem cell and growth factor administration.[52] A linear correlation between the distribution of the viscoelastic moduli and those of the volume fractions of the HA was found in porcine vocal fold tissue (regarding the similar structures between human and porcine vocal folds), concluding that HA may control regulate the nanoscale viscoelasticity of human vocal folds, supporting the use of HA-based implants for the restoration of vocal fold viscoelasticity.[64]

Method of vocal folds implantation with HA has been applied in medical centers as a voice improving in several indications such as presbyphonia, vocal folds paralysis (abduction), glottic insufficiency, vocal fold scars, or sulcus vocalis.[53] All of these illnesses prevent the patients to obtain full glottis closure and produce good quality voice. Voice disorders resulting from glottic insufficiency are a significant clinical problem in everyday phoniatric practice. A method of treatment is injection laryngoplasty. Some studies aimed to assess the voice quality of patients treated with HA injection into the vocal fold, which improves phonatory functions of the larynx and the quality of voice in patients with glottic insufficiency. This method for treatment and improvement of voice disorders proves to be simple, easy, effective, and fast for the restoration of good voice quality, and can be a safe alternative to the conservative methods of treatment used in rehabilitation of voice disorders, such as surgery method in the rehabilitation of voice.[53]

The results indicate that application of HA augmentation to the vocal fold in cases having insufficient glottic closure provides good and very good voice quality and speeds rehabilitation of the voice. The satisfactory glottic closure was obtained after surgery in all cases. The outcomes were confirmed by videostroboscopic assessment and objective acoustic examination of the voice.[53] Augmentation of the vocal fold with HA is an effective and voice-improving method, restoring speech production. Medialization laryngoplasty is also an option for elderly patients with a vocal handicap, because it is a safe and viable procedure. With few exceptions, patients achieve a clinically meaningful improvement in their voice-related quality of life with a low complication rate.

16.3   ANALYSIS, CONCLUSIONS, AND CHALLENGES

The understanding of these complex systems and their characterization is very important to understand their structure and to model them to practical applications, such as pharmaceutical and medicinal applications.[65,66] Studies which have been taken into account their transport properties,[23,24,67–69] provided a direct measure of the molecular mobility, an important factor in the preservation of biological materials in sugar matrices. Data of the mutual diffusion coefficients and behavior of these systems, at therapeutic dosage, are available in the literature for aqueous systems containing this polysaccharide,[23] using the Taylor technique in aqueous solutions. The studied molecular mechanics gave additional information concerning the probable interactions in this system containing Na-HA, and helped in obtaining a better understanding of the diffusion in these systems and helped to better understand the factors governing the formation of these structures.

Berkland et al. have reviewed the physical and chemical characteristics of HA as applied to tissue engineering, dermal filling, and viscosupplementation.[37] Difficulties such as potential toxicity of cross-linking techniques, high viscosity of HA solutions, and rapid elimination have been raised as limitations to improve biomedical products derived from HA, thus, current and emerging strategies to modify HA were reviewed as potential approaches.[38] Other biomedical applications of HA are also emerging (e.g., drug conjugation and delivery, therapeutic and/or immune modulation effects, etc.). New chemical modifications performed on HA are also appearing in the literature, for safely cross-linking HA in situ as an important paradigm for tissue engineering and viscosupplementation. Formulations that improve product performance will also utilize physical interactions among components (e.g., interactions between HA nanoparticles) and utilizing particle/polymer or particle/particle interactions may provide new approaches to tune the viscoelastic properties of HA formulations as applied to tissue engineering, dermal filling, and osteoarthritis treatment.[38]

Recent studies of new polymeric materials for use in modified drug delivery systems, whose main potential application is the coating of solid oral dosage forms, reveal evidences that HA-based responsive films (cross-linking of HA with trisodium trimetaphosphate), applied as coating material of oral solid dosage forms, can prevent the premature release of drugs in harsh acidic stomach conditions, but control the release in the alkaline gastrointestinal tract distal portion, assuring safety to intestinal mucosa.[70]

There are limitations including the in vivo loss of HA limiting the duration of effect, however, new HA materials are being synthesized, and can be used in future. A new cross-linked HA-gelatin (HA-Ge) hydrogel is injectable biomaterial to be used in vocal fold tissue engineering. With a remarkable viscoelastic and tuned mechanical—biological properties, these HA-Ge hydrogels are similar to those of human vocal fold tissue (storage, loss moduli, and frequency-dependency), thus, this could improve the tissue regenerationin vocal fold tissue treatment.[54]

Hyaluronic acid preparations have a short half-life, therefore, the long-term effects cannot solely be attributed to the substitution of the molecule itself. The artificial biomaterials investigated to date mainly have not recapitulated the highly complex multilayer architecture or the aerodynamic-to-acoustic energy transfer and high frequency vibration of native lamina propria to completely restore scarred or atrophied vocal fold. The growing body of literature in vocal fold physiology and developmental biology will inspire the creation of more complex, fully functional synthetic matrices that allow the optimization of scaffold-cell interaction, correlation of the relationship between vocal fold protein constituents and rheological outcome, as well as standardization of in vitro protocols and in vivo vocal fold animal models to generate safe and clinically useful biomaterials for patients with voicedisorders.[52]

16.4   ACKNOWLEDGMENT

The authors are grateful for funding from "The Coimbra Chemistry Centre" which is supported by the Fundaçãopara a Ciência e a Tecnologia (FCT), Portuguese Agency for Scientific Research, through the programs UID/QUI/UI0313/2013 and COMPETE. The authors (A. Mráček, L. Grundělová, and A. Minařík) are grateful for the support of the Operational Program Research and Development for Innovations cofounded by the European Regional Development Fund (ERDF) and the national budget of the Czech Republic within the framework of the project Centre of Polymer Systems (reg. number: CZ.1.05/2.1.00/03.0111). This work was supported by the Ministry of Education, Youth and Sports of the Czech Republic, Program NPU I (LO1504), and by the European Regional Development Fund (Grant CZ.1.05/2.1.00/19.0409).

KEYWORDS

- hyaluronic acid (HA)
- mutual diffusion
- aqueous solutions
- transport properties
- viscoelastic properties
- hyaluronan hydrogels
- injectable biomaterials
- vocal fold paralysis
- acoustic analysis
- injection laryngoplasty
- videostroboscopy
- glottic insufficiency
- synovial joint fluid

REFERENCES

1. Meyer, K.; Palmer, J. W. The Polysaccharide of the Vitreous Humor. *J. Biol. Chem.* **1934**, *107*, 629–640.
2. Meyer, K.; Dubos, R.; Smyth, E. M. The Hydrolysis of the Polysaccharide Acids of the Vitreous Humor, of Umbilical Cord, and of Streptococcus by the Autolytic Enzyme of Pneumococcus. *J. Biol. Chem.* **1937**, *118*, 71–78.

3. Meyer, K.; Smyth, E. M.; Gallardo, E. On the Nature of the Ocular Fluids II. The Hexosamine Content. *Am. J. Ophthalmol.* **1938,** *21,* 1083–1090.
4. Meyer, K.; Hobby, G. L.; Chaffee, E.; Dawson, M. H. The Hydrolysis of Hyaluronic Acid by Bacterial Enzymes. *J. Exp. Med.* **1940,** *71,* 137–146.
5. Meyer, K.; Chaffee, E. Hyaluronic Acid in the Pleural Fluid Associated with a Malignant Tumor Involving the Pleura and Peritoneum. *J. Biol. Chem.* **1940,** *133,* 83–91.
6. Meyer, K.; Chaffee, E. The Mucopolysaccharides of Skin. *J. Biol. Chem.* **1941,** *138,* 491–499.
7. Meyer, K.; Chaffee, E.; Hobby, G. L.; Dawson, M. H. Hyaluronidases of Bacterial and Animal Origin. *J. Exp. Med.* **1941,** *73,* 309.
8. Mörner, C. T. Untersuchung der Protein substanzen in denlichtbrechendenMedien des Auges. *Z. physiol. Chem.* **1894,** *18,* 233.
9. Toole, B.P. Hyaluronan is not Just a Goo! *J. Clin. Invest.* **2000,** *106,* 335–336.
10. Necas, J.; Bartosikova, L.; Brauner, P.; Kolar, J. Hyaluronic Acid (Hyaluronan): A Review. *Vet. Med.* **2008,** *53,* 397–411.
11. Garg, H.; Hales, C. *Chemistry and Biology of Hyaluronan;* Elsevier Science 2004, ISBN: 9780080443829.
12. Whelan, J. *The Biology of Hyaluronan;* John Wiley & Sons 1989, ISBN: 0471923052.
13. Laurent, T.C. Biochemistry of Hyaluronan. *Acta Oto-Laryngol. Suppl.* **1987,** *442,* 7–24.
14. Smejkalova, D.; Hernannova, M.; Buffa, R.; Cozikova, D.; Vistejnova, L.; Matulkova, Z.; Hrabica, J.; Velebny, V. Structural Characterization and Biological Properties of Degradation Byproducts from Hyaluronan After Acid Hydrolysis. *Carbohydr. Polym.* **2012,** *88,* 1425–1434.
15. Laurent, T. C.; Fraser, J. R. Hyaluronan. *FASEB J.* **1992,** *6,* 2397–2404.
16. Laurent, T. C.;Laurent, U. B. G.; Fraser, J. R. The Structure and Function of Hyaluronan: An Overview. *Immunol. Cell Biol.* **1996,** *74,* A1–7.
17. Scott, J. E.; Cummings, C.; Brass, A.; Chen, Y. Secondary and Tertiary Structures of Hyaluronan in Aqueous Solution, Investigated by Rotary Shadowing Electron Microscopy and Computer Simulation. Hyaluronan is a Very Efficient Network-Forming Polymer. *Biochem. J.* **1991,** *274*(3), 699–705.
18. Gřundělová, L.; Mráček, A.; Kašpárková, V.; Minařík, A.; Smolka, P. The Influence of Quarternary Salt on Hyaluronan Conformation and Particle Size in Solution. *Carbohydr. Polym.* **2013,** *98,* 1039–1044.
19. Heatley, F.; Scott, J. E. A Water Molecule Participates in the Secondary Structure of Hyaluronan. *Biochem. J.* **1988,** *254*(2), 489–493.
20. Haxaire, K.; Bracciny, I.; Milas, M.; Rinaudo, M.; Pérez, S. Conformation Behaviour of Hyaluronan in Relation to its Physical Properties as Probed by Molecular Modeling. *Glycobiology* **2000,** *10*(6), 587–594.
21. Cowman, M. K.; Spagnoli, C.; Kudasheva, D.; Li, M.; Dyall, A.; Kanai, S.; Balazs, E. A. Extended, Relaxed, and Condensed Conformations of Hyaluronan Observed by Atomic Force Microscopy. *Biophys. J.* **2005,** *88,* 590–602.
22. Sheehan, J.; Arundel, C.; Phelps, C. Effect of the Cations Sodium, Potassium and Calcium on the Interaction of Hyaluronate Chains: A Light Scattering and Viscometric Study. *Int. J. Biol. Macromol.* **1983,** *5,* 222–228.

23. Veríssimo, L. M. P.; Valada, T. I. C.; Sobral, A. J. F. N.; Azevedo, E. F. G.; Azevedo, M. L. G.; Ribeiro, A. C. F. Mutual Diffusion of Sodium Hyaluranate in Aqueous Solutions. *J. Chem. Thermodyn.* **2014**, *71*, 14–18.

24. Mráček, A.; Gřundělová, L.; Minařík, A.; Veríssimo, L. M. P.; Barros, M. C. F.; Ribeiro, A. C. F. Characterization at 25°C of Sodium Hyaluronate in Aqueous Solutions Obtained by Transport Techniques. *Molecules* **2015**, *20*, 5812–5824.

25. Mracek, A.; Varhanikova, J.; Lehocky, M.; Grundelova, L.; Pokopcova, A.; Velebny, V. The Influence of Hofmeister Series Ions on Hyaluronan Swelling and Viscosity. *Molecules* **2008**, *13*, 1025–1034.

26. Mracek, A. The Measurement of Polymer Swelling Processes by an Interferometric Method and Evaluation of Diffusion Coefficients. *Int. J. Mol. Sci.* **2010**, *11*, 532–543.

27. Mráček, A.; Benešová, K.; Minařík, T.; Urban, P.; Lapcik, L. The Diffusion Process of Sodium Hyaluronate (Na-HA) and Na-HA-n-alkyl Derivatives Films Swelling. *J. Biomed. Mater. Res.* **2007**, *83*A, 184–190.

28. Fraser, J. R. E.; Laurent, T. C.; Laurent, U. B. G. Hyaluronan: Its Nature, Distribution, Functions and Turnover. *J. Intern. Med.* **1997**, *242*, 27–33.

29. Weigel, P. H.; Hascall, V. C.; Tammi, M. Hyaluronan Synthases. *J. Biol. Chem.* **1997**, *272*, 13997–14000.

30. Evanko, S. P.; Tammi, M. I.; Tammi, R. H.; Wight, T.N. Hyaluronan-Dependent Pericellular Matrix. *Adv. Drug Delivery Rev.* **2007**, *59*, 1351–1365.

31. Yang, C.; Cao, M.; Liu, H.; He, Y.; Xu, J.; Du, Y.; Liu, Y.; Wang, W.; Cui, L.; Hu, J.; Gao, F. The High and Low Molecular Weight Forms of Hyaluronan Have Distinct Effects on CD44 Clustering. *J. Biol. Chem.* **2012**, *287*, 43094–43107.

32. Camenisch, T. D.; Spicer, A. P.; Brehm-Gibson, T.; Biesterfeldt, J.; Augustine, M. L.; Calabro, A., Jr.; Kubalak, S.; Klewer, S. E.; McDonald, J. A. Disruption of Hyaluronan Synthase-2 Abrogates Normal Cardiac Morphogenesis and Hyaluronan-Mediated Transformation of Epithelium to Mesenchyme. *J. Clin. Invest.* **2000**, *106*, 349–360.

33. Choi, K. Y.; Saravanakumar, G.; Park, J. H.; Park, K. Hyaluronic Acid-Based Nanocarriers for Intracellular Targeting: Interfacial Interactions with Proteins in Cancer. *Colloids Surf. B* **2012**, *99*, 82–94.

34. Tezel, A.; Fredrickson, G.H. The Science of Hyaluronic Acid Dermal Fillers. *J. Cosmet. Laser Ther.* **2008**, *10*, 35–42.

35. Volpi, N.; Schiller, J.; Stern, R.; Soltes, L. Role, Metabolism, Chemical Modifications and Applications of Hyaluronan. *Curr. Med. Chem.* **2009**, *16*, 1718–1745.

36. Laurent, T. C.; Laurent, U. B. G.; Fraser, J. R. E. Serum Hyaluronan as a Disease Marker. *Ann. Med.* **1996**, *28*, 241–253.

37. Bernanke, D. H.; Markwald, R. R. Effects of Hyaluronic Acid on Cardiac Cushion Tissue Cells in Collagen Matrix Cultures. *Tex. Rep. Biol. Med.* **1979**, *39*, 271–285.

38. Fakhari, A.; Berkland, C. Applications and Emerging Trends of Hyaluronic Acid in Tissue Engineering, as a Dermal Filler, and in Osteoarthritis Treatment. *Acta Biomater.* **2013**, *9*, 7081–7092.

39. Tamer, T. M. Hyaluronan and Synovial Joint: Function, Distribution and Healing. *Interdiscip. Toxicol.* **2013**, *6*, 111–125.

40. McDonald, J. N.; Levick, J. R. Effect of Intra-Articular Hyaluronan on Pressure-Flow Relation Across Synovium in Anaesthetized Rabbits. *J. Physiol.* **1995,** *485,* 179–193.

41. Coleman, P. J.; Scott, D.; Mason, R. M.; Levick, J. R. Characterization of the Effect of High Molecular Weight Hyaluronan on Trans-Synovial Flow in Rabbit Knees. *J. Physiol.* **1999,** *514,* 265–282.

42. Van den Bekerom, M. P. J.; Mylle, G.; Rys, B.; Mulier, M. Viscosupplementation in Symptomatic Severe Hip Osteoarthritis: A Review of the Literature and Report on 60 Patients. *Acta Orthop. Belg.* **2006,** *72,* 560–568.

43. Gigante, A.; Callegari, L. The Role of Intra-Articular Hyaluronan (Sinovial) in the Treatment of Osteoarthritis. *Rheumatol. Int.* **2011,** *31,* 427–444.

44. Hiraoka, N.; Takahashi, K. A.; Arai, Y.; Sakao, K.; Mazda, O.; Kishida, T.; Honjo, K.; Tanaka, S.; Kubo, T. Intra-Articular Injection of Hyaluronan Restores the Aberrant Expression of Matrix Metalloproteinase-13 in Osteoarthritic SubchondralBone. *J. Orthop. Res.* **2011,** *29,* 354–360.

45. Kumahashi, N.; Naitou, K.; Nishi, H.; Oae, K.; Watanabe, Y.; Kuwata, S.; Ochi, M.; Ikeda, M.; Uchio, Y. Correlation of Changes in Pain Intensity with Synovial Fluid Adenosine Triphosphate Levels After Treatment of Patients with Osteoarthritis of the Knee with High-Molecular-Weight Hyaluronic Acid. *The Knee.* **2011,** *18,* 160–164.

46. Bannuru, R. R.; Natov, N. S.; Dasi, U. R.; Schamid, C. H.; McAlindon, T. E. Therapeutic Trajectory Following Intra-Articular Hyaluronic Acid Injection in Knee Osteoarthritis—Meta-Analysis. *Osteoarthr. Cartil.* **2011,** *19,* 611–619.

47. Muzzarellia, R. A. A.; Grecoa, F.; Busilacchia, A.; Sollazzob, V.; Gigantea, A. Chitosan, Hyaluronan and Chondroitin Sulfate in Tissue Engineering for Cartilage Regeneration: A Review. *Carbohydr. Polym.* **2012,** *89,* 723–739.

48. Lùrati, A.; Laria, A.; Mazzocchi, D.; Re, K. A.; Marrazza, M.; Scarpellini, M. Effects of Hyaluronic Acid (HA) Viscosupplementation on Peripheral Th Cells in Knee and Hip Osteoarthritis. *Osteoarthr. Cartil.* **2015,** *23,* 88–93.

49. Anniko, M.; Hellström, S.; Laurent, C. Reversible Changes in Inner Ear Function Following Hyaluronan Application in the Middle Ear. *Acta Oto-Laryngol.* **1987,** *104,* 72–75.

50. El Kechai, N.; Mamelle, E.; Nguyen, Y.; Huang, N.; Nicolas, V.; Chaminade, P.; Yen-Nicolaÿ, S.; Gueutin, C.; Granger, B.; Ferrary, E.; Agnely, F.; Bochot, A. Hyaluronic Acid Liposomal Gel Sustains Delivery of a Corticoid to the Inner Ear. *J. Controlled Release* **2016,** *226,* 248–257.

51. Stenfors, L. E. Repair of Traumatically Ruptured Tympanic Membrane Using Hyaluronan. *Acta Oto-Laryngol.* **1987,** *104,* 88–91.

52. Li, L.; Stiadle, J. M.; Lau, H. K.; Zerdoum, A. B.; Jia, X.; Thibeault, S. L.; Kiick, K. L. Tissue Engineering-Based Therapeutic Strategies for Vocal Fold Repair and Regeneration. *Biomaterials* **2016,** *108,* 91–110.

53. Szkiełkowska, A.; Miaśkiewicz, B.; Remacle, M.; Krasnodębska, P.; Skarżyński, H. Quality of the Voice After Injection of Hyaluronic Acid into the Vocal Fold. *Med. Sci. Monit.* **2013,** *19,* 276–282.

54. Kazemirad, S.; Heris, H. K.; Mongeau, L. Viscoelasticity of Hyaluronic Acid-Gelatin Hydrogels for Vocal Fold Tissue Engineering. *J. Biomed. Mater. Res. Part B.* **2016,** *104,* 283–290.

55. Stern, R.; Kogan, G.; Jedrzejas, M. J.; Soltés, L. The Many ways to Cleave Hyaluronan. *Biotechnol. Adv.* **2007,** *25,* 537–557.

56. Šoltés, L.; Mendichi, R.; Kogan, G.; Schiller, J.; Stankovska, M.; Arnhold, J. Degradative Action of Reactive Oxygen Species on Hyaluronan. *Biomacromolecules* **2006,** *7,* 659–668.

57. Toole, B. Hyaluronan in Morphogenesis. *J. Intern. Med.* **1997,** *242,* 35–40.

58. Laurent, T. C.; Dahl, I. M. S.; Dahl, L. B.; Engström-Laurent, A.; Eriksson, S.; Fraser, J. R. E. The Catabolic Fate of Hyaluronic Acid. *Connect. Tissue Res.* **1986,** *15,* 33–41.

59. Rodgers, L. S.; Lalani, S.; Hardy, K. M.; Xiang, X.; Broka, D.; Antin, P. B.; Camenisch, T. D. Depolymerized Hyaluronan Induces Vascular Endothelial Growth Factor, a Negative Regulator of Developmental Epithelial-to-Mesenchymal Transformation. *Circ. Res.* **2006,** *99,* 583–589.

60. Goa, K. L.; Benfield, P. Hyaluronic Acid. *Drugs* **1994,** *47,* 536–566.

61. Kikuchi, T.; Yamada, H.; Fujikawa, K. Effects of High Molecular Weight Hyaluronan on the Distribution and Movement of Proteoglycan Around Chondrocytes Cultured in Alginate Beads. *Osteoarthr. Cartil.* **2001,** *9,* 351–356.

62. Julovi, S. M.; Ito, H.; Nishitani, K.; Jackson, C. J.; Nakamura, T. Hyaluronan Inhibits Matrix Metalloproteinase-13 in Human Arthritic Chondrocytes via CD44 and P38. *J. Orthop. Res.* **2011,** *29,* 258–264.

63. Mo, Y.; Takaya, T.; Nishinari, K.; Kubota, K.; Okamoto, A. Effects of Sodium Chloride, Guanidine Hydrochloride, and Sucrose on the Viscoelastic Properties of Sodium Hyaluronate Solutions. *Biopolymers* **1999,** *50,* 23–34.

64. Mirin, A. K.; Heris, H. K.; Mongeau, L.; Javid, F. Nanoscale Viscoelasticity of Extracellular Matrix Proteins in Soft Tissues: A MultiscaleApproach. *J. Mech. Behav. Biomed. Mater.* **2014,** *30,* 196–204.

65. Loh, W. A técnica de dispersão de taylor para estudos de difusão em líquidos e suas aplicações" *Quím. Nova* **1997,** *20,* 541–545.

66. Barthel, J.; Gores, H. J.; Lohr, C. M.; Seidl, J. J. Taylor Dispersion Measurements at Low Electrolyte Concentrations. I. Tetraalkylammonium Perchlorate Aqueous Solutions. *J. Solution Chem.* **1996,** *25,* 921–935.

67. Callendar, R.; Leist, D. G. Diffusion Coefficients for Binary, Ternary, and Polydisperse Solutions from Peak-Width Analysis of Taylor Dispersion Profiles. *J. Solution Chem.* **2006,** *35,* 353–379.

68. Lampreia, I. M. S.; Santos, A. F. S.; Barbas, M. J. A.; Santos, F. J. V.; Lopes, M. L. S. M. Changes in Aggregation Patterns Detected by Diffusion, Viscosity, and Surface Tension in Water + 2-(Diethylamino)Ethanol Mixtures at Different Temperatures. *J. Chem. Eng. Data* **2007,** *52,* 2388–2394.

69. Ribeiro, A. C. F.; Lobo, V. M. M.; Leist, D. G.; Natividade, J. J. S.; Veríssimo, L. P.; Barros, M. C. F. Cabral, A. M. T. D. P. V. Binary Diffusion Coefficients for Aqueous Solutions of Lactic Acid. *J. Solution. Chem.* **2005,** *34,* 1009–1016.

70. Sgorla, D.; Almeida, A.; Azevedo, C.; Bunhak, É. J.; Sarmento, B.; Cavalcanti, O. A. Development and Characterization of Crosslinked Hyaluronic Acid Polymeric Films for Use in Coating Processes. *Int. J. Pharm.* **2016,** *511,* 380–389.

INDEX